BUILDING EARTH OBSERVATION CAMERAS

BUILDING EARTH OBSERVATION CAMERAS

GEORGE JOSEPH

CRC Press
Taylor & Francis Group
Boca Raton London New York

CRC Press is an imprint of the
Taylor & Francis Group, an **informa** business

CRC Press
Taylor & Francis Group
6000 Broken Sound Parkway NW, Suite 300
Boca Raton, FL 33487-2742

First issued in paperback 2021

© 2015 by Taylor & Francis Group, LLC
CRC Press is an imprint of Taylor & Francis Group, an Informa business

No claim to original U.S. Government works

ISBN 13: 978-1-138-74819-4 (pbk)
ISBN 13: 978-1-4665-6647-7 (hbk)

Dedicated to

all my colleagues in SAC

and

other centers of ISRO

who contributed to build world-class

Earth observation cameras

Contents

Preface...xiii
Author..xvii

1. Introduction 1
   1.1 Remote Sensing 2
   1.2 Civilian Earth Imaging System...3
   1.3 Indian Earth Observation Program: The Evolution.........................5
   1.4 Earth Observation System: The Paradigm Shift...............................6

2. Image Formation...9
   2.1 Introduction 9
   2.2 Electromagnetic Radiation 10
        2.2.1 Quantum Nature of Electromagnetic Radiation...............11
        2.2.2 Thermal Radiation...12
        2.2.3 Propagation of Electromagnetic Radiation from One
               Medium to Another..12
        2.2.4 Diffraction..13
   2.3 Some Useful Terminologies of the Imaging Systems 14
   2.4 Aberrations 17
        2.4.1 Spherical Aberration..18
        2.4.2 Coma 19
        2.4.3 Astigmatism 20
        2.4.4 Distortion 20
        2.4.5 Curvature of the Field 20
        2.4.6 Chromatic Aberration 20
   2.5 Wave Optics 22
   2.6 Image Quality Evaluation...25
   2.7 Modulation Transfer Function 27
   2.8 Source of Electromagnetic Radiation for Imaging.........................31
   2.9 Radiometric Consideration...31
   References 34

3. Imaging Optics...35
   3.1 Introduction 35
   3.2 Refractive Optics 36
        3.2.1 Telecentric Lenses 40

3.3   Reflective and Catadioptric Systems..43
      3.3.1   Types of Reflective Telescope Systems.................................43
      3.3.2   Increasing Field of View of Telescopes.............................46
           3.3.2.1   Catadioptric System...47
           3.3.2.2   All Reflective Wide Field of View
                     Telescope..48
3.4   Stray Light Control and Baffling...52
3.5   Building a Reflective Telescope..54
      3.5.1   Selection of Mirror Material..55
      3.5.2   Mirror Fabrication...58
           3.5.2.1   Lightweighting of Mirror...................................59
           3.5.2.2   Optimizing Lightweight Mirror
                     Structure..63
      3.5.3   Mirror Mounts..64
           3.5.3.1   Bipod Mounts...65
      3.5.4   Alignment of Mirrors...67
References 71

4. Earth Observation Cameras: An Overview 75
4.1   Introduction..75
4.2   Spatial Resolution..76
4.3   Spectral Resolution 82
      4.3.1   Interference Filter..86
4.4   Radiometric Resolution..87
      4.4.1   Radiometric Quality...88
4.5   Temporal Resolution...90
4.6   Performance Specification 91
4.7   Imaging Modes...92
4.8   On-Orbit Performance Evaluation..93
References 98

5. Optomechanical Scanners 101
5.1   Introduction..101
5.2   Principle of Operation 101
5.3   Scanning Systems 103
      5.3.1   Scan Geometry and Distortion 110
5.4   Collecting Optics..112
5.5   Dispersive System and Focal Plane Layout...................................113
5.6   Detectors..117
      5.6.1   Detector Figure of Merit 117
      5.6.2   Thermal Detector 120
      5.6.3   Photon Detectors..121
           5.6.3.1   Photoemissive Detectors 122
           5.6.3.2   Photoconductive Detector................................123
           5.6.3.3   Photovoltaic Detector.......................................123

        5.6.4   Quantum Well Infrared Photodetectors 126
        5.6.5   Operating Temperature 126
        5.6.6   Signal Processing 130
    5.7   System Design Considerations 133
    5.8   Enhanced Thematic Mapper Plus 138
    References 142

6. Pushbroom Imagers 145
    6.1   Introduction 145
    6.2   Principle of Operation 145
    6.3   Linear Array for Pushbroom Scanning 147
        6.3.1   Charge-Coupled Devices 147
        6.3.2   CMOS Photon Detector Array 150
        6.3.3   Hybrid Arrays 152
    6.4   CCD Signal Generation and Processing 153
        6.4.1   CCD Output Signal 154
        6.4.2   Off-the-Chip Signal Processing 155
    6.5   Spaceborne Pushbroom Cameras 159
    6.6   IRS Cameras: LISS-1 and -2 159
        6.6.1   Focal Plane Layout 162
                6.6.1.1   Single Collecting Optics Scheme 162
                6.6.1.2   Multiple Lens Option 165
        6.6.2   Mechanical Design 166
        6.6.3   Electronics 168
        6.6.4   Alignment and Characterization 168
                6.6.4.1   Focusing of the Camera System 169
                6.6.4.2   Image Format Matching and BBR 172
                6.6.4.3   Flat Field Correction 174
        6.6.5   Qualification 179
    6.7   IRS-1C/D Camera 180
        6.7.1   LISS-3 Design 181
        6.7.2   Wide Field Sensor 182
        6.7.3   PAN Camera 184
                6.7.3.1   Payload Steering Mechanism 187
    6.8   RESOURCESAT Series 188
        6.8.1   RESOURCESAT LISS-3 189
        6.8.2   Advanced Wide Field Sensor (AWiFS) 190
        6.8.3   LISS-IV Multispectral Camera 190
    6.9   SPOT Earth Observation Camera 194
    6.10 Landsat 8: Landsat Data Continuity Mission 198
        6.10.1   Operational Land Imager 199
        6.10.2   Thermal Infrared Sensor 202
    6.11 Hybrid Scanner 205
        6.11.1   GEO High-Resolution Imaging Systems: Technology
                 Challenges 208

      6.11.2   GEO-Resource Survey Systems 209
               6.11.2.1 ESA GEO-HR System 209
               6.11.2.2 Geostationary Ocean Color Imager 212
               6.11.2.3 ISRO Geostationary Imaging
                            Satellite (GISAT) 213
               6.11.2.4 Geostationary Hyperspectral Imaging
                            Radiometer (NASA) 214
  References 214

7. Submeter Imaging 219
  7.1   Introduction 219
  7.2   Considerations for Realizing a High-Resolution Imaging
       System 220
  7.3   Increasing the Integration Time 221
       7.3.1   Time Delay and Integration 222
       7.3.2   Asynchronous Imaging 223
       7.3.3   Staggered Array Configuration 225
  7.4   Choosing Faster Optics 227
       7.4.1   Choosing Charge-Coupled Device Pixel
              Dimension 227
       7.4.2   Increasing Collecting Optics Diameter 229
  7.5   Data Transmission 229
       7.5.1   Data Compression 232
  7.6   Constraints on the Satellite 234
  7.7   Imaging Cameras with Submeter Resolution 237
       7.7.1   IKONOS 237
       7.7.2   QuickBird-2 239
       7.7.3   GeoEye-1 240
       7.7.4   WorldView Imaging Systems 240
       7.7.5   Indian Remote Sensing Satellite High-Resolution
              Imaging Systems 243
  7.8   What Limits the Spatial Resolution? 244
  References 247

8. Hyperspectral Imaging 251
  8.1   Introduction 251
  8.2   Hyperspectral Imaging Configuration 254
       8.2.1   Scanner Approach 255
       8.2.2   Pushbroom Approach 256
  8.3   Spectrometers: An Overview 257
       8.3.1   Dispersive Spectrometers 258
       8.3.2   Fourier Transform Spectrometers 261
              8.3.2.1 Michelson Interferometers 263
              8.3.2.2 Sagnac Interferometer 266
       8.3.3   Filter-Based Systems 270

8.4 Distortions: "Smile" and "Keystone" Effects 273
8.5 Hyperspectral Imaging Instruments.. 274
    8.5.1 MightySat II: Fourier Transform
        Hyperspectral Imager 275
    8.5.2 National Aeronautic and Space Administration Earth
        Observation 1: Hyperion 275
    8.5.3 NASA EO-1 Linear Etalon Imaging Spectral Array
        Atmospheric Corrector 276
    8.5.4 Indian Space Research Organization Hyperspectral
        Imager... 277
References.. 278

9. **Adding the Third Dimension: Stereo Imaging** 281
9.1 Introduction... 281
9.2 Stereo Pair Generation Geometries.. 284
    9.2.1 Across-Track Stereoscopy 285
    9.2.2 Along-Track Stereoscopy 287
9.3 Along-Track Stereo Using Multiple Telescopes......................... 289
    9.3.1 IRS CARTOSAT 1.. 289
    9.3.2 SPOT High-Resolution Stereo Camera 292
    9.3.3 Advanced Land Observation Satellite Stereo
        Mapping Camera: PRISM 294
9.4 Stereo Pairs with Single Optics 295
    9.4.1 Monocular Electro-Optical Stereo Scanner 295
    9.4.2 Chandrayaan1 Terrain Mapping Camera 298
    9.4.3 Along Track Stereo by Satellite Tilt 299
References.. 300

10. **Journey from Ground to Space** 303
10.1 Introduction... 303
10.2 Launch Environment 303
10.3 Space Environment 304
    10.3.1 Thermal Environment 304
    10.3.2 Vacuum 305
    10.3.3 Radiation Environment 306
10.4 Space Hardware Realization Approach.. 309
    10.4.1 Model Philosophy.. 313
10.5 Environmental Tests 314
    10.5.1 Mechanical Tests... 314
    10.5.2 Thermovacuum Test.. 316
    10.5.3 Electromagnetic Interference/Compatibility
        Tests 317
    10.5.4 Environmental Test Levels 317
    10.5.5 Ground Support Equipment/Facilities............................ 318
    10.5.6 Contamination Control... 319

10.6 Reviews 320
       10.6.1 Baseline Design Review... 320
       10.6.2 Preliminary Design Review (PDR) 320
       10.6.3 Critical Design Review 320
       10.6.4 Preshipment Review 321
       10.6.5 Configuration Change Control and Nonconformance
              Management... 321
       10.6.6 Failure Review Committee... 322
10.7 Parts/Components Procurement 323
10.8 Reliability and Quality Assurance 324
References... 325

Appendix: Representative Imageries 327
Index 335

Preface

Humankind has always marveled at viewing the Earth from great heights, because such perspectives provide a broader, holistic view of the neighborhood and the Earth's surface. Founded on this basic human trait, the ability to image any part of the globe from space and obtain a seamless perspective of the Earth's objects has become an important modern-day technological achievement. Today, there are many satellites that orbit the Earth (or, for that matter, are even in outer space) and provide images of the Earth using advanced imaging technology, bringing far-reaching impacts to humanity. Even though the use of imagery from space was initially for reconnaissance and military purposes, soon scientists recognized the potential of the images for many civilian and public-good applications and also for scientific understanding of the Earth as a total system, and a new discipline called *remote sensing* has emerged. Remote sensing has now paved the way for a large number of societal, commercial, and research applications in almost every nation of the world.

An important component of any remote sensing activity is the *imaging system*—the *eye in the sky*—that is not just a technological marvel in space but is a system of excellent engineering, based on sound physical principles. This then is the subject matter of this book—looking at imaging system technology and providing insights into different aspects of building an imaging system for a space platform. In principle, any part of the electromagnetic radiation within the *atmospheric window* can be the basis of remote sensing; the present book specifically deals with electro-optical sensors that operate in the optical–infrared region. Design, development, and characterization of such an imaging system and finally qualifying it for operations in space require intense knowledge in various disciplines of engineering and science. I believe that an overview of the total system is essential for any imaging system project manager. Such a broad understanding, generally, is difficult to find in one source textbook—though information is available in various specialized books, journals, and so on. I have tried to bridge this gap through this book.

I have written this book with manifold purposes. First, the book should be able to help practicing imaging system project managers and scholars and researchers of electro-optical sensor systems obtain fundamental information and a broad overview of various entities of imaging systems—thus, it should help them to look deeper into the systems that they are developing. Second, the application scientists who use satellite imagery should also find the book very helpful in understanding various technical aspects and the terminology used in defining the performance of the image system so that they can choose the most appropriate dataset for meeting their needs. Third,

I foresee that this book will also serve as a guideline to the new entrants to this field to understand the concepts and challenges of building space-based Earth observation systems.

Writing this book has been a personal challenge for me. From my deep association with the development of electro-optical sensors, I find the subject to be so vast that each subsystem of an Earth observation camera deserves to be a book by itself—the challenge before me has been to *cover everything possible but keep it minimalistic*. Another challenge I foresaw was that because of the multidisciplinary nature of the imaging systems, the reader community would be quite varied across science and engineering disciplines. Thus, maintaining the structure of the book and presenting the content in a manner that can be appreciated by a range of readers has been always at the back of my mind. Therefore, I decided on three things: (1) the book must serve as much as possible as a single source of all technical aspects of imaging systems, (2) there should be broad coverage and so minimalistic depth, and (3) it should be interesting for the wide-ranging reader community. Of course, I give a long list of references at the end of each chapter—if anyone wants to deep dive, then that too would be possible.

In the opening chapter, the book traces the historical development of imaging systems and reviews the evolution of Earth observation systems in the world and the trends in the technology and end utilization. The second chapter provides the basic concepts and fundamental principles of image formation and the physical laws and principles. The next seven chapters cover various aspects of design, system trade-offs, realization, and characterization of various types of imaging systems—specific examples from Indian Remote Sensing (IRS) Satellite systems illustrate the design and development issues. To maintain continuity and facilitate nonspecialist in following the content easily, I have consciously repeated some definitions and explanations in later chapters though they are also explained earlier. There are a number of Earth observation systems—past and present. It is not possible to cover all of them. However, I have made a concerted effort to cover representative imaging systems from other agencies so that one can get an understanding and appreciate various engineering challenges that have been addressed in different manners. The last chapter gives a broad framework of the tasks involved in qualifying a payload for space use. I must make it abundantly clear that any references that I make to the Indian space program and the IRS are my personal views and have no endorsement of the Indian Space Research Organisation (ISRO) as an organization.

I am grateful to Dr. K. Radhakrishnan, the present chairman of ISRO, for giving me an honorary position in ISRO, without which I could not have accomplished this task. I am thankful to Dr. R. R. Navalgund and Prof. A. S. Kiran Kumar, the past and present directors of Space Applications Centre (SAC) who extended to me the necessary facilities in the course of preparation for the book. Many of my colleagues at ISRO/Department of Space (DOS) helped me in different manner—by giving valuable input, generating tables and figures, critically going through the manuscript, and so on. There are so many of these wonderful colleagues that have helped me, but it is practically impossible to mention each of them by name. I am deeply indebted to all of them. My special thanks to A. V. Rajesh for generating many of the line drawings and the contributing to the cover page design and for the final formatting of the text.

My wife, Mercy, who, because of my professional involvement and pursuits has made many sacrifices over the years, deserves special thanks for her constant support and encouragement, without which the book would not have been finished. Our grandchildren—Nishita, Reshawn, and Riana—are always a source of inspiration to take up new challenges.

George Joseph

Author

Dr. George Joseph started his research career in 1962 at the Tata Institute of Fundamental Research (TIFR), Mumbai, where he was involved in the study of cosmic rays. For his pioneering work at TIFR on detection of emission of neutrons from the sun, he was awarded a PhD in 1971. In 1973, he was invited to join the Space Applications Centre (SAC), one of the major centers of the Indian Space Research Organization (ISRO), primarily for developing Earth observation systems. Under his leadership, a variety of Earth observation cameras were developed for the Indian Remote Sensing (IRS) Satellite and the Indian National Satellite (INSAT). Apart from being the guiding force for the development of Earth observation remote sensors developed by ISRO, Dr. Joseph has made substantial contributions toward the realization of various other remote-sensing-related activities. He served SAC in various capacities including as its director from 1994 to 1998.

Dr. Joseph has served in a number of national and international committees/organizations including president of Technical Commission–1 of the International Society for Photogrammetry and Remote Sensing (ISPRS) during 1996–2000. During 2006–2009, he was the director of the Centre for Space Science and Technology Education in Asia and the Pacific (CSSTE-AP), affiliated to the United Nations. He is a fellow of a number of national academies such as the Indian Academy of Sciences; the National Academy of Sciences, India; and the Indian National Academy of Engineering. In recognition of his outstanding contributions to electro-optical sensor development and for his distinguished achievements in furthering remote sensing utilization in the country, the Government of India honored him in 1999 with a civilian award—the Padma Bhushan.

1

Introduction

From the humble beginnings of an invention of a pinhole camera around 1000 AD to highly sophisticated data gathering from space, the history of imaging has been a captivating series of technological advances and their applications. The most ancient pinhole camera, also known as the camera obscura, could just project an image of a scene outside, upside-down, onto a viewing surface. It was only around 1816 that Nicéphore Niépce became the first man to capture a camera image on a paper medium coated with silver chloride, which darkened where it was exposed to light. Shortly after, George Eastman created an epoch by making commercial cameras that could be used by nonspecialists, and the first Kodak camera entered the market in 1888 preloaded with enough film for 100 exposures.

Early humans used to climb up high mountains or treetops to get a better view of their surroundings to identify greener pastures or the threat of approaching hostile situations. In other words, a synoptic view was important for their survival. It is known that by 1858 humans could capture an aerial image, when Gaspard-Félix Tournachon (also known as Félix Nadar) took the first known aerial photograph by placing a camera on a balloon. Other methods to take aerial photographs included the use of kites and pigeons attached with cameras. Homing pigeons were used extensively during World War I. Also, aircraft platforms became indispensable means for reconnaissance during the world wars.

It was indeed the war needs that gave the maximum impetus for advancements in camera and photographic film. During world wars, obtaining intelligence about the enemy and their activities was one of the key needs for each warring side. The development of films capable of recording images in near infrared (NIR) was a boon to detect camouflaged objects. A green paint or even cut tree branches give different responses compared to a living tree canopy when imaged in NIR. Even after the end of World War II, during the ensuing Cold War, reconnaissance of others' territory was found essential. According to the international laws that were in force after World War II, each country had sovereign rights over its airspace—the portion of the atmosphere above its territory and up to a height that any aircraft could fly. Thus, any violation of air space entitles the affected country the right of self-defense and accordingly the right even to shoot down the intruders. A new dimension for reconnaissance began with the launch of Sputnik in 1957, which marked the beginning of space age, which soon granted the world nations a legal regime for activities in the space environment, which was declared as a province of all mankind and hence imaging on any territory from space was no bar. In the beginning, indeed, imaging of Earth from space was mostly used for snooping into others' territory!

The initial operational photoreconnaissance satellites of the United States, called CORONA, had photographic cameras, and after the mission the exposed film was ejected from the satellite and collected in midair for processing and exploitation. The camera system was nicknamed KH (key-hole). Beginning in 1960, several Corona systems were launched with cameras developed by Itek using Eastman Kodak film; early flights had spatial resolution in the rage of 10–12 m in early flights, and by 1972, the resolution improved to about 2–3 m. Soon after the first U.S. Corona mission, the Soviets also built successful photoreconnaissance satellites (Zenit). These photographic systems are broadly similar to the KH cameras. As technology advanced, KH became digital. The KH-11, also named CRYSTAL, launched in 1976 and was the first American spy satellite to use electro-optical digital imaging, using an 800 by 800 pixel charge-coupled device (CCD) array for imaging; hence it had a real-time observation capability. The 2.4 m diameter telescope of KH-11 could provide a theoretical ground resolution of about 15 cm, though the actual realized spatial resolution from orbit will be much poorer due to atmospheric and other effects. Because of the high secrecy involved, the technical details and capabilities of these satellites are not available in the public domain. The latest reconnaissance system, KH-12 (improved CRYSTAL), operates in the visible, NIR, and thermal infrared (TIR) regions of the electromagnetic (EM) spectrum. These sensors probably have low-light-level CCD image intensifiers, which can provide nighttime images. Thus, during the early years of imaging from space, military applications were responsible for most of the advances.

Scientists soon realized the potential of space imagery for varied civilian applications in fields such as geology, forestry, agriculture, cartography, and so on. After the world wars, the scientists pursued more vigorously the use of space imagery for the public good. A new term, "remote sensing," was added to the technical lexicon.

1.1   Remote Sensing

The term remote sensing literally means making observations about an object, where the sensor used is not in physical contact with the object under consideration. This is in contrast to in-situ measurement, for example measuring the body temperature using doctor's thermometer. Though any observation from a distance can be termed remote sensing, United Nations Resolution (41/65) dealing with "The Principles Relating to Remote Sensing of the Earth from Outer Space" adopted on December 3, 1986 defines remote sensing as follows:

> The term "remote sensing" means the sensing of the Earth's surface from space by making use of the properties of electromagnetic waves emitted, reflected or diffracted by the sensed objects, for the purpose of improving natural resources management, land use and the protection of the environment.

From a scientific standpoint in remote sensing, we study various properties of the EM radiation, primarily reflected or emitted by an object, to make inferences about the object under study. Thus, remote sensing requires a source that produces the EM radiation, a sensor that has the capability of measuring the desired properties of the EM radiation emanating from the object being investigated, and a platform to carry the sensor. When the source of EM radiation is naturally occurring, such as solar radiation or the self-emission from the object, it is called passive remote sensing. It is also possible that the sensor carries its own EM radiation to illuminate the target, as in the case of radar, and such a system is called active remote sensing.

When we observe the Earth from space, the intervening atmosphere does not transmit all parts of EM spectrum. Thus, the atmosphere is transparent only to certain portions of the EM radiation, which are called "atmospheric window" regions. Earth observation is carried out in these windows of the EM spectrum. The EM energy transmitted from the source (sun, transmitter carried by the sensor) interacts with the intervening atmosphere and is spectrally and spatially modified before reaching the Earth's surface. The EM radiation falling on the Earth interacts with the Earth surface resulting in part of the energy being sent back to the atmosphere. In addition, self-emission from the surface due to its temperature also takes place. This is the "signature" of the target, which is detected by sensors placed on suitable platforms, such as aircraft, balloons, rockets, satellites, or even ground-based sensor-supporting stands. The sensor output is suitably manipulated and transported back to the Earth; it may be telemetered as in the case of unmanned spacecraft, or brought back through films, magnetic tapes, and so on, as in aircraft or manned spacecraft systems. The data are reformatted and processed on the ground to produce photographs or stored in a computer compatible digital data storage medium. The photographs/digital data are interpreted visually/digitally to produce thematic maps and other information. The interpreted data so generated need to be used along with other data/information, and so on, to arrive at a management plan. This is generally carried out using a geographic information system.

1.2 Civilian Earth Imaging System

The beginning of space-based imaging of the Earth can be dated back to 1891, when Germans were developing rocket-propelled camera systems, and by 1907 gyro-stabilization had been added to improve picture quality. In August 1959, less than 2 years after the first man-made satellite was launched, the

United States' Explorer 6 transmitted the first picture of the Earth to be taken from a satellite. Soon meteorologists realized that satellite-based Earth observation could provide a synoptic view and global coverage, which are important to understand weather phenomenon. Thus, the systematic Earth observation from space started with the launch of Television Infrared Observation Satellite (TIROS-1) in April 1, 1960 by the United States, which was designed primarily for meteorological observations. The camera on board TIROS-1 was a half-inch vidicon. Since then, each successive spacecraft has carried increasingly advanced instruments and technology. These low Earth orbit observation systems and later, imaging from geosynchronous orbit, have substantially improved weather forecasting capabilities.

The birth of a civilian satellite to survey and manage Earth resources has a different story. In the 1960s, photographs of the Earth taken from the Mercury, Gemini, and Apollo missions were available to geologists of the U.S. Geological Survey and they realized that these photographs with large spatial coverage in a single frame, compared to aerial imagery, are of immense value for geological studies and mapping. After a number of deliberations regarding the security and legal aspects, finally in 1970 the National Aeronautics and Space Administration received approval to develop an Earth imaging satellite initially named ERTS (Earth Resources Technology Satellite), and currently known as Landsat. The Landsat 1 launched in 1972 carried two imaging systems—the Return Beam Vidicon (RBV) in three spectral bands with a spatial resolution of about 30 m and a four spectral band multispectral scanner system (MSS) with a spatial resolution of about 80 m. The RBV was supposed to be the primary instrument and the MSS, an experimental secondary instrument. However, the MSS imagery was found to be superior to the RBV imagery. The RBV was discontinued after Landsat 3, the MSS sensor was the workhorse for remote sensing data, and an improved instrument, thematic mapper (TM), was flown on Landsat 4 and 5.

The data from the Landsat MSS system were available to a number of interested countries outside the United States also. It is a matter of credit to the United States that uninterrupted Landsat data were available across the world. The data were analyzed by government and private sector agencies, and it was realized that the Earth's features and landscapes can be identified, categorized, and mapped on the basis of their spectral reflectance and emission; this has created new insights into geologic, agricultural, and land-use surveys, and allowed for resource exploration and exploitation in a scientific manner. Many countries recognized the potential of satellite imagery for their resource management and realized that once the data become a part of their planning process the data need to be available on an assured basis. Thus, other countries also planned their own satellite-based imaging system. France was the next to launch a series of multispectral Earth observation systems starting with SPOT-1 in 1986. Japan launched its first Marine Observation Satellite MOS-I in 1987 to gather data about the ocean's surface and the Japanese Earth Resources Satellite JERS-I in 1992, which carried both a visible and infrared instrument and a synthetic aperture radar

(SAR). India launched Indian Remote Sensing (IRS) Satellite IRS-1A in 1988, which was the first of a series of operational Earth observational systems. India has one of the best constellations of Earth observation systems and the global community uses the data. More importantly, the remote sensing data is operationally used in various sectors in India for planning, implementation, and monitoring of various national resources.

1.3   Indian Earth Observation Program: The Evolution

The satellite program in India started with the design and development of a scientific satellite, which was launched using a Soviet rocket from a cosmodrome in the erstwhile USSR on April 19, 1975. The satellite after launch was named Aryabhata, after the fifth century Indian astronomer. Encouraged by the success of Aryabhata, it was decided to develop an application satellite that could be a forerunner to meet the goal of providing services for national development. Since the Aryabhata was a low Earth orbiting satellite, developing a remote sensing satellite was the most appropriate choice. In view of the complexities of realizing an operational remote sensing satellite in the first mission, it was decided to go for an experimental system that can provide the necessary experience in the design, development, and management of a satellite-based remote sensing system for an Earth resources survey. The satellite carried a two-band TV camera system with a spatial resolution of about 1 km and a multiband microwave radiometer. The satellite was launched on June 7, 1979 from the erstwhile USSR. The satellite was named Bhaskara, after the twelfth century Indian mathematician Bhaskaracharya. The follow-on satellite Bhaskara 2 was also launched from the USSR on November 20, 1981.

The Bhaskara missions provided valuable experience in a number of interrelated aspects of satellite-based remote sensing systems for resource survey and management. The next logical move was to embark on a state-of-the-art operational Earth observation system. Way beyond the 1 km camera system used in the learning phase, India has successfully launched a series of remote sensing satellites—the IRS satellites—starting with IRS-1A in 1988. When IRS-1C was launched in 1995, with its PAN camera providing 5.8 m resolution, it had the highest spatial resolution among civilian Earth observation satellites in the world and retained its number one position until the advent of IKONOS in 1999, with 1 m resolution capability. Since then a number of Earth observation satellites have been launched for land, ocean, and meteorological studies including a series of cartographic satellites, which can be used for updating of topographic maps. Another milestone is the launch of the Radar Imaging Satellite in 2012—a state-of–the-art microwave remote sensing satellite carrying a SAR payload operating in the C-band (5.35 GHz). These are low Earth orbiting satellites. In addition, the Indian Space Research Organization (ISRO) has a constellation of geostationary satellites carrying Earth observation cameras to provide information for meteorological studies and forecast.

Although IRS satellites were developed strongly on the basis of meeting national development needs, the excellence that they promoted in technology, applications, and capacity building enabled them to play a significant role in the international arena. The data from the IRS satellites are available globally through Antrix Corporation Limited (Antrix), the marketing arm of the ISRO.

1.4   Earth Observation System: The Paradigm Shift

As mentioned earlier, systematic observation of the Earth from space began in 1960 when the TIROS satellite was launched mainly for meteorological observations and applications. But it was not till 1972 when the Landsat 1 was launched that formal "remote sensing" gained a great impetus with the availability of systematic images of the whole globe; this opened up an era of applications of remote sensing data for a wide variety of themes. Since this epoch event of 1972, Earth observation has seen a number of improvements in terms of technology, applications, and management. After the launch of Landsat 1 in 1972, the United States was the only player in space-based remote sensing and it took 14 years for another remote sensing satellite to emerge in the French SPOT-1 system, which was launched in 1986. Soon after, India launched its first IRS-1A system in 1988 and entered the operational space-based remote sensing era and soon many others followed. Though there are now a number of nations that have Earth observation systems, today it is mainly the U.S. program of Earth observation that dominates the scene of remote sensing—both in the public sector and commercial sector, providing valuable data for resource management and scientific understanding of the Earth system processes.

The initial concept of Earth observation was to mainly use specialized sensors on orbiting satellite systems to obtain images or data—either of land, ocean, or atmosphere separately, and missions such as Landsat, Seasat, Upper Atmosphere Research Satellite, and so on contributed massive volumes of such images and data. But it was soon realized that land, ocean, and atmosphere are a "coupled system" and that it is necessary to have simultaneous observation for different parameters of land, oceans, and atmosphere and that too at global scales. Such a "synchronized strategy" was felt essential to understand Earth processes and the interlinkages of the "coupled system"—thereby better understanding our Earth system. Thus, the approach shifted to observing the Earth as an integrated system and concepts of global observation strategy emerged. However, such a multidisciplinary global approach requires a variety of sensors operating in various regions of the EM spectrum and in different modes—thereby, sensor and instrumentation development got a great boost and a number of unique and specialized sensors have been developed. Accommodating a variety of sensors and instrumentation on a satellite platform made the satellites heavier, demanded more power, and increased complexity of operations—thereby such Earth observations satellites became obviously more expensive and required large budgets and timelines for development. For example, TERRA (formerly known as EOS AM-1) supports five instruments to study various aspects of land, ocean, and atmosphere covering a broad spectrum of EM radiation. The satellite weighs about 4600 kg, needs 3 kW of power in orbit, and took almost a decade of planning and development to become operational in December 1999. Similarly, the ENVISAT of the European Space Agency carries sensors covering the visible–IR–NIR–TIR–microwave region to make measurements of various processes of atmosphere, land, and ocean. ENVISAT weighs 8200 kg and generates 6.6 kW power in orbit and the satellite also took more than 10 years to develop before launch in 2002. No doubt, the technology of building satellites also saw tremendous growth—advanced concepts of bus design, payload operations, power generation, data storage, and transmission were realized and became core to such large Earth observation systems.

With technological advancement in electronics, miniaturization, materials, and computer science, Earth observation satellite design is also seeing rapid changes. The current thinking is to develop smaller and agile Earth observation satellites having highly sensitive sensors and instrumentation, faster electronics, and faster response to mission needs, thereby bringing more efficiency in cost and schedule but also improved performance. In order to have "concurrent measurements," the concept of distributed satellite systems has developed—each carrying different sensors and instruments, but "flying in formation" such that each satellite crosses the equator within a few minutes of others and thereby achieving multidisciplinary observation capability "as in a single big satellite." Such a coordinated approach reduces risks of total loss of observational capability due to mission failures. However, such an approach poses a tremendous challenge in mission operations and maintenance, thereby requiring the complex technology of autonomous mission operations on the ground. The EOS A-Train system consists of six satellites combining a full suite of instruments for observing clouds and aerosols from passive radiometers to active LIDAR and radar sounders, flying in close proximity, is an example of formation flying. The satellites cross the equator within a few minutes of one another at around 1:30 p.m. local time.

Another development in the Earth observation scenario came about in the form of commercial Earth observation satellites—with the private sector building–owning–operating Earth observation satellites. This shift was a result of a 1992 commercialization effort of the United States—which licensed private agencies to develop, own, and operate imaging systems—thereby quickly opening up commercial distribution and the availability of meter/submeter resolution images. However, the policy posed "U.S. government control" in the form of a restriction referred to as "Shutter Control," by which the U.S. government reserved its right to interrupt or limit commercial operations, should extraordinary circumstances (not clearly defined) warrant it. The advancement of focal plane charge-coupled device (CCD) arrays with time delay and integration (TDI) capability made it possible to realize Earth observation cameras with spatial resolution of 1 m and less. The first such commercial Earth observation system was IKONOS, launched in 1999. This has been followed by a number of other commercial ventures to provide sub-meter image data. The commercial viability of these high-resolution imaging systems is due to the large market for mapping, especially for urban areas, change detection, disaster management, infrastructure, and also defense and security requirements. However, it is important to note that these high-resolution imaging systems would not have systematic, global coverage, which is the cornerstone of remote sensing as provided by the Landsat/IRS/SPOT type of systems. Although the high resolution of imaging proves to be a great advantage in "seeing details" and reaches "near aerial-image" quality, the limited swath and poor temporal resolutions make them less suited for applications requiring systematic repeat coverage. However, these high-resolution imaging satellites are today meeting a large market demand, even as traditional systematic systems, like Landsat, SPOT, IRS, and so on are still very essential and meet various needs, such as for natural resources management (as in land use, crop production estimation, watershed management, and on), for scientific studies (as in global change), and for many others uses.

Earth observation technology has come a long way from its early beginnings in 1960s—it is no longer an issue of "what the use of the images is," but more an issue of "meeting market demand, creating commercial viability, and maintaining security." The technology and systems of Earth observation are fast changing—detailed, rapid, and large volumes of data collection are being distributed efficiently across the globe and enabling a large suite of market economic, driven by government–private sector partnerships. With a large suite of technological capability in Earth observation (resolutions, bands, coverage, sophistication of measurements, etc.), the concept of user segment has also seen a dramatic change. Today, a user of an Earth observation satellite is no more the scientist/analyst in a laboratory but instead a large community of professionals in government, private sector, academia, and even citizens—who seek a solution to their problem. Thus, today an end user is interested in getting a solution for his problem and this may require analysis, fusion, and integration of various images, maps, location data, social media data, complex tables, and so on in an "analysis model" that can be customized as per user needs. Such a change in perspective and demand has brought about yet another trend in the form of a large value-addition and customization market with Earth observation images forming one big data element, and the making of "tailored solutions." Thus, a large market exists where users are ready to pay a cost for obtaining their solution needs—rather than just obtaining images and data.

Remote sensing has become a part of everyday life, from weather forecasts to military surveillance. There is a continued demand to get images with enhanced performance to meet newer applications. The Earth observation camera designers have to rise to the occasion to meet this challenge. The following chapters will help you to face this challenge.

2

Image Formation

2.1 Introduction

An image is a projection of a three-dimensional scene in the object space to a two-dimensional plane in the image space. An ideal imaging system should map every point in the object space to a defined point in the image plane, keeping the relative distances between the points in the image plane the same as those in the object space. An extended object can be regarded as an array of point sources. The image so formed should be a faithful reproduction of the features (size, location, orientation, etc.) of the targets in the object space to the image space, except for a reduction in the size; that is, the image should have geometric fidelity. The imaging optics does this transformation from object space to image space.

The three basic conditions that an imaging system should satisfy to have a geometrically perfect image are (Wetherell 1980) as follows:

1. All rays from an object point (x, y) that traverse through the imaging system should pass through the image point (x', y'). That is, all rays from an object point converge precisely to a point in the image plane. The imaging is then said to be *stigmatic*.

2. Every element in the object space that lies on a plane normal to the optical axis must be imaged as an element on a plane normal to the optical axis in the image space. This implies that an object that lies in a plane normal to the optical axis will be imaged on a plane normal to the optical axis in the image space.

3. The image height h must be a constant multiple of the object height, no matter where the object (x, y) is located in the object plane.

The violation of the first condition causes image degradations, which are termed *aberrations*. The violation of the second condition produces *field curvature*, and the violation of the third condition introduces *distortions*. Consequences of these deviations will be explained in Section 2.4. In addition, the image should faithfully reproduce the relative radiance distribution of the object space, that is, radiometric fidelity.

2.2 Electromagnetic Radiation

As electromagnetic (EM) energy forms the basic source for Earth observation
cameras, we shall describe here some of the basic properties of EM radiation,
which help us to understand image formation. Various properties of EM radiation can be deduced mathematically using four differential equations, generally referred to as Maxwell's equations. Readers interested in the mathematical
formulation of EM theory may refer to the work by Born and Wolf (1964). Using
Maxwell's equations, it is possible to arrive at a relationship between the velocity
of an EM wave and the properties of the medium. The velocity c_m in a medium
with an electric permittivity ε and magnetic permeability μ is given as follows:

$$c_m = \frac{1}{\sqrt{\varepsilon\mu}} \tag{2.1}$$

In vacuum,

$$\varepsilon = \varepsilon_0 \simeq 8.85 \times 10^{-12} \text{ farad/m}$$

$$\mu = \mu_0 \simeq 4\pi \times 10^{-7} \text{ henry/m}$$

Thus, the velocity of the EM radiation in vacuum, c, is given by

$$c = \frac{1}{\sqrt{\varepsilon_0\mu_0}} \sim 3 \times 10^8 \text{ ms}^{-1}$$

This value is familiar to the readers as the velocity of light. The wavelength λ, frequency ν, and the velocity of the EM wave c are related such that

$$c = \nu\lambda \tag{2.2}$$

Other quantities generally associated with wave motion are the period
T $(1/\nu)$, wave number k $(2\pi/\lambda)$, and angular frequency $\omega(2\pi\nu)$.

The terms ε and μ in a medium can be written as $\varepsilon = \varepsilon_r\,\varepsilon_0$ and $\mu = \mu_r\mu_0$,
where ε_r is the relative permittivity (called the dielectric constant) and μ_r is
the relative permeability. Therefore the velocity of EM radiation in a medium
c_m can be expressed as,

$$c_m = \frac{1}{\sqrt{\varepsilon_r\varepsilon_0\mu_r\mu_0}} = \frac{1}{\sqrt{\varepsilon_r\mu_r}}\frac{1}{\sqrt{\varepsilon_0\mu_0}}$$

$$c_m = \frac{c}{\sqrt{\varepsilon_r\mu_r}} = \frac{c}{n} \tag{2.3}$$

That is, the velocity in a dielectric medium is reduced by a factor $\sqrt{\varepsilon_r\mu_r}$ compared to that in vacuum (frequency remains constant but the wavelength is reduced). This factor is referred to as refractive index (RI), n. That is,

Refractive index, n, of the medium $= \dfrac{\text{velocity of EM radiation in vacuum}}{\text{velocity of EM radiation in the medium}}$

The media we consider are normally nonmagnetic, and hence $\mu_r = 1$. Therefore,

$$n = \sqrt{\varepsilon_r}$$

In general, ε_r can be complex (and hence n), which indicates a lossy medium, that is, a portion of the wave is absorbed. The imaginary part is responsible for the absorption of the wave; hence, when ε_r is a real positive number the medium does not absorb EM waves. The refractive index of a medium depends on the wavelength of the radiation. The variation of refractive index with wavelength is called *dispersion*. The splitting of colors of a white light when it passes through a prism is due to this phenomenon.

2.2.1 Quantum Nature of Electromagnetic Radiation

The phenomenon of interference and polarization requires the EM radiation to behave like a wave, whereas for some interactions, like in the photoelectric effect, the radiation behaves like a particle. Thus, the EM radiation has a dual nature—wave and particle. The particulate nature of the EM radiation is explained in terms of quantum theory. According to quantum theory, EM radiation moves in space as discrete packets or quanta of energy propagating with the same speed and direction as defined by the wave theory. Each quantum of radiation—called a photon—has an energy e_ν related to the frequency such that

$$e_\nu = h\nu \tag{2.4}$$

where h is Planck's constant (6.63×10^{-34} Ws^2)
 Since $c = \nu\lambda$, the energy of a photon e_λ in terms of wavelength is given by

$$e_\lambda = \frac{ch}{\lambda}$$

In classical EM radiation the radiant energy is dependent on the square of wave amplitude, whereas in the quantum concept it depends on the number of photons and energy of each photon.

2.2.2 Thermal Radiation

Any object above absolute zero emits EM radiation. Thus, the objects we see around us, including ourselves, are thermal radiators! An ideal thermal radiator is called a black body, which emits radiation as per Planck's law:

$$M_\lambda = \frac{2\pi hc^2}{\lambda^5 \left[\exp\left(\frac{ch}{\lambda kT} \right) - 1 \right]} \qquad (2.5)$$

M_λ = Spectral radiant exitance in $Wm^{-2}\mu m^{-1}$
h = Planck's constant—$6.6256 \times 10^{-34} Ws^2$
c = Velocity of light—$2.9979 \times 10^8 \ ms^{-1}$
k = Boltzmann's constant—$1.3805 \times 10^{-23} \ JK^{-1}$
T = Absolute temperature in kelvin
λ = Wavelength in microns

A black body is an ideal surface that absorbs all the incident radiation regardless of the wavelength or direction of the incident radiation. For a given temperature and wavelength, no object can emit more energy than a black body. However, a real surface does not emit at this maximum rate. The emission from a real surface is characterized with respect to a black body through the term emissivity. Emissivity (ε) is defined as the ratio of radiant exitance of the material of interest (M_m) to the radiant exitance of a black body (M_b) at the same temperature. That is,

$$\text{Emissivity} = \frac{M_m}{M_b} \qquad (2.6)$$

Emissivity is a dimensionless quantity whose value lies between 0 and 1. In general, emissivity varies with wavelength. When emissivity of a body is less than 1 but remains constant throughout the EM spectrum, it is called a gray body.

2.2.3 Propagation of Electromagnetic Radiation from One Medium to Another

EM radiation travels in a straight line in any homogenous medium and may be referred to as a ray. A homogeneous medium is one in which the refractive index is constant everywhere. When an EM wave falls on a boundary between two lossless homogeneous media with different refractive indexes, a part of the wave is reflected back to the incident medium (Fresnel reflection) and the rest is transmitted to the second medium; however, it changes its direction (Figure 2.1). This change in direction or the "bending" of light rays as they pass from one medium to another is referred to as refraction. The ray propagating to the second medium is called the

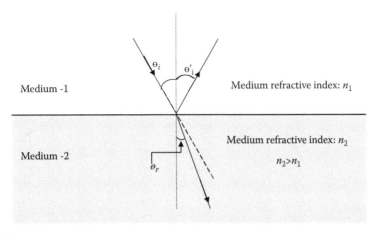

FIGURE 2.1

Schematics showing reflection and refraction of electromagnetic radiation. When the reflection takes place at a smooth surface, the angle of reflection θ_i' is equal to the angle of incidence θ_i. The angle of refraction θ_r follows Snell's law.

refracted ray. The angle of reflection θ_i' is equal to the angle of incidence θ_i, and the angle of refraction θ_r is given by Snell's law: $n_1 \sin \theta_i = n_2 \sin \theta_r$

$$\sin \theta_r = \frac{n_1}{n_2} \sin \theta_i \qquad (2.7)$$

where θ_i is the angle between the incident ray and the normal to the interface and θ_r is the angle between the refracted ray and the normal to the interface. The angle of refraction θ_r and the reflected and transmitted power depend on the angle of incidence θ_i and the refractive indexes n_1 and n_2 of the two media. In Equation 2.7, when

$$\frac{n_1}{n_2} \sin \theta_i = 1$$

then $\theta_r = \pi/2$, that is, the refracted ray emerges tangent to the boundary of separation as shown by ray b in Figure 2.2. The incidence angle at which this happens is called the *critical angle* (θ_c). Any further increase in incidence angle reflects back the total radiation to the incident medium (ray c). This phenomenon is called *total internal reflection*. As $\sin \theta_i$ cannot have a value greater than 1, for this condition to exist n_1 should be greater than n_2. That is, to have total internal reflection the ray has to pass from a denser to a rarer medium. (The subscript for the refractive index n is 1 for the incident medium and 2 for the medium to which the ray is refracted.)

2.2.4 Diffraction

We generally think of light as traveling in a straight line. Therefore, we expect a sharp geometric shadow when light encounters an obstacle in its

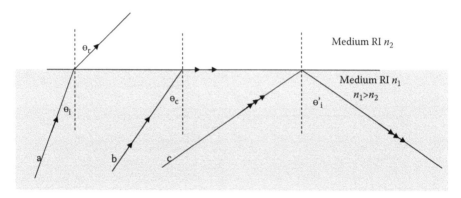

FIGURE 2.2
Schematics showing total internal reflection. θ_c, critical angle. Ray c undergoes total internal reflection.

path. However, one of the characteristics of waves is to bend around the "edges" they encounter. As the EM radiation also behaves like a wave, it bends around an obstacle—an edge, slit, hole, and so on—producing some spreading of energy in the geometrical shadow. Thus, the shadow produced by EM radiation is fuzzy and not sharp. This bending of waves around an obstacle is termed *diffraction*. Diffraction is an important physical aspect inherent in all wave phenomena. Diffraction by an aperture depends on λ/d, where λ is the wavelength and d is the aperture width (diameter). The effect is prominent when λ/d is large. The diffraction pattern for a slit or hole consists of a bright central maxima flanked on either side by secondary maxima and minima, with the intensity of each succeeding secondary maximum decreasing as the distance from the center increases (see Section 2.6). Diffraction plays an important role in limiting the resolving power of imaging instruments.

2.3   Some Useful Terminologies of the Imaging Systems

In this section, we shall explain some of the terms usually associated with imaging systems so that the reader will find it easier to follow later discussions. The basic concepts can be explained with reference to a simple lens. A simple lens is formed by two curved surfaces.

In a lens element with two spherical surfaces, the *optical axis*—also referred to as the *principal axis*—is a straight line that passes through both the centers of curvature of the lens surfaces. An imaging lens assembly usually has a number of centered lens elements. In a centered optical system, the centers of curvature of all the surfaces lie on a common line, which is referred to as the optical axis. For an axially symmetrical reflecting surface, the optical axis is the line joining the center of curvature and the vertex of the mirror.

The rays close and parallel to the optical axis (paraxial rays) converge at a point on the optical axis called the *focal point*. A plane at right angles to the optical axis and passing through the focal point is called the *focal plane*. *Focal length* is a fundamental parameter of an imaging system that defines the scale of the image, that is, the size of an object in the image plane with respect to the object size (of course it also depends on the distance between the object and the imaging system). For a thin lens, the focal length is the distance between the center of the lens and the focal point on the optical axis. However, a practical lens consists of a number of individual lens elements. Then, how do we define the focal length? For a multielement lens (referred to as a thick lens compared to a single thin lens), a ray parallel to the optical axis entering the optical system emerges from the system passing through the focal point. Now if the rays entering the system and those emerging from the system are extended until they intersect, the points of intersection will define a surface, called a principal surface. In a well-corrected optical system, the principal surfaces are spheres, centered on the object and the image. In the paraxial region, the surfaces can be treated as if they are planes and referred to as *principal planes* (Figure 2.3). We can consider that all refraction is taking place at the principal plane. There can be two such planes within the lens—one for the rays traveling from the left and another for the rays traveling from the right of the lens. The intersections of these surfaces with the optical axis are the *principal points*. The *effective focal length* (EFL) of a system is the distance from the

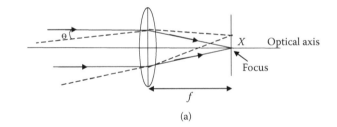

(a)

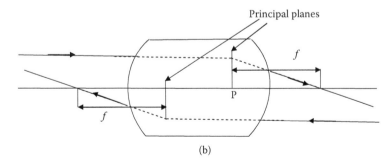

(b)

FIGURE 2.3
Concept of focal length f. (a) For a thin lens, the rays parallel and close to the optical axis converge to a point on the optical axis called the focal point. The distance between the center of the thin lens and the focal point is the focal length. (b) For a multielement lens, the distance between the principal point P and the focal point is the focal length f.

principal point to the focal point. However, locating the principal plane accurately in a multielement lens system is not easy. The focal length can be measured accurately from the displacement of image in the focal plane when a paraxial ray is tilted by a small angle. If x is the shift in the image point in the focal plane for a tilt in the ray incident on the lens through an angle $\theta°$ (Figure 2.3a), the focal length f is given by

$$f = \frac{x}{\tan \theta} \qquad (2.8)$$

The radiation incident on an imaging system passes through a number of refracting/reflecting surfaces. The finite size of these elements limits the amount of radiation reaching the image plane. Consider a single lens imaging a star. The radiation from the star falls on the lens and beyond. As the lens has a finite diameter, only those rays that are falling on the clear aperture of the lens are allowed to pass through. Here, the rim of the lens limits the irradiance at the image plane. In every optical system, there is one aperture that limits the passage of radiation through the system. The aperture that limits the bundle of rays finally reaching the image plane is called the *aperture stop*. Thus, the aperture stop controls the amount of radiance from the scene reaching the film or the detector located in the focal plane. The aperture can be a physical restraint (mechanical component) whose sole purpose is to limit the bundle of rays that passes through the optical system, or it could be the edge of an optical element itself. The aperture is usually centered on axis. In a multielement lens assembly, it is usually positioned between the lens elements. In a Cassegrain telescope, the clear aperture of the primary mirror itself acts as an aperture stop. The image of the aperture stop formed by all lenses preceding the stop is called the *entrance pupil*. The image of the aperture stop formed by all lenses following the stop is called the *exit pupil*. If the aperture stop is at, or in front of, the first optical system, then the entrance pupil is at the same location as the aperture stop. In a single lens element or a simple telescope, like the Newtonian, the clear entrance aperture that allows the light rays itself forms the entrance pupil. A ray from the object that passes through the center of the aperture stop and appears to pass through the centers of the entrance and exit pupils is called the *chief ray* or principal ray. A ray from the object passing through the edge of the aperture stop is called a *marginal ray* (Figure 2.4).

Another stop associated with an imaging system is the *fieldstop*. The aperture that limits the extent of the scene that is imaged (detected) is called the field stop. In other words, it decides the field of view (FOV) of the optical system. In Earth imaging cameras, this is usually the detecting system (film, CCD, etc.) placed in the image plane. In such cases, the projection of the detector at the object space defines the extent of the area imaged by the camera.

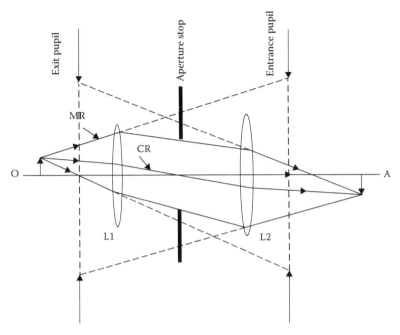

FIGURE 2.4
Schematics showing the pupils and chief and marginal rays. OA, optical axis; CR, chief ray; and MR, marginal ray. The image of the aperture stop (AS) by lens L1 is the entrance pupil. The image of AS by lens L2 is the exit pupil.

Another important parameter for an imaging system is the *f-number*, also referred to as f-stop, represented as f/no or F/#:

$$F_{/\#} = \frac{f}{D}$$

where f is the EFL of a focusing optical system and D the system's circular entrance pupil diameter. It is usually written like F/5, which means the entrance pupil diameter is 1/5th of the EFL. The irradiance at the image plane is inversely proportional to the square of the f/no. The systems that have a large f/no are called slow systems (as the light-gathering capability is low), and one with a smaller f/no is called a faster system.

2.4 Aberrations

In a perfect imaging system, all the rays originating from an object point cross at a single image point in the image plane. For a real optical system, this does not happen due to aberrations. Aberrations are described in terms of the amount by which a geometrically traced ray departs from a *desired* (ideal) location in the image formed by an optical system (Smith 1966). The ideal location of the image points are based on *first-order optics*. The refraction of light follows Snell's law. That is, in the equation $n_1 \sin \theta_1 = n_2 \sin \theta_2$, $\sin \theta$ can be expanded in terms of θ (expressed in radians) as an infinite Taylor series as follows:

$$\sin \theta = \theta - \frac{\theta^3}{3!} + \frac{\theta^5}{5!} - \frac{\theta^7}{7!} + \ldots$$

For a small value of θ, the higher order terms can be neglected and we can write $\sin \theta = \theta$. Therefore, Snell's equation can be written as follows:

$$n_1 \theta_1 = n_2 \theta_2 \tag{2.9}$$

When rays are traced through an optical system using the aforementioned approximation, it is termed first-order optics. Sin $\theta = \theta$ is strictly valid only for an infinitesimally small value of θ. That is, all the angles of incidence, refraction, and reflection should be such that the rays are close to the optical axis. Therefore, this is also called *paraxial optics*, also referred to as *Gaussian optics*. For aberration measurement, we can consider the location of the image as given by the first order or paraxial law of image formation as the reference. Hence, the aberrations are measured by the amount by which rays (nonparaxial) deviate from the paraxial image point.

To account for nonparaxial rays, we have to include higher order terms in the expansion of $\sin \theta$. If we also include the θ^3 term (i.e., $\sin \theta = \theta - \frac{\theta^3}{3!}$), we have *third-order optics*. The third-order theory gives five monochromatic aberrations: spherical aberration, coma, astigmatism, field curvature, and distortion. It may be noted that in actual practice these aberrations could occur in combinations rather than alone. These defects are also referred to as *Seidel aberrations*. Without getting into the analytical details, we shall try to understand the behavior of each of these aberrations.

2.4.1 Spherical Aberration

A spherical lens or mirror does not bring all the rays parallel to the optical axis at a single point in the image plane. The rays that are farther away from the optical axis are brought to focus closer to the lens than those closer to the optical axis (paraxial focus). As the ray height increases from the optical axis, the focus shifts farther and farther away from the paraxial focus. *Longitudinal spherical aberration* is the distance between paraxial and marginal focal points. *Transverse spherical aberration* is the radius of the circle produced by the marginal ray at the paraxial focal plane (Figure 2.5a). The spot size is minimum at a certain location between the paraxial focus and the marginal focus and is called the *circle of least confusion*. The image of a point

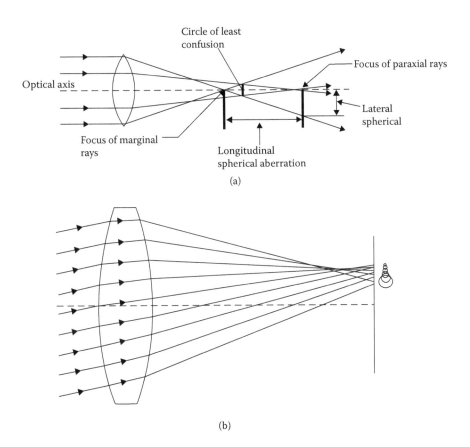

FIGURE 2.5
Schematics showing (a) spherical aberration and (b) coma.

formed by a lens having spherical aberration is usually a bright dot surrounded by diffuse light. The effect of spherical aberration on an extended image is to reduce the contrast of the image, thereby producing a blurred effect on edges.

2.4.2 Coma

Comatic aberration appears when a light beam passes through the optics obliquely. When a bundle of oblique rays is incident on an optical system with coma, the rays passing through the outer portions of the lens/mirror are imaged at different heights than those passing through the center portion. In addition, rays that pass through the periphery (outer zone) form a larger image than the rays that are closer to the axis. That is, the optical system has different degrees of magnification for the bundle of rays passing through different zones (ring-shaped areas concentric with the optical axis) of the optics. Each concentric zone of the optics forms a ring-shaped image (comatic circle) whose sum produces a cone-shaped (comet-like flare) blurred image (Figure 2.5b).

An optical system corrected for both spherical aberration and coma is called an *aplanatic* system.

2.4.3 Astigmatism

For an optical system with astigmatism, the rays from an off-axis point lying in two orthogonal planes produce different line images at different distances from the lens. The planes considered are the *tangent plane,* which contains the optical axis and the object point (also called the meridional plane), and the plane perpendicular to the tangent plane, containing the chief ray, referred to as the *sagittal plane* or radial plane. Thus, in the presence of astigmatism the image of an off-axis point is not imaged as a point but as two separate lines (Figure 2.6a). As one moves from tangential to sagittal focus, the image is an ellipse that progressively reduces its size and approaches a circle of least area and further expands to an ellipse. The circle of least confusion is located halfway between the two foci. The separation between the two foci gives the magnitude of astigmatism.

2.4.4 Distortion

Even if a perfect off-axis image point is formed, its location may not be as predicted by the first-order theory. This happens because the focal length of the lens for various points on the image plane (and hence the magnification) varies. Because of this, the image of an object that has a straight side is seen curved in (pincushion) or curved out (barrel). Here, every point may be in focus, but points on the image will not be in proper locations (Figure 2.6b). The distortion is expressed either as a percentage of the paraxial image height or as an absolute value in lens units.

2.4.5 Curvature of the Field

A planar object normal to the axis will be imaged into a curved surface instead of a plane (Petzval field curvature) (Figure 2.6c). That is, if the detector is a plane, as in the case of a film, a charge-coupled device array, and so on, a plane object cannot be in perfect focus at the center and edge simultaneously.

The aforementioned five aberrations, which are applicable to both refractive and reflective systems, are called monochromatic aberrations. In addition to these, when a refractive element is used there is an additional aberration called chromatic aberration.

2.4.6 Chromatic Aberration

The refractive index of a material is a function of wavelength. Therefore, every property of the optical system using a refractive element is dependent on the refractive index. Generally, for most of the glass materials

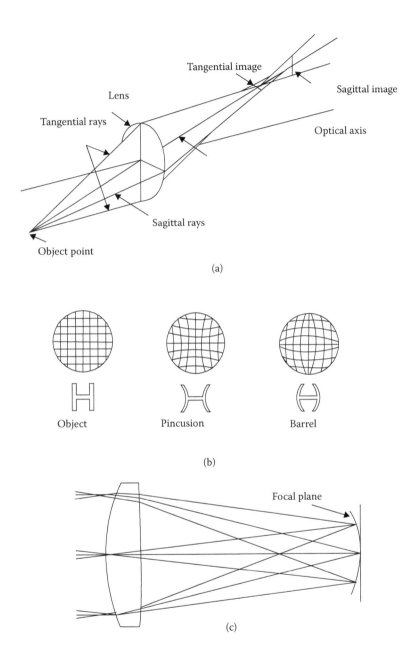

FIGURE 2.6
Schematics showing (a) astigmatism, (b) distortion, and (c) field curvature.

used in an optical system the refractive index is higher for shorter wavelengths. Consider the example of visible light where $n_{red} < n_{yellow} < n_{blue}$. This makes the shorter blue wavelength bend more compared to red wavelength at each surface. Therefore, for a simple convex (positive) lens

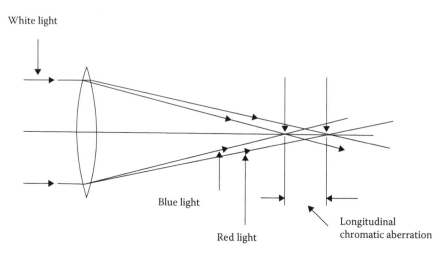

FIGURE 2.7
Schematics showing chromatic aberration in a lens.

blue light is focused closer to the lens compared to red light (any two wavelengths can be considered; blue and red are taken as examples). The distance between the two focal points is called *axial* (longitudinal) *chromatic aberration* (Figure 2.7). The magnification also depends on the refractive index. The difference in image height between red and blue is called *lateral chromatic aberration*.

2.5 Wave Optics

What we discussed in Section 2.4 considers geometrical optics, wherein EM radiation is considered as a ray that travels in a straight line. We shall now consider how aberrations are represented in wave optics. Consider a point source in an isotropic medium. The point source emits light uniformly in all directions. The locus of the points having the same phase, called the wave front, lies on a sphere and we have a spherical wave. Thus, in a three-dimensional space the spherical wave front lies on the surface of a sphere. In two dimensions, we may represent the spherical waves as a series of circular arcs concentric with the source, as shown in Figure 2.8. Each arc represents a surface over which the phase of the wave is constant. In general, we may say that the wave front emerging from (or converging to) a point is spherical. The direction normal to the wave front indicates the direction of motion of waves; that is the ray we refer to in geometrical optics. Far away from the source, the spherical wave tends to be a plane wave, that is, wave fronts are parallel to each other; in geometrical optics these are called parallel rays.

Thus, when we observe a star we have plane waves entering the telescope and in an ideal aberration-free system the radiation emerges from the exit pupil

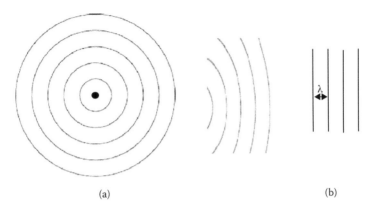

(a)   (b)

FIGURE 2.8
Schematics showing wave fronts. The directions of propagation of the rays are at right angles to the wave fronts. As one moves further from the source (a), the wave front appears planar (b), which is what we call parallel rays.

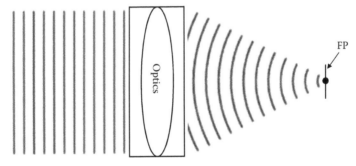

FIGURE 2.9
Image formation by an ideal imaging system, with a plane wave being transformed to a perfect spherical wave.

as a spherical wave with its center at the image point (Figure 2.9). However, in a practical optical system the wave front emerging from the exit pupil deviates from an ideal spherical surface. Wave front aberration is defined as the optical path difference (OPD) between the actual wave front and the ideal wave front (Figure 2.10). Wave front aberration gives the aggregate effect of the optical system on image quality. This is a very useful parameter for studying the design, fabrication, and testing of optical systems.

One way of expressing wave front error is in terms of OPD, which is the maximum departure from a perfect reference sphere, referred to as peak to valley (P-V) wave front error. This correlates well with image quality only when the shape of the wave front is relatively smooth, but this will not represent the true nature of image degradation at all times. For example, if we have a perfect reflecting surface with a bump or valley in a small localized area, its effect on the overall image quality will be negligible even if the P-V OPD of the valley/bump is large. A better way to evaluate the extent of image degradation

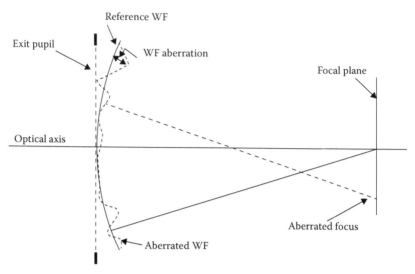

FIGURE 2.10
Schematics showing wave front aberration.

caused by wave front deformations is to express the OPD by its deviation from the reference sphere averaged over the entire wave front as root mean square (RMS) wave front error, usually expressed in units of the wavelength of light.

The wave front aberration values can be represented in the plane of the exit pupil, normalized to unit circle with zero mean, defined as a surface for which the sum of deviations on either side—opposite in sign one to another—equals zero (Vladimir 2006). This can then be expressed as a weighted sum of power series terms (polynomial) that are functions of the pupil coordinates. Zernike polynomials are normally used for describing the shape of an aberrated wave front. The Zernike polynomials are usually defined in polar coordinates (ρ,θ), where ρ is the radial coordinate ranging from 0 to 1 and θ is the azimuthal component ranging from 0 to 2π. Without getting into the mathematical details, we may state that Zernike polynomials are made up of terms that are of the same form as the types of aberrations often observed in optical tests, and hence the terms can be correlated to the type of aberration.

Optical system design aims at minimizing various aberrations. This is accomplished by using different lens materials, curvature, or even different surface profiles (such as aspherics). In actual practice, the designer chooses to optimize one of the parameters that represent the image quality of the optical system, at the same time producing a design that can be manufactured and is consistent with the available budget and schedule. This is a major challenge for an optical designer. With modern computer-aided optical design tools, it is possible to achieve a design such that most of the aberrations are minimized and the performance is close to the theoretical limit due to diffraction, that is, a diffraction-limited design. In Section 2.6, we discuss some of the performance parameters usually considered for an imaging system.

2.6 Image Quality Evaluation

Here, we shall consider the merit functions usually used to compare the quality of images from different imaging systems. Assume a hypothetical case of an imaging optics with no aberrations. The image of a point source is decided by the diffraction effect. A plane wave incident at the circular entrance pupil produces a bright disk (*airy disk*) followed by a number of concentric sharp dark rings alternating with broader bright rings of decreasing brightness called *airy pattern* (Figure 2.11). This is the Fraunhofer diffraction at a circular aperture. The airy pattern is the point spread functions (PSFs) for a perfect optical system, based on circular elements, that is, the two-dimensional distribution of light intensity in the telescope focal plane for a point source. The diameter d of the first minima is given by

$$d = 2.44 \frac{\lambda}{D} f = 2.44\lambda \left(F_{\#} \right) \qquad (2.10)$$

where D is the diameter of the entrance pupil, f the focal length, and λ the wavelength of observation. The airy disk diameter is dependent on F/# and not on D alone. The next two dark rings occur at radii of 2.23 and 3.24 times λ(F/#), respectively. The central disk of the airy pattern carries about 84% of the intensity and 91% is within the first bright ring, whereas the rest 9% is distributed in the successive bright rings in decreasing order.

What are the ways of assessing the effect of wave front aberrations on image quality? The Fraunhofer diffraction is the fundamental limit of performance of an imaging system with a circular aperture. Let us now understand what happens to the airy pattern when the system has aberrations. As the amount of radiance forming a point image is the same with and without aberrations, the irradiance at the center of the images has to decrease and spill over to adjacent rings when the image size increases. Therefore, one way of evaluating the image quality is to assess how far the PSF of the telescope under test deviates from the Fraunhofer diffraction pattern. One of the parameters that decide the departure from an ideal PSF is *encircled energy*, which is the ratio of the energy inside a circle of radius, say, d, and centered on the diffraction pattern to the total energy in the PSF produced by the system. For a given radius this will be maximum for the airy pattern, and hence the aberration always tends to decrease the factor from the value obtained with an airy pattern.

Another way of expressing the effect of wave front aberrations on image quality is the *Strehl ratio*. The Strehl ratio is defined as the on-axis irradiance of the berated system divided by the on-axis irradiance that would be formed by a diffraction-limited system of the same f-number (Figure 2.12). The ratio indicates the level of image quality in the presence of wave front aberrations. A system is regarded as well corrected if the Strehl ratio is greater than or equal to 0.8, which corresponds to an RMS wave front error less than or

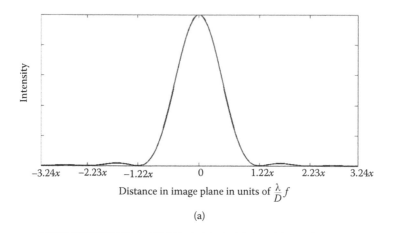

Distance in image plane in units of $\frac{\lambda}{D} f$

(a)

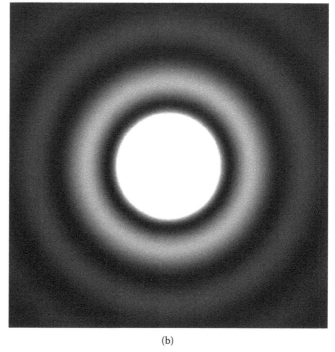

(b)

FIGURE 2.11
(a) Intensity distribution of airy pattern of a circular aperture. (b) Two-dimensional intensity distribution. In the figure, 0 corresponds to optical axis location in the image plane.

equal to $\lambda/14$ (Wyant and Katherine 1992). This is not a good measure of image quality for optical systems with large wave front errors.

The PSF gives the system performance in the spatial domain, that is, one finds out the radiance distribution in two-dimensional space. Another way of

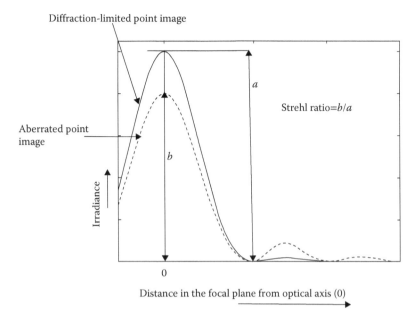

FIGURE 2.12
Schematics defining the Strehl ratio.

looking at the image quality is to assess how the system responds to the spatial variation of targets in object space. We shall discuss this aspect in Section 2.7.

2.7 Modulation Transfer Function

Fourier transform of the PSF describes how the optical system behaves in the frequency domain. Here the frequency should not be confused with the terms used in the communication/electrical engineering discipline, wherein we talk about the amplitude variation of an oscillation in the time domain represented as cycles per second. Here the frequency refers to the irradiance distribution of targets in the image space, usually referred to as line pairs (cycles) per unit distance. Consider alternate black and white targets (Figure 2.13); a line pair is one black and one bright target and if the combined width is d mm the spatial frequency is $1/d$ line pair (or cycles) per millimeter. The Fourier transform of the PSF is called the optical transfer function (OTF), which gives the irradiance as a function of spatial frequency. Without getting into the mathematical details, we shall explain how the OTF is used in understanding the performance of an imaging system. In general, the OTF is a complex number and thus it has a real term and an imaginary term. The real term is the modulation transfer function (MTF) and the imaginary part is the phase transfer function (PTF) or the change in phase position as a function of spatial frequency. If the

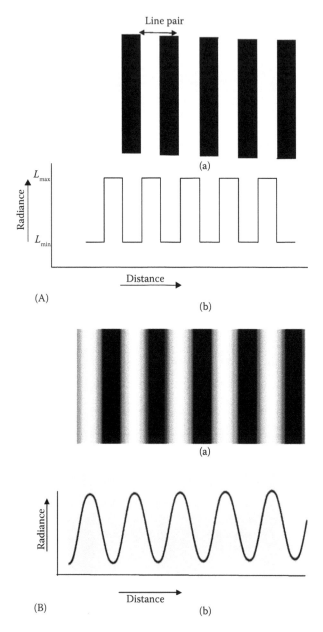

FIGURE 2.13
Schematics showing spatial frequency patterns: (A) square wave and (B) sine wave. In both cases, part (a) represents the spatial pattern and part (b) the radiance distribution.

PTF is linear with frequency, it represents a simple lateral displacement of the image and in the worst case a phase shift of 180° producing a reversal of image contrast, wherein bright and dark patterns are inverted.

Before we proceed further, it is useful to familiarize with the concept of contrast ratio. The difference in radiance of the objects in a scene plays an important role in their detection.

The contrast is usually referred with respect to two adjacent areas. If L_{max} and L_{min} are the maximum and minimum radiance between adjacent areas (Figure 2.13), the contrast ratio C_R is given by

$$C_R = \frac{L_{max}}{L_{min}} \tag{2.11}$$

The contrast modulation (modulation depth) C_M is given by

$$C_M = \frac{L_{max} - L_{min}}{L_{max} + L_{min}} \tag{2.12}$$

where L_{max} and L_{min} are the maximum and minimum values of the exitance (irradiance) of the white and black targets at the object (image) space. The contrast in the image will be less than that of the scene (object space). This is because the imaging system reduces the contrast. The degradation of contrast by the imaging system is represented by the MTF. That is, the MTF is the ratio of the output to input modulation depth as a function of spatial frequency v.

The MTF describes how the amplitudes of different spatial frequency components are modified by the optical system, that is, the contribution of the imaging system to reduce the contrast from the object space to image space. Objects with low levels of contrast are difficult to discern in the presence of other degradations like noise and so on. The designer tries to get the best possible MTF at the highest spatial frequency of interest. The MTF is usually normalized to unity at zero spatial frequency. Theoretically, the MTF can vary from 0 to 1. For a circular aperture, the MTF at low spatial frequencies is close to 1 (i.e., 100%) and generally falls as the spatial frequency increases until it reaches zero at the cutoff frequency. Taking an analogy from electrical engineering, we may consider that the imaging optics behaves like a "low pass filter" for spatial frequencies. When the contrast value reaches zero, the image does not show any features and is seen as a uniform shade of gray. An optical system cannot perform better than its diffraction-limited MTF; any aberrations will lower the MTF.

Mathematically, the MTF applies to a sinusoidal input target. That is, the intensity of the bright and dark bars changes gradually in a sinusoidal fashion. The system MTF is then the product of the MTF of individual components like optics, detector, image motion, and even the atmosphere. Thus, the overall MTF of an imaging system is given by

$$MTF_{System} = MTF_{Optics} \times MTF_{Detector} \times MTF_{Image\ motion} \times MTF_{Atmosphere} \times \cdots$$

For a perfect optical system, the frequency at which MTF = 0, called the cutoff frequency v_c, is given by

$$\frac{1}{\lambda(F/\#)}$$

So, as the F/# decreases, the cutoff frequency increases. Therefore, theoretically a faster system (low F/#) gives a better MTF compared to a slower system. However, in practice as the fabrication complexity and aberrations are relatively less for a slower system the realizable MTF of a system may not strictly follow the F/# advantage as stated earlier.

For MTF measurement in the laboratory, one generally uses a square wave target (i.e., the maximum radiance is the same for all the bright bars [L_{max}], and the minimum radiance L_{min} is the same for all the dark bars), as it is relatively easy to fabricate compared to a sine wave target. When a square wave target is used, the transfer function is generally referred to as the *contrast transfer function* (CTF). However, the square wave–derived CTF and the sine wave–derived MTF are not equal. As the camera specifications are normally given in terms of sine wave MTF, we should be able to convert the CTF measured with a square wave target to its equivalent sine wave MTF. Coltman (1954) has given an analytical relationship between the CTF of a square wave pattern of infinite extent and a sine wave MTF as

$$\text{MTF}(v) = \frac{\pi}{4}\left[\text{CTF}(v) + \frac{\text{CTF}(3v)}{3} - \frac{\text{CTF}(5v)}{5} + \dots \right] \qquad (2.13)$$

Thus, the CTF is generally higher than the MTF for the same spatial frequency. The Coltman formula is strictly not applicable to discrete sampling systems, as it assumes that the imaging system is analog (e.g., photographic film camera), and the number of target bars is infinite (Nill 2001). Nevertheless, this is a good approximation to convert measured CTF to MTF. The values of the second and higher terms in Equation 2.13 are generally very low as the CTF falls as frequency is increased. For evaluation frequencies greater than one-third of the cutoff frequency, only the first term needs to be considered, resulting in a simple relationship:

$$\text{MTF}(v) = \frac{\pi}{4}\text{CTF}(v) \text{ for } v > \frac{v_c}{3}$$

To compare the quality of various imaging optics, the MTF is the best practical tool. The MTF also plays a role in the radiometric accuracy of the final data reaching the users. We shall deal with this aspect in Section 4.4.1.

2.8 Source of Electromagnetic Radiation for Imaging

Solar radiation is the primary source of energy for imaging from space in the optical and near IR wavelength region (about 0.4–2.5 µm). The Sun's radiation covers the ultraviolet, visible, IR, and radio frequency regions and the maximum exitance occurs at around 0.55 µm, which is in the visible region. However, in passing through the atmosphere the solar radiation is scattered and absorbed by gases and particulates. There are certain spectral regions where the EM radiation is transmitted through the atmosphere without much attenuation, and these are called *atmospheric windows*. Imaging of the Earth's surface is generally confined to these wavelength regions, that is, the 0.4–1.3, 1.5–1.8, 2.0–2.26, 3.0–3.6, 4.2–5.0, and 7.0–15.0 µm and 1–30 cm wavelength regions of the EM spectrum. When observing the Earth at wavelengths beyond a few micrometers, emission of the Earth becomes a dominant source for passive remote sensing. That is, the observation made beyond a few micrometers is essentially on the basis of temperature distribution and/or emissivity variation.

2.9 Radiometric Consideration

The data from an Earth observation camera is more than just an image. The camera is essentially a radiometer, that is, an instrument that gives quantitative information on the radiation given out by emission and/or reflectance by targets. There are five quantities that are the basis of radiometry, which are given in Table 2.1. From the standpoint of radiometry, the optical system transfers the radiance from the target to the focal plane. We shall now present the relationship between the target radiance and the instrument parameters. Consider an optical system with diameter D transferring radiance from the target to the detector at the focal plane (Figure 2.14).

The radiant flux (power) delivered to the detector is given by

$$\phi_d = \frac{\pi}{4} O_e \Delta\lambda L_\lambda \beta^2 \ D^2 \text{watts}$$

(2.14)

where
L_λ is the target radiance (W·m^{-2}·sr^{-1}·µm^{-1})
$\Delta\lambda$ is the spectral bandwidth of the radiation to be measured (µm)
O_e is the optical efficiency—transmittance of the optical system including atmosphere (<1)
β is the system geometric FOV; in remote sensing, usually referred to as instantaneous FOV (radians)

TABLE 2.1

Radiometric Quantities and Units

	Quantity	Symbol	Unit	Definition
1.	Radiant energy	Q	Joule	Quantity of energy carried. A measure of the radiation to do work.
2.	Radiant flux (radiant power)	Φ	Watts	Radiant energy emitted or incident on a surface per unit time. This is the rate of flow of energy.
3.	Irradiance Exitance	E M	Watts m^{-2} Watts m^{-2}	Radiant flux falling per unit area of the surface. Symbol M is used for radiant flux emitted by unit area.
4.	Radiant intensity	I	Watts sr^{-1}	Radiant flux leaving per unit solid angle, in a specified direction.
5.	Radiance	L	Watts m^{-2}sr^{-1}	Radiant flux per unit projected area and per unit solid angle

Source: Joseph, G., *Fundamentals of Remote Sensing,* Universities Press (India) Pvt Ltd., Hyderabad, India, 2005. With permission.

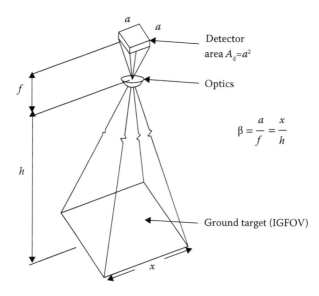

$$\beta = \frac{a}{f} = \frac{x}{h}$$

FIGURE 2.14

Imaging geometry of an optical system. (Reproduced from Joseph, G., *Fundamentals of Remote Sensing,* Universities Press (India) Pvt Ltd., Hyderabad, India, 2005. With permission.)

It may be noted that the radiant flux at the detector (i.e., the detector output) is independent of height, h, of the camera from the target. This is because as the height h increases the area of the target covered increases for the same β (assuming the target size projected totally fills the detector) and hence compensates for the increased distance.

If A_d is the detector area, $\beta^2 = A_d/f^2$, where f is the focal length; the aforementioned equation can be rewritten as follows:

$$\phi_d = \frac{\pi}{4} O_e \Delta\lambda L_\lambda \frac{A_d}{f^2} D^2 \text{watts} \qquad (2.15)$$

Since f/D is the f-number (F/#) of the system, the irradiance at the detector $E_d(\Phi_d/A_d)$ is given by

$$E_d = \frac{\pi}{4} \frac{O_e \Delta\lambda}{\left(F/_{\#}\right)^2} L_\lambda \ \text{W} \cdot \text{m}^{-2} \qquad (2.16)$$

The aforementioned expression gives on-axis image irradiance. A more exact expression for the irradiance in the image plane is given by McCluney (1994):

$$E_d(\theta) = \pi O_e \Delta\lambda L_\lambda \left[\frac{1}{1 + (2\ F/_{\#}\ )^2}\right] \cos^4\theta \ \text{W} \cdot \text{m}^{-2} \qquad (2.17)$$

where θ is the angle of image location with respect to the optical axis. For the image along the optical axis, $\theta = 0$.

When F/# is large, which usually is the case with Earth observation systems, Equation 2.17 can be approximated to

$$E_d(\theta) = \frac{\pi O_e \Delta\lambda L_\lambda}{4(F/_{\#})^2} \cos^4\theta \ \text{W} \cdot \text{m}^{-2} \qquad (2.18)$$

If we make observation of the target for a time τ seconds (integration time), then the energy per unit area at the detector is given by

$$Q_d(\theta) = \frac{\pi O_e \Delta\lambda L_\lambda}{4(F/_{\#})^2} \tau \cos^4\theta \ \text{J} \cdot \text{m}^{-2} \qquad (2.19)$$

This expression is convenient to evaluate the output of the detector, as the detector responsivity is usually given in V/(μJ/cm²).

References

Born, M. and E. Wolf. 1964. *Principles of Optics*. Pergamon Press, Oxford, United Kingdom.

Coltman, J. W. 1954. The specification of imaging properties by response to a sine wave target. *Journal of the Optical Society of America*. 44: 468–469.

Joseph, G. 2005. *Fundamentals of Remote Sensing*. Universities Press, Hyderabad, India.

McCluney, W. R. 1994. *Introduction to Radiometry and Photometry*. Artech House, Boston.

Nill, N. B. 2001. Conversion between Sine Wave and Square Wave Spatial Frequency Response of an Imaging System. http://www.dtic.mil/dtic/tr/fulltext/u2/a460454.pdf (accessed on January 28, 2014).

Smith, W. J. 1966. *Modern Optical Engineering*. McGraw-Hill, NY.

Vladimir, S. 2006. http://www.telescope-optics.net/aberrations.htm (accessed on January 26, 2014).

Wetherell, W. B. 1980. *Applied Optics and Optical Engineering*. Vol. VIII. Academic Press, NY.

Wyant, J. C. and K. Creath. 1992. Basic wavefront aberration theory for optical metrology. In *Applied Optics and Optical Engineering*. Vol. Xl. ed. Robert R. S. and J. C. Wyant. Academic Press, NY.

3

Imaging Optics

3.1 Introduction

The quality of the imagery reaching the end user from an Earth observation camera depends primarily on the imaging optics, the focal plane detector array and associated electronics, and the processing (applying corrections, map projection, etc.) of the raw data. The function of imaging optics is to transfer the radiance from the object space to image plane. In doing so, the optical system should ensure geometric and radiometric fidelity of the image, that is, the ability to maintain the shape, orientation, relative dimension, and relative radiance values as in the object space. The collecting optics for any imaging system is chosen from one of three basic configurations, which are (1) all refractive; (2) all reflective; and (3) a combination of reflective and refractive, called catadioptric. All three configurations have been used in spaceborne Earth imaging cameras. The selection of a particular configuration for a mission primarily depends on (1) the spectral range to be covered, (2) total field of view (FOV), and (3) aperture size. A variety of optical design software is available to optimize the configuration of imaging optics. A designer generally tries to get the best image quality within the FOV of interest and at the same time does a tolerance analysis, that is, how the image performance varies with changes in mechanical parameters and temperature change, to ensure practical realizability of such a system. For a spaceborne optical system, the design and system realization should take into account a wide range of environmental loads during fabrication, launch, and in space. These include mechanical and thermal stresses, and the radiation environment in space. It is required to pay utmost attention to optomechanical design to ensure that the shape and position of the functional elements of the system are unaffected under these loads, so that the system performance requirements are satisfied. The stringent mechanical and optical stability requirements to ensure high-quality imagery make the design and realization of the spaceborne imaging system a real challenge. In this chapter, we shall discuss some of these aspects.

3.2   Refractive Optics

Refractive optics, usually referred to as the lens, has the advantage of covering a very wide FOV. For example, the large-format camera (LFC) flown on board the space shuttle with a single lens covered a field angle of 80°. Such a large coverage with a single collecting optics is not possible with reflective optics and difficult with catadioptric systems. The major limitation of a lens is the spectral coverage. It is impossible with a single refractive optics to cover right from the visible to thermal infrared (TIR) region. Apart from the difficulty of correcting aberrations for a large spectral region, there is no material that has good transmission over such a broad wavelength region.

An imaging lens assembly is made of a number of lens elements. The lens elements allow lens designers to reduce various aberrations by suitably choosing the material, shape, and curvature. In other words, the total power of the collecting optics is suitably divided among the lens elements to get the best performance. Whereas this is the general philosophy of realizing an imaging camera lens, a lens for a spaceborne camera demands much more detailed attention for the selection of various parts and the fabrication methods.

Space radiation has a major effect on the glass elements and the coatings. The space radiation environment includes electromagnetic radiation extending from ultraviolet (UV) to X-rays, primarily from the Sun, charged particles, mainly electrons and protons that are trapped by the Earth's magnetic field in the Van Allen Belt, and cosmic radiation covering protons to heavier nuclei. This radiation in turn can produce secondary radiation by interacting with the spacecraft material. The UV radiation affects only those components that are directly exposed to space, because inside the assembly they are absorbed by the intervening material. The amount of radiation encountered depends on the orbit. We shall discuss more details on the space environment in Chapter 10. The interaction of this radiation with optical materials can cause various types of radiation damage such as transmission degradation (solarization), refractive index change, dielectric breakdown, stress buildup, and radio luminescence (Czichy 1994). The damage depends on the glass composition and the types of radiation and the duration over which the material is exposed to the radiation. The damage either can be superficial or may extend to the inside.

The ionizing radiation induces the formation of color centers that give rise to absorption bands causing reduced transmission of the lens system, which in turn decreases the radiance throughput of the optics and hence the signal to noise ratio. The effect is more prominent in the lower wavelength spectrum of the visible region, as can be seen in Figure 3.1. The changes in refractive index produce changes in the focus of optical systems (depending on how sensitive the optical design is with respect to variation in the refractive index), thereby reducing the resolution of the image generated. The front element of the lens, which is exposed to the outside, will have the maximum

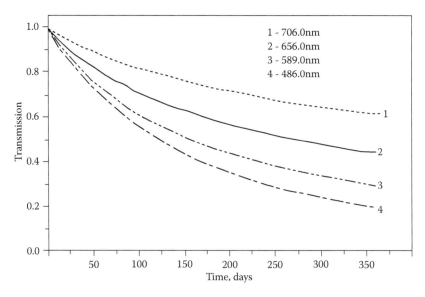

FIGURE 3.1
Model simulation of transmission degradation of a 5-cm thick BK7 block in space under 100 krad/year radiation for four wavelengths. (From Fruit et al., Measuring space radiation impact on the characteristics of optical glasses, measurement results and recommendations from testing a selected set of materials, *Proceedings of SPIE*, 4823, 132–141, 2002. Reproduced with permission from SPIE.)

impact due to radiation, whereas for other lens elements the effect is less severe due to the shielding of the preceding lens elements and the lens housing material as well as other packages surrounding the camera. An early National Aeronautics and Space Administration (NASA)-sponsored study shows that sapphire has the least effect from solar radiation followed by fuse silica (Firestone and Harada 1979). Boies et al. (1994) studied the effect of electron radiation on different glasses and concluded that fused silica (Corning 7958) samples did not discolor, and it was selected as the material of choice for the TOPEX LRA retro reflectors. They also found that MgF_2 antireflective coating of about 133-nm thickness was not susceptible to any degradation due to electron radiation.

Another important aspect to be taken care of in spaceborne lenses is the effect of air on vacuum change. The refractive index of air is greater than that of vacuum. Due to this finite difference in refractive index, when used in the high vacuum environment prevalent in space, optical systems using refractive components will show a change in focal length, compared to air. Therefore, on ground the detectors have to be offset in the focal plane so that in the operating environment of high vacuum the detectors are at the best focus. Another possibility is to fill the lens assembly with helium gas during alignment and measurements of various parameters in the laboratory. Helium has a refractive index of 1.000033, very close to the refractive index of vacuum, as against the refractive index of 1.000273 for air. The lens should have provision to vent out the inside air to space when it is in orbit. This can be achieved by having a special provision on the lens barrel, such as carrying a valve with a diaphragm that bursts at a predetermined differential pressure.

Refractive optics has been used in a number of Earth observation missions starting from TIROS1. The Landsat 1 BBV camera also used a lens as the imaging optics. These were the systems where one exposure covered one picture frame. The first operational multispectral camera system, from Sunsynchronous orbit, using refractive optics with charge-coupled device (CCD) arrays, is the IRS LISS 1&2 cameras. Although the LISS cameras will be discussed in Chapter 6, we shall describe here the design considerations for an imaging lens for spaceborne cameras taking the LISS lenses as an example. The four spectral bands of the LISS 1/2/3 cameras use separate refractive collecting optics for each band. In such a scheme, the lenses can be designed to get optimized performance for each band covering a narrow spectral range. In any multispectral imaging camera, one of the basic requirements is to ensure that all the bands look at the same ground pixels concurrently or view the ground pixels with a fixed known bias; this setting should be maintained at all environmental conditions. This demands that the four lenses imaging in four spectral bands should have the same focal length and distortion characteristics and also should have the same behavior when the lenses are subjected to various environmental conditions. In addition, the lenses have to be as athermal as possible so that temperature control of the lens on board is realizable to a high degree of tolerance. This is a great challenge for the lens manufacturer. Matra Defense—DOD/UAO—have developed, produced, and tested a number of LISS 1/2/3 lenses along with their interferential filters meeting these requirements (Lepretre 1994). We shall briefly discuss the details of the LISS lens realized by Matra.

The LISS lenses are derived from a double Gauss concept, consisting of nearly symmetric arrangements of elements about a central stop. Figure 3.2 gives the optical schematics of a typical double Gauss lens. The Matra design consists of eight lens elements. The last element, which is closer to

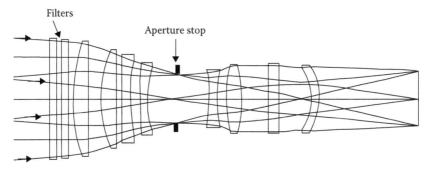

FIGURE 3.2
Schematics showing optical layout of a typical double Gauss lens with spectral band limiting filter.

the image plane, has a low power and its position can be adjusted to match the focal length of the four lenses in a camera. For example, for the IRS LISS 3 lens of about 350 mm effective focal length (EFL), a 100 μm variation of L7–L8 airspace produces a 20 micron change in EFL (Lepretre 1994). The spectral selection is carried out by an interference filter kept in front of the lens assembly. To protect the lens assembly against radiation and thermal effect, a parallel plate optical window of suitable thickness fabricated from synthetic fused silica optical glass material (Suprasil) is placed just before the interference filter, thus forming the first element of the lens assembly. The overall optical design is optimized such that all four lenses in a group have similar sensitivity of optical axis change due to individual lens element decenter. Although fabricating LISS lenses having relatively narrow spectral bandwidth (~0.1 μm) and modest FOV (less than 10°) is not a great effort, the real challenge is to make the four lenses of the camera such that all of them have the same image format and the collinearity is maintained under all environmental conditions with a high degree of tolerance (~1 micron). Here, the term image format refers to the image size for a fixed FOV. For all the lenses to have the same format, the EFL is required to be the same. The image format has to be matched for any angle within the designed FOV. This necessitates a distortion variation for a set of four matched lens assemblies within 0.005%. Because the CCD array location is fixed with respect to the flange, the variation of flange focal distance with temperature has to be within a specified limit. The next challenge is to maintain collinearity, that is, the optical axis with respect to some mechanical reference of the lens barrel, usually the flange, should remain fixed against various environmental conditions. Collinearity change is caused by mechanical (shock and vibration) and thermal stresses producing displacements of the lens elements within the barrel, which support them beyond the allowed tolerance. That is, under all operating conditions, each optical component should be constrained so that the decenter, tilt, and axial spacing and the stress-induced surface deformations are within the design tolerance limits. To achieve this, both lateral and axial constraints are needed for each component.

To achieve all these stringent requirements, proper mechanical design of the individual lens element assemblies in the barrel is required. One method of mounting the lens elements is to fix each lens in its own "subcell" using suitable adhesive, and they in turn are assembled inside the lens barrel with "interference fit." However, interference fit can cause stress on the lens elements (Richard and Valente 1991), which can worsen under temperature change. The assembly techniques usually adopted for the lenses meant for terrestrial applications are not sufficient enough for spaceborne lens assemblies, which have to undergo the stress levels specified for space hardware. In order for the lens assembly to be stable against the environmental stresses it is subjected to, a suitable flexure is preferred for mounting optical components for high-precision optical systems (Vukobratovich and Richard 1988; Yoder 2008). A flexure is an elastic element that allows small controlled relative motion of components in the desired direction, while restraining their motion in other directions. In the present case, the lens mount consists of a barrel with three blades symmetrically placed as a flexure. A three blade flexure is stiff in the direction of the length of the blade but flexes in the radial direction. The lens is glued to the blade using suitable adhesive. Then, differential expansion of materials caused by uniform temperature changes will not produce tilt or decenter. The length, width, and thickness are selected based on analysis to have minimum stress on the lens elements, and the eigenfrequencies of the elementary lens mount and barrel are different. The barrel is then fine machined to align the mechanical axis of the barrel with respect to the lens optical axis. The individual lens elements are assembled with shrink fit (thermal fit) in the barrel with a negative play to produce the final lens assembly. All IRS cameras such as LISS, AWiFS, and Ocean Color Monitor (OCM) have refractive collecting optics. Figure 3.3 shows the mechanical assembly of a typical LISS lens in the barrel.

3.2.1 Telecentric Lenses

In many applications, wider swath coverage is required in a single pass. A wider swath enables improved temporal resolution, that is, the frequency of observation under the same viewing angle increases as the swath is increased. For example, with a swath of about 2900 km the NOAA AVHRR instrument can visit any location on the Earth every day. Such large swath

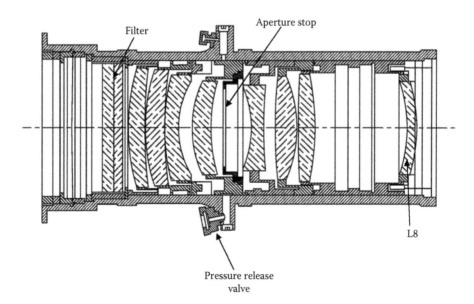

FIGURE 3.3
Mechanical assembly of a typical LISS lens assembly with barrel. (Reproduced with permission from REOSC [SAFRAN Group, France].)

systems were realized earlier by optomechanical scanners, which can collect data from horizon to horizon. Here, the optics needs to be corrected only for a narrow field, usually less than 0.1° depending on the instantaneous field of view. However, when linear arrays, such as CCDs, are used in a pushbroom mode, the imaging optics ought to have a FOV corresponding to the swath. For example, the POLDER radiometer flown on board ADEOS has a total FOV of 114°, the vegetation instrument on board SPOT4 has a total FOV of 101°, and the OCM instrument on board IRS P4 satellite has a FOV of 86°. In these instruments, spectral selection is achieved by interference filters. The central wavelength of an interference filter shifts to shorter wavelengths with increasing angle of incidence. (This property can be utilized to fine-tune filter characteristics.) If a lens with conventional design is used as the imaging optics, the filter will experience varying angles of incidence along the CCD array, whether the filter is placed at the entrance of the lens or at the focal plane. Therefore, the spectral characteristics of the imagery will not be the same for all pixels in a picture frame, which is not acceptable. Let us consider a scheme in which the spectral selection filter is kept at the focal plane. To take care of the issue discussed earlier we need to have an optical system, such that irrespective of the angle with which the rays enter the optical system the rays approaching the focal plane should be near normal. Such a configuration is possible and is referred to as a telecentric lens.

In a telecentric lens, the chief rays are parallel to the optical axis in image and/or object space. A telecentric lens can be designed with three types of telecentricity: object space telecentricity, image space telecentricity, and bilateral telecentricity. Object space telecentric lenses have constant magnification over a specific range of object distances (depth of field), which helps capture accurate measurements of three-dimensional parts and components of differing heights. Such telecentric lenses are used extensively in machine vision applications, such as optical metrology. In the case of an imaging system, what is relevant is image space telecentricity, where the rays incident on the focal plane are near normal so that the angular effect on the interference filter does not happen. Such a design also reduces off-axis irradiance falloff (i.e., $\cos^4\theta$ effect) in large-format imagery (Bai and Sadoulet 2007). Figure 3.4 shows the difference of angle of incidence at the focal plane between a conventional lens and a telecentric lens. As a typical example of a telecentric lens used in a spaceborne camera for Earth observation, we shall discuss the lenses for OCM launched on board IRS-P4 (OCEANSAT) in 1999. OCM has eight spectral bands covering visible and near IR regions distributed from 412 to 865 nm having a spectral bandwidth of 20 nm except for the two longer wavelengths, which have 40 nm. The swath coverage of 1420 km is achieved by telecentric lenses manufactured by REOSC (SAGEM Group), France (Thépaut et al. 1999). Each lens has 20.0 mm EFL and an f-number of 4.3 with a total FOV of 86°. As the imageries from all the bands are required to have inherent registration at instrument level, as in the case of LISS lenses discussed in Section 3.2, the lenses have stringent requirements on EFL and

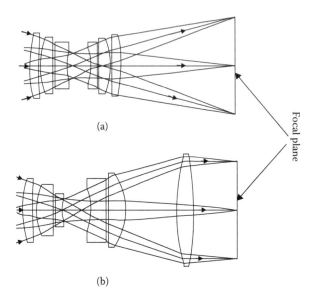

(a)

(b)

Focal plane

FIGURE 3.4
Schematics showing angle of incidence at the image plane for a (a) conventional lens and an (b) image plane telecentric lens. For a telecentric lens, all rays fall close to normal onto the image plane.

distortion variation among the eight lenses and optical axis (collinearity) stability against specified environmental conditions. For one set of matched lenses, the EFL variation should be within a tolerance of ±0.01% (i.e., ±2 µ), the variation in distortion less than ±0.3 µ for each point in the field, and the change in collinearity less than ±1 µ. These are in addition to other specifications of MTF, uniformity of illumination within the image format, veiling glare, and so on. All these have to be met under various environmental conditions. We shall briefly present here how the OCM lenses are realized (Thépaut et al. 1999).

Each lens assembly consists of 10 lens elements (Figure 3.5). To correct the distortion level over the field, the first powered element has a parabolic concave contour on the rear surface. By judicious design, the distortion could be controlled point by point in the field. Because of the large spectral range between the eight bands, each lens assembly uses different glasses according to each spectral band to get the best performance. All lenses are coated with antireflection coatings, which have a residual reflectivity (in the working band) lower than 0.4% per surface at the maximum incidence angle. A parallel plate optical window fabricated out of fused silica is located at the front to reduce thermal gradients within the lens assembly. The interference filter (IF) is placed at the exit of the lens as the last element. The lens design limits the maximum angle of incidence on to the IF to about ±7° instead of ±43°, if the IF were mounted at the entrance of the lens as in the case of LISS lenses. To match EFL

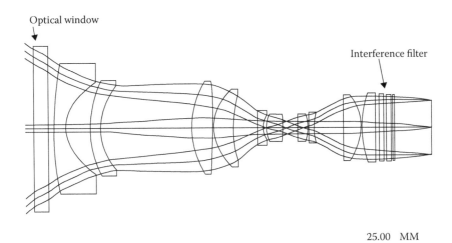

FIGURE 3.5
Optical schematics of a typical OCM lens assembly. (Reproduced with permission from REOSC [SAFRAN Group, France].)

and distortion, two spacers are included in the design so that they can be adjusted with little influence on the image quality (MTF). The overall design is fairly athermal.

3.3 Reflective and Catadioptric Systems

The lens optics has limitations when it is required to cover an extended wavelength region from visible to TIR. The lens materials that are transparent in the TIR region do not transmit in the very near infrared (VNIR) region and vice versa. For example, germanium, one of the common lens materials for realizing lenses in the TIR region, is opaque in the VNIR region, whereas fused silica and a host of glass materials transparent to visible radiation are opaque to the TIR region. In addition, for spaceborne cameras when the aperture and focal length of collecting optics increase, the weight and volume of the imaging optics using refractive elements are prohibitively high. Cameras generating high spatial resolution imagery invariably require long focal length and large aperture optics. For these reasons, a reflective telescope is preferred as the collecting optics for spaceborne high spatial resolution electro-optical cameras.

3.3.1 Types of Reflective Telescope Systems

Let us consider a concave reflecting surface (Figure 3.6a) on which the radiation is incident parallel to the optical axis. The rays converge to its focal point. This is the most basic form of a reflective telescope. In this case, the focus is in

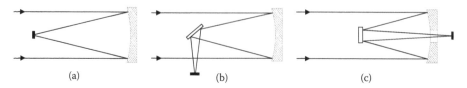

FIGURE 3.6
Schematics of different configurations of telescopes using a single powered reflective surface.

the path of the incoming beam and hence there is a constraint on the size of the focal plane assembly that can be put in the prime focus to limit the obscuration of the incoming beam. This concern can be taken care of by placing a plane mirror at an angle before the prime focus so that the converging beam is deflected and the focus can be located outside the incoming beam (Figure 3.6b). Such a system is called a Newtonian telescope. The classical Newtonian telescope uses a concave paraboloidal mirror as primary. Access to the focus to place instruments can be greatly improved by folding the converging beam so that the focus is brought behind the primary mirror. Therefore, another possible configuration is to take the beam through a central hole in the primary mirror (Figure 3.6c). The Newtonian telescope is not attractive for space application because of the overall size is larger and there is a very limited FOV. In addition, a single-mirror system like a Newtonian telescope can correct only spherical aberration. However, significant coma and astigmatism aberrations remain, both of which produce off-axis blur. These limitations can be alleviated by introducing a powered secondary mirror instead of a plane mirror in the optical path. In addition, two powered elements allow three variables to optimize the performance of the telescope, that is, surface profile of two mirrors and the distance between primary and secondary mirrors. A two-mirror system can be designed to be free of both spherical aberration and coma, leaving only astigmatism affecting the image quality (Korsch 1977).

There are two basic arrangements to realize a two-mirror telescope system. The Gregorian telescope has a concave secondary mirror along the optical axis of a concave primary mirror so that the rays reflect back toward the primary mirror and pass through a small hole in its center (Figure 3.7a). The classical Gregorian telescope uses a parabolic primary mirror and an elliptical secondary. There can be a number of variants of the same configuration depending on the surface profile chosen for the primary and secondary mirrors. In the other configuration, the incident light is reflected from a concave mirror onto a convex secondary mirror placed before the focus of the primary and then back through a hole in the concave mirror to form the image; this is called a Cassegrain telescope (Figure 3.7b). The Cassegrain secondary mirror slows the convergence of the light cone from the primary mirror, thereby increasing the system focal length. That is, the secondary mirror "magnifies" the focal length of the primary mirror. In the classical configuration, the primary mirror is a concave paraboloid and the secondary mirror is a convex hyperboloid. The classical Cassegrain telescope is free from spherical

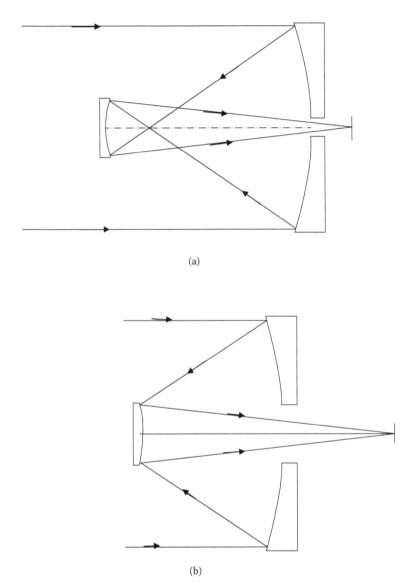

(a)

(b)

FIGURE 3.7
Optical layout of two-mirror telescopes: (a) Gregorian and (b) Cassegrain.

aberration, and the major aberration is coma. Depending on the primary and secondary surface profiles, a number of variants of the Cassegrain telescope are possible. Of all the three configurations, that is, Newtonian, Gregorian, and Cassegrain, the Cassegrain telescope has the shortest tube length (distance between the primary and secondary) keeping the other parameters constant. Therefore, most Earth observation cameras the use Cassegrain telescope or one of its variants as the collecting optics.

A variant of Cassegrain that is widely used in imaging systems is the Ritchey-Chretien (RC) telescope. It has a hyperbolic primary mirror and convex hyperbolic secondary mirror and is aplanatic, that is, it is free of spherical aberration and coma. Astigmatism and field curvature limit the useful FOV though the usable FOV is larger compared to a conventional Cassegrain configuration for equivalent f-number. In another configuration called Dall-Kirkham (DK), the primary mirror is ellipsoidal and the secondary mirror is spherical convex. Although this comes with a penalty of having a higher coma compared to classical Cassegrain, its mirrors are easier to fabricate and its alignment sensitivity is considerably lower than in classical Cassegrain (Sacek 2006).

In Earth observation cameras, two-mirror systems are normally used for optomechanical scanning imagers wherein the optics needs to be corrected for less than a fraction of a degree. For example, the Landsat TM telescope primary mirror is a 41-cm-diameter RC system, which needs to be corrected for only about 0.04°. The AVHRR sensor on board the NOAA satellite is an optomechanical scanner with a DK telescope whose field is corrected for 0.075° only. As mentioned earlier in the case of pushbroom scanners, the camera FOV has to cover the full swath (across track direction) and therefore requires a much larger FOV than that required for optomechanical scanners. We shall now discuss how to realize a large FOV telescope.

3.3.2 Increasing Field of View of Telescopes

Advances in telescope technology have evolved primarily to get better astronomical observations. Though telescopes have been used to study celestial objects since Galileo's time, the early reflective astronomical telescopes could provide only narrow FOV observation. However, astronomers realized the need for telescopes with wider FOVs to survey galactic and extragalactic objects. The first practical wide field astronomical telescope was based on a design by Bernhard Schmidt in 1928, referred to as the Schmidt telescope. There are two basic design concepts in realizing a wider FOV telescope system. In one configuration, the light collection is primarily by a mirror and the correction to achieve a larger field/flat field is achieved by placing refractive elements in an appropriate location in the optical path. As this scheme uses both reflective and refractive components, it is referred to as a catadioptric system. We have seen that two-mirror systems can provide a good image over a narrow field. One could add more reflective surfaces so that one has more parameters (like surface profile and its location), which can be varied to compensate for aberrations over a wider FOV, thereby making an all reflective optical system. There are a number of variants in both the schemes. However, we shall deal with only those telescope configurations that are commonly used in spaceborne Earth observation cameras. Interested readers may refer to Ackermann et al. (2010) to get a comprehensive review of a wide FOV telescope.

3.3.2.1 Catadioptric System

As mentioned in Section 3.3.2, one of the early designs for an astronomical telescope with a wide FOV for sky survey is the Schmidt telescope. A concave spherical mirror with aperture stop at its center of curvature is free of coma astigmatism and distortion but has spherical aberration (Mahajan 1991). To compensate for the spherical aberration in the Schmidt telescope, a corrector plate known as a Schmidt corrector is introduced in the optical path. The Schmidt corrector is designed with spherical aberration that is equal but opposite to the spherical aberration of the primary mirror, thus compensating for the spherical aberration produced by the mirror (Figure 3.8a). Such telescopes for ground use have been created with FOVs as large as 9 × 9 degree (Ackermann et al. 2010).

In the classical Schmidt telescope, as the corrector plate is placed at the center of curvature the tube length is twice the focal length. A variant of the Schmidt corrector is Schmidt–Cassegrain, which combines a Cassegrain reflector's optical path with a Schmidt corrector plate to make a compact telescope. Light enters through the aspheric Schmidt correcting refractor and strikes the spherical concave primary mirror and is reflected to a convex spherical secondary mirror, which reflects the light. The light gets out through an opening in the primary mirror to form the image behind the primary mirror as in the case of Cassegrain telescope (Figure 3.8b). The basic Schmidt and its many variants are still in use today for inground-based observatory, though limited to apertures less than about 1.25 m (Ackermann et al. 2010). Schmidt telescopes are not completely aberration-free as the refractor introduces some chromatic aberration, though small. It has a curved focal plane and requires the use of a field-flattening optics. A good example of a spaceborne camera for Earth observation using a Schmidt telescope is the HRV telescope of the SPOT satellite (Chevrel et al. 1981). We discuss the details of the SPOT telescope in Chapter 6.

Because the Schmidt corrector is at the entrance of the telescope, its size has to be same as the reflector entrance aperture. This poses many problems such as weight, deformation due to temperature gradient, and so on, especially for a

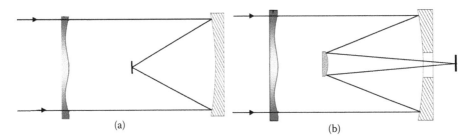

(a)   (b)

FIGURE 3.8
Optical layout of the Schmidt telescope: (a) classical Schmidt and (b) Schmidt-Cassegrain.

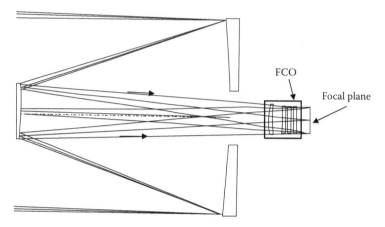

FIGURE 3.9
Optical layout of a typical catadioptric telescope with focal plane corrector (FCO). The FCO has three powered elements and an optical window.

spaceborne system. The problem can be alleviated if the corrector is used near the focal plane. A number of optical designs have been carried out to get larger FOVs using refractive elements near the focal plane on Cassegrain-type telescopes, which can correct off-axis aberrations and also helps to flatten the field. These field correctors usually have low power so that they do not have much effect on the f-number of the telescope. Early correctors used two lenses with spherical surfaces, but later to improve the performance field correctors having three or four lenses with some including at least one aspheric surface have been designed (Wilson 1968; Wynne 1968; Faulde and Wilson 1973; Epps and DiVittorio 2003). One generally optimizes the parameters of the mirrors and the lens as a single system to get the best performance. Ackermann et al. (2008) have realized a corrector with five spherical lenses, which can be used when the mirror positions are already fixed, as in the case of an existing telescope, producing extremely high image quality with near zero optical distortion. Many existing Cassegrain-type telescopes can incorporate this five-element corrector to extend the field without changing the mirror configuration. The examples given here are for large aperture ground-based astronomical telescopes, though these designs are general in nature. The IRS satellite CARTOSAT-2A launched in 2008 has an RC telescope with a 700-mm-diameter primary mirror with f-number 8. To achieve a flat field over ±0.5°, a focal plane corrector optics consisting of three lens elements is placed before the focal plane. The optical layout of the telescope assembly is given in Figure 3.9.

3.3.2.2 *All Reflective Wide Field of View Telescope*

As discussed in Section 3.3.1, any practical two-mirror telescope configuration can only be corrected for maximally two aberrations, usually spherical aberration and coma. It has long been recognized that the FOV of two-mirror telescope configurations can be improved by going to optical configurations that use three powered mirrors (Lampton and Sholl 2007). This is because with three mirrors there are nine parameters available for design optimization: the curvature and asphericity of the three mirrors, their relative distances, and the distance of the focus from the tertiary. Three-mirror telescopes can be designed to correct four aberrations, that is, spherical aberration, coma, astigmatism, and field curvature (Korsch 1977). The three-mirror anastigmatic (TMA) telescope offers a diffraction-limited wide FOV and superior stray light baffling. Two configurations of "obscured TMA" (OTMA) telescope were proposed by Korsch with good geometric spot size over a flat field of 1.5° with excellent stray light suppression. With three aspheric mirrors, all third-order aberrations can be corrected simultaneously and higher order aberrations can be adequately controlled. One of the configurations proposed by Korsch is shown in Figure 3.10. Here, the combination of primary and secondary mirrors acts like a Cassegrain configuration forming a real image just behind the primary mirror. (Instead, it is also possible that the image is formed between the primary and the secondary.) This image is directed to the tertiary mirror via a fold mirror placed diagonally between the primary and the tertiary mirrors. The reflected rays from the tertiary mirror pass through the hole in the fold mirror. The tertiary reimages the image formed by the primary–secondary mirror combination at approximately unit magnification, forming the focal plane behind the fold mirror. The intermediate image plane provides an ideal location for a baffle to reduce stray light. The tertiary mirror can also magnify the first focus of the Cassegrain telescope to the final focus producing the required final EFL as in the case of the High Resolution Imaging Science Experiment (HiRISE) camera on board NASA's

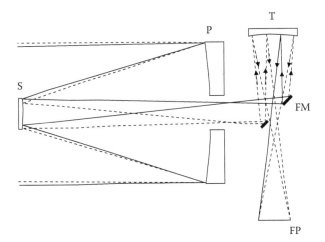

FIGURE 3.10
Typical obscured three-mirror anastigmatic optical layout. P, primary mirror; S, secondary mirror; T, tertiary mirror; FM, fold mirror; and FP, focal plane.

Mars Reconnaissance Orbiter spacecraft (Gallagher et al. 2005). A number of variants of the Korsch TMA configuration have been successfully implemented, depending on how the tertiary is placed in the beam (Gallagher et al. 2005; Lampton and Sholl 2007). OTMA telescopes have been used in many Earth imaging cameras such as IKNOS and Pleiades (Bicknell et al. 1999; Fappani and Ducollet 2007). As a typical example of OTMA, we shall briefly describe the Pleiades telescope. The telescope has an entrance aperture of 650 mm, with an EFL of 13 m. The optical layout is similar to that shown in Figure 3.10. It has an on-axis Cassegrain-type telescope with a primary elliptical concave mirror and a hyperbolic convex secondary mirror. A fold mirror directs the beam to the tertiary mirror, which is an off-axis elliptical concave. All the mirrors are made of Zerodur and lightweighted up to 70%–80%. The mirror fixing devices (MFDs) are made of invar. The mirrors are attached to MFDs using a space qualified structural glue and finally the mirror–MFD assembly is attached to the telescope structure with titanium screws. The telescope provides good performance over the designed FOV of about 1.7°.

The major drawback of the reflecting telescopes we have hitherto discussed is the central obscuration due to one of the reflective elements. The central obscuration reduces the effective collecting aperture, and its diffraction effect reduces the MTF of the system. Figure 3.11 shows the MTF reduction for various obscuration ratios. For a centrally obstructed telescope, an enlarged secondary mirror baffle further reduces the MTF. The secondary

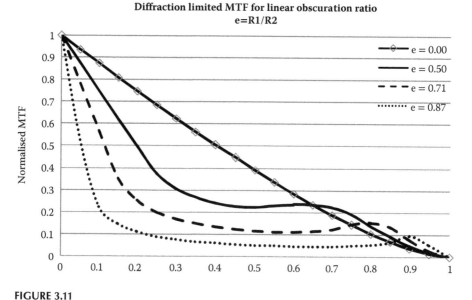

FIGURE 3.11

Effect of central obscuration on diffraction-limited MTF for Cassegrain-type telescope. R1, radius of central obscuration; R2, radius of entrance aperture. (Calculated from Wetherell, E., and B. William, *Applied Optics and Optical Engineering*, Academic Press, New York, 1980; Adalka Dheeraj, 2014, personal communication.)

mirror mechanical support connecting the mirror to the rest of the telescope also affects the diffraction pattern. In addition, if the angular field is very wide the obscuration can block a large portion of the incoming beam (Marsh and Sissel 1987). The obscuration issue can be circumvented by adopting the unobscured three-mirror anastigmatic (UTMA) design. Here, the first mirror surface is off-axis so that the focus is taken out of the incoming beam, and hence the secondary can be placed such that it does not obstruct the incoming beam. As it is easy to fabricate a symmetric mirror to the specifications, the off-axis portion is usually cut from a centrally symmetric mirror. For example, QuickBird has an UTMA telescope with an unobscured 0.6-m primary mirror, which is a subaperture of an approximately 1.8-m parent optic (Sholl et al. 2008). Figure 3.12 gives the optical layout of a typical UTMA telescope. For a review of unobscured multiple-mirror imaging designs ranging from two to many mirrors, readers may refer to the work by Rodgers (2002). In the UTMA design, one can reduce the FOV in the direction of the mirror offset to have a larger FOV in the orthogonal direction while maintaining the same overall wave front error (WFE). For example, in a particular system viewing a distant object, angular fields of $2° × 3°$, $1° × 6°$, and $0.5° × 8°$ can all give similar worst-case WFE (Rodgers 2002). This is of particular interest for pushbroom imaging cameras, since the along track field requirement is very small compared to across track FOV, which defines the swath. A number of spaceborne imaging sytems have the UTMA telescope as the imaging optics. These include IRS-PAN, QuickBird, EO1-ALI, and Topsat, to name a few (Joseph et al. 1996; Figoski 1999; Bicknell et al. 1999; Greenway et al. 2004). The design details of the IRS-PAN UTMA configuration are described in Chapter 6. The PROBA-Vegetation payload, a follow-on to the SPOT vegetation camera, carries an UTMA, with an FOV of $34.5° × 5.5°$ with a focal ratio of 7, and is telecentric in the image

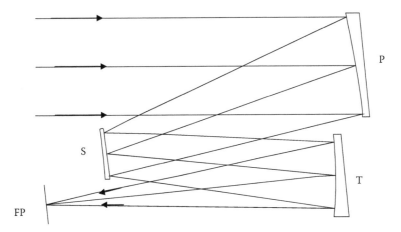

FIGURE 3.12
Typical optical layout of an unobscured three-mirror anastigmat. P, primary mirror; S, secondary mirror; T, tertiary mirror; and FP, focal plane.

space. Though the focal length is only 110 mm, it may be the largest FOV UTMA realized for Earth observation (Grabarnik et al. 2010). Silny et al. (2011) describe the design and realization of a five-mirror unobscured system with a 36° cross track FOV and about 90 mm EFL operating at f-number 3.0.

One of the factors that impacts cost and schedule is the fabrication of off-axis aspheric surfaces. Shafer (1978) has shown that by the proper configuration of four spherical mirrors it is possible to devise telescope designs that are unobscured, with no spherical aberration, coma, or astigmatism, at the third-order level. Some variants are also corrected for Petzval curvature and distortion as well. However, as the number of reflecting surfaces increase, the alignment becomes more complex.

3.4 Stray Light Control and Baffling

An Earth imaging camera is essentially a radiometer. The digital number given by any pixel should be a faithful reproduction of the radiance from an area covered by the projection of the focal plane detector through the optics onto the ground. However, this ideal situation does not happen due to various reasons, resulting in radiometric error. One of the factors that produce radiometric error is stray light, if the optical design has not taken care of it. When there are high reflecting objects such as snow, cloud, and so on close to the out of field region, the effect on image quality can be quite detrimental.

Any radiation falling on the image plane other than that expected as per the optical design is a stray light. The radiation that originates from outside the sensor's FOV is the major source of stray light, which may be termed *out-of-field stray light*. In a two-mirror system like Cassegrain telescope, an out-of-field ray can directly reach the detector without striking the primary or secondary through the hole in the primary mirror. These unwanted "direct hit" rays do not attenuate and hence reduce the image quality seriously. In addition to direct hit rays, there can be single/multiple scattering from mirror edges or other support structures giving a diffused radiation in the focal plane (Figure 3.13). There is also a possibility that the radiation inside the FOV may get scattered by the reflecting/refracting elements and the scattered light reaches the image plane forming a diffused background, though because of attenuation these indirect hits have lesser radiance compared to direct hit rays. This in-field scattering depends primarily on the roughness and contamination of the telescope mirrors. The diffuse stray light reaching the image plane of an optical system reduces image contrast and is termed *veiling glare*.

The contribution of out-of-field stray light to the focal plane can be avoided by using suitable baffles. For a two-mirror system like an RC telescope, usually two truncated baffles are incorporated, one baffle around the central hole of the primary mirror and a secondary baffle around the secondary mirror. These baffles if properly designed substantially prevent not only the

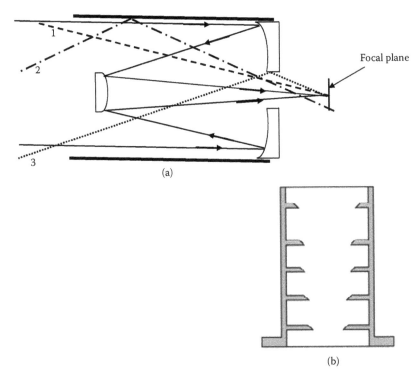

(a)

(b)

FIGURE 3.13
(a) Stray light paths in a Cassegrain telescope. 1, Direct ray; 2, ray reflected from telescope metering tube; and 3, ray scattered from primary mirror hole edge. (b) Sectional view of baffle with vanes.

direct light from any incident angle but also the light reflecting from the tube between the primary and the secondary used as the metering structure from reaching the focal plane.

The baffles should be designed such that they provide minimum obscuration to the within-field rays but have maximum blocking for stray light. The goal of baffle design is to find the length and diameter of baffles that satisfy the aforementioned criterion. However, there can be specular and scattered rays at the baffle, which may reach the detector. To get the best stray light suppression, we have to block such radiation entering the detector. This is achieved by positioning vanes along the baffles, the action of which is as follows. The space between the vanes acts like a "cavity." Once radiation enters a vane cavity, it makes multiple reflections before it can exit. The multiple reflections absorb and diffuse the radiation incident on the baffle. The vane cavity is covered with a black coating, which absorbs some of the radiation falling on it and scatters the rest in all directions. When scattering takes place, the reflected radiation from the vane cavity spreads out in all directions, so the intensity falls off as the distance increases from the scattering surface. However, one should try to optimize the number of vanes because fewer vanes not only reduce mass and manufacturing costs but also provide fewer vane edges that can scatter light toward the detector.

Blackened vane cavities increase the thermal load onto the baffles increasing the temperature. This is not a welcome situation for imagers in the TIR region as self-emission from the baffle will add to the background radiation. In thermal imaging systems, specular vane cavities have been suggested (Bergener et al. 1985). However, specular vane cavities can work only when the off-axis source angle is fixed.

Baffles for stray light suppression have been designed with graphical and analytical methods (Young 1967; Hales 1992). Kumar et al. (2013) developed a baffle design method based on a combination of the results of optical design software and analytical relations formulated herein. The method finds the exact solution for baffle parameters by iteratively solving the analytical relations using the actual ray coordinates of the telescope computed with the aid of optical design software. Estimation of the contribution of stray light at the focal plane is more complex, and requires a detailed knowledge of optical design, mechanical structure surface roughness, coating material properties, scatter models, and so on. Many of these parameters are to be modeled, whose validity is difficult to establish and hence baffle design should be extensively tested experimentally in the laboratory to confirm adequacy of the design.

When the telescope has an intermediate focus as in the Korsch TMA, a stop can be placed at this focus, which substantially reduces contributions from stray light to the focal plane.

3.5 Building a Reflective Telescope

Once the telescope configuration is chosen one has to work out the specification of each mirror, mirror separations, mirror orientation, and so on to get optimum optical performance within the total FOV of interest, which is carried out using optical design software. There is a variety of optical design software available, both commercial and open source. The telescope design should minimize the tube length, for lower instrument volume, reduce the secondary size to keep obscuration low, and have a flat focal field. However, these requirements are interdependent. For example, configuration with a fast primary (low f-number) has a shorter tube length and a small secondary mirror but a higher field curvature (JODAS 1999). In addition, the misalignment tolerance of the secondary varies as the cube of the primary focal ratio (Bely 2003), which makes alignment of a telescope with a fast primary more difficult. A fast mirror is also more difficult to manufacture and hence more expensive. The design should also take into account that the specifications arrived at should be realizable in practice. One can have an excellent performance on paper, but if the tolerable limits on mirror surface contour, separation, tilt, and so on are very tight and cannot be realized in practice then the design is not practicable. Putting tight tolerances in the optical design, even if it is achievable, increases the cost of fabrication. The designer should also look for ease of fabrication of the element, availability of the material, and its cost. Therefore, a judicial compromise has to be made between optical performance and other aspects such as material choice, limits on tolerances, and so on so as to realize the optical system meeting the laid down specifications within the schedule and budgetary constraints. Once the design is frozen, the next task is to realize the hardware that meets the optical performance requirements and stability requirements against various specified environmental conditions.

3.5.1 Selection of Mirror Material

It is important to choose the right material for fabricating mirrors, which meets the necessary optical quality through fabrication, mounting, and alignment while operating through various stipulated environmental specifications. Traditionally, glass ceramic with very low coefficient of thermal expansion (CTE) like Cervit and Zerodur are widely used as mirror substrate. (Cervit is manufactured by Corning, Inc., and Zerodur by Schott AG Company). The CTE is generally specified within a range of values around zero, and the actual value of a lot will be within the range. Reflecting optics of IRS telescopes, as well as in many other missions like SPOT, Landsat, and so on, use conventional glass material, either Cervit or Zerodur. However, newer materials are increasingly finding applications as space telescope mirror materials. We shall review the general requirements of a material to realize space telescope mirrors and then present some examples of systems that have been developed using materials other than the conventional glass material. The mechanical property of importance is lower density (ρ), to reduce mass, and higher Young's modulus of elasticity, E, which is the measure of stiffness or rigidity of a material. The desirable property taking into account both the aforementioned material properties is specific stiffness, which is defined as the ratio of elastic modulus to density (E/ρ). The higher the specific stiffness, the lower the distortion caused during the polishing process and assembly and the better the ability of the mirror to withstand vibration and shock loads. When comparing components with the same mass, a parameter considered is the mass-deflection proportionality factor $\left[\left(\dfrac{\rho^3}{E} \right)^{1/2} \right]$, which gives the relative deflection of a component with fixed mass, that is, when comparing components with the same mass this factor gives the relative deflections (Paquin 1995). Another important parameter is the thermal behavior of the mirror material. The material should have low CTE, α, to minimize dimensional changes due to temperature variation, but high thermal conductivity, k, to reduce thermal gradient. Distortions due to thermal gradients can be minimized by selecting materials with high ratios of thermal conductivity to CTE, k/α. Use of a material with higher k/α value greatly increases the allowable temperature envelope and still minimizes the distortions in mirror shape. Table 3.1 gives these values for some of the commonly used mirror materials.

Apart from the aforementioned properties, one should take into account the dimensional stability of the material, both the inherent stability and the instabilities that can be induced by the fabrication process. To maintain dimensional stability, the mirror material should have isotropic behavior in mechanical and thermal properties. Of course, the material must be able to be machined and polished to the required optical tolerances and amenable for stable optical coating to reflect the optical wavelength region under consideration. Besides the technical requirements, other aspects such as schedule and cost also influence the choice of mirror material.

As seen from Table 3.1, beryllium (Be) has the best mechanical performance, whereas silicon carbide (SiC) gives the best thermal behavior. However, Be machining requires special precautions due to its toxic nature. Beryllium is not usually used for telescope mirrors in Earth imaging cameras. However, due to its low mass beryllium has been used as a scan mirror in optomechanical scanners such as Landsat MSS/TM, INSAT VHRR, NOAA AVHRR, and so on. As beryllium substrate is a powder metallurgy product, there are limits to the finish that can be achieved by direct polishing of the beryllium substrate. So the beryllium surface is generally plated with electroless nickel to have a surface that can be polished with mirror finish. The polished surface is coated with a reflective metal like gold, silver, or aluminum, depending on the wavelength region of interest to get the desired reflectance. Finally, a protective overcoat, usually silicon oxide or magnesium fluoride, is

TABLE 3.1

Room Temperature Properties of Selected Materials Used for Mirror Fabrication

	Density ρ (g/cm³)	Young's Modulus E (GPa)	CTE $\propto$ $(10^{-36}/$ K)	Thermal Conductivity K (W/mK)	Specific Stiffness E/ρ	K/α	Mass Deflection Proportionality Factor $\left(\dfrac{\rho^3}{E}\right)^{1/2}$
Desirable Material	Small	Large	Small	Large	Large	Large	Small
Zerodur	2.53	90.3	0.02	1.46	35.7	73	0.42
ULE Corning 7972	2.21	67.6	0.03	1.3	30.6	43.3	0.39
CVD SiC	3.21	465	2.2	280	145.1	127	0.27
Beryllium	1.85	287	11.3	216	155	19	0.15
Aluminum	2.7	68	22.5	167	25	7.4	0.55

Note: The values are indicative only; the exact values for a batch can be different.

deposited. However, nickel coating has some issues under changing temperature. Nickel and beryllium differ in thermal expansion and thermal conductivity, so a nickel coating will introduce bimetallic problems. Therefore, depending on the thickness of the nickel overcoat, the temperature exertion could change the mirror surface profile. The temperature effect can be minimized by using an electroless nickel with a CTE matched closely with beryllium by adjusting the phosphorus content of the nickel coating (photonics). The Landsat MSS scan mirror was formed by electric discharge machining from a solid beryllium billet along with integral connecting ribs. To give a good optical finish, the front surface was electroless nickel plated and for stress balancing the rear side was also plated with the same material. For uniform spectral reflectance across the MSS spectral range, a silver coating was vacuum deposited on the front surface and a protective coating was then applied to prevent tarnishing (Lansing et al. 1975).

Beryllium is particularly suited for cryogenic systems because thermal expansion drops drastically below room temperature and approaches 0 below about 80 K and the thermal conductivity increases below about 150 K (Paquin 1995). The best example of the use of beryllium is NASA's James Webb Space Telescope (JWST), operating below 40 K, for astronomical observation. JWST uses beryllium mirror segments for the primary mirror, each segment being 1.3 m. The mirror billets are generated from beryllium powder using a hot isostatic press process. Each blank is machined to lightweight the mirror, in a process that removes nearly 92% of its original mass, and gold coated.

As seen from Table 3.1, SiC has many attractive properties such as low thermal expansion coefficient, high thermal conductivity, and very high specific stiffness. However, the properties of SiC depend on the process of fabrication. SiC made by sintering or hot pressing of SiC powders or by the reaction-bonding process is usually porous and when polished does not produce a good optical surface. To realize an optical surface that can be polished, these forms of SiC require an overcoat of chemical-vapor-deposited (CVD) SiC, or any other suitable material (Han Yuan-yuan et al. 2006). As CVD-coated SiC has the same thermal properties of SiC, there is no bimetallic problem as is encountered in nickel-coated Be mirrors. The support structure and interconnecting metering structure can also be made of SiC, thereby avoiding differential thermal expansion, and hence such a configuration will have minimal thermal distortion and will have some saving in the overall mass of the telescope.

SiC mirrors have been successfully used in astronomical space telescopes and Earth observation cameras. The Herschel IR telescope, which was launched in 2009 by the European Space Agency, has a 3.5-m-diameter SiC primary mirror, lightweighted to 21.8 kg/m^2 aerial density. The 3.5-m diameter was achieved by brazing together 12 segments, which are formed by cold isostatic pressing and sintering of silicon carbide. After machining and polishing to the required surface contour and finish, a 10-nm thick adhesive layer of nickel-chrome is coated on the surface followed by deposition of a reflective layer of aluminum. The aluminum surface is protected by a layer of silicon-based polymer, which prevents corrosion of the aluminum surface by water vapor during ground operations. All the major telescope components—the primary and secondary mirrors and the hexapod that supports the secondary mirror—are made from silicon carbide, allowing the telescope mass to be reduced to 300 kg rather than the 1.5 t that would have resulted from using conventional materials (ESA 2004).

Another method of realizing a lightweight SiC mirror is to have a sandwich structure composed of a porous silicon carbide core sandwiched between two SiC face sheets having a chemical-vapor-deposited coat of SiC on the surface. The primary and secondary mirrors of the IR astronomy satellite (AKARI) telescope developed by the Japan Aerospace Exploration Agency are of this design (Onaka and Salama 2009). The primary mirror of the telescope, which has a physical diameter of 71 cm, weighs only 11 kg.

A number of Earth observation cameras are now using SiC for telescope optics. Following are examples of Earth observation cameras using SiC mirrors for the telescope (however, the list is not exhaustive). The advanced Land Imager launched on NASA's EO1 has telescope mirrors made of SiC. ROCSat-2, a Taiwanese Earth imaging satellite, carries a Korsch telescope, which has mirrors with the structure and the focal plane support assembly fabricated out of SiC. THEOS, the Earth observation satellite of Thailand, carries a panchromatic camera with a primary mirror made of SiC. The Sentinel-2 satellite being developed by the European Space Agency for the Earth observation mission Global Monitoring for Environment and Security carries Earth imaging cameras whose telescope mirrors are also made of SiC (Spoto et al. 2012). Most of these are satellites that are required to have high agility either for faster revisit or for acquiring data of a specific region of interest. All SiC telescope helps to achieve overall satellite mass low and hence will have lower inertia, thereby making the satellite more agile.

Aluminum mirrors are used in the Earth observation camera REIS (RapidEye Earth Imaging System), designed and developed by Jena-Optronik GmbH, Germany, and referred to as JSS-56. The collector optics is a 145 mm TMA design based on an all-aluminum construction (Risse et al. 2008) The aluminum mirrors are Ni coated to achieve a suitable surface polishing quality (Kirschsteina et al. 2005). The PROBA V telescope, which is to replace the vegetation camera of SPOT, also has an all-aluminum TMA telescope.

3.5.2 Mirror Fabrication

Irrespective of the type of telescope used, the first element, the primary optics, has the maximum size. As the spatial resolution increases generally the telescope primary mirror size also increases. Larger apertures bring about a number of challenges for realizing the mirror and associated mechanical system. In this section, we discuss some of the critical issues in realizing mirrors for a spaceborne optical system.

The thickness of the mirror is chosen so as to have adequate stiffness to reduce the deflection due to gravity. As a rule of thumb, the classical way is to have the thickness of the mirror such that the ratio of diameter to thickness (referred to as the aspect ratio) is between 6 and 8. For a given aspect ratio, the mass of the mirror increases as the cube of the radius. As the mirror mass increases, the distortion caused by the mirror self-weight also increases. The root mean square (RMS) surface deformation, δ, due to self-weight of the mirror depends on the density, ρ, and the elasticity modulus, E, of the mirror material, and the diameter, D, and thickness, h, of the mirror. When the mirror axis is vertical, δ is given by the following equation (Yong Yan et al. 2007):

$$\delta = k \frac{\rho}{E} \left(\frac{D^2}{h} \right)^2 \tag{3.1}$$

where k is a proportionality parameter related to the support system of the mirror. From this equation, it is seen that when the diameter of the mirror is doubled the thickness has to increase four times so that there is no change in δ. But this increases the mass of the mirror 16 times.

The increase in mirror thickness has other implications. In general, the front and back surfaces will have different temperatures. Thermal management of the spacecraft tries to minimize the temperature nonuniformity. However, as the diameter of the mirror increases it is more difficult to control the temperature gradient. The temperature gradient will produce an additional curvature radius (ΔR) to the mirror surface as per the following relation (Yong Yan et al. 2007):

$$\Delta R = h\alpha^{-1}\Delta T^{-1} \tag{3.2}$$

where ΔT is the temperature gradient in the direction of the thickness of the mirror, α is the thermal expansion coefficient of the mirror, and h is the thickness of the mirror. From the aforementioned equation, it is clear that as the mirror thickness increases the thermal gradient–induced curvature also increases. Therefore, this additional curvature can defocus the image unless the detector assembly is repositioned to the new focal plane location.

3.5.2.1 Lightweighting of Mirror

When the primary mirror mass increases, the mass of the structure supporting the mirror also increases. Thus, the overall mass of the telescope depends strongly on the primary mirror mass. Therefore, it is necessary to optimize the mirror weight without affecting the optical performance under the environmental conditions encountered by the telescope. Lower weight also increases the eigenfrequency. Thus, lightweight optics is of crucial importance to improve the overall performance of the telescope.

The reflecting property of the mirror is carried out only by the front surface. The lightweighting is essentially achieved by removing material from the back of the reflecting surface, but still ensuring structural stiffness to meet the mechanical and optical requirements at a lower mass. The schemes used to produce lightweight mirror can be broadly classified into three categories: contoured back mirrors, sandwich mirrors, and open back/semi–open back mirrors.

We shall briefly describe them, without getting into design details. Contoured back mirrors are the simplest scheme to reduce the weight of mirrors. In this configuration, the back is contoured to reduce the overall weight. Figure 3.14 shows two types of contoured mirrors: single arch and double arch. The parameters that can be varied for optimizing stiffness include height, edge thickness, and shape of contour (parabolic, convex, and straight). Weight reduction up to about 25% is feasible in comparison with a right circular solid mirror of aspect ratio 6. As seen in the figure, the mirror thickness varies across the mirror surface. This will cause different portions of the mirror to reach thermal equilibrium at different times when the temperature changes and could produce surface distortion when temperature variations are fast. A contoured back mirror produces maximum thermal distortion compared to other types (Vukobratovich 1997).

Sandwich mirrors typically have an optical plate (the reflecting surface) and a backing plate separated by a lightweight core (Figure 3.15). The core geometry generally used is triangular, square, or hexagonal. The core is attached to the optical and back plate by bonding or fusing. The parameters that can be adjusted for best performance include the plate thickness, core structure including the rib thickness and spacing, and distance between the outer plates. An edge band may be used, which provides additional stiffness in the direction of the circumference and also can be used for mirror mounting. A foam core of the same material has also been used, resulting in extremely low mass with high rigidity (Novi et al. 2001). Sandwich mirrors provide the highest stiffness to weight ratio compared to any lightweight mirrors, that is, contoured, semi–open back, or open back, providing a weight reduction of up to 85% in comparison with a right circular solid cylindrical mirror of 6:1 aspect ratio. However, they are comparatively more difficult to fabricate (and

(a)   (b)

FIGURE 3.14

Schematics showing contoured back mirrors: (a) single arch and (b) double arch. (Adapted from Vukobratovich, D., *Handbook of Opto-Mechanical Engineering*, CRC Press, Boca Raton, Florida, 1997.)

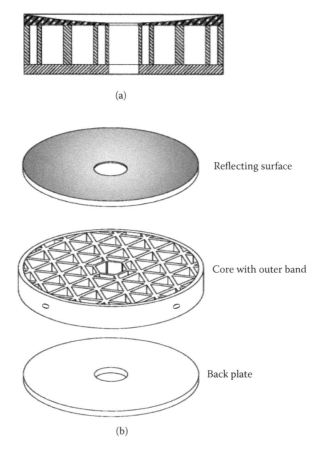

(a)

Reflecting surface

Core with outer band

Back plate

(b)

FIGURE 3.15
Schematics showing sandwich mirror: (a) cross-sectional view and (b) exploded view.

hence more expensive), and designing of mirror mounts against dynamic loads is technically challenging.

An open back/semi–open back mirror configuration consists of a thin face sheet, which is the reflecting surface, and a shear core at the back to form pockets. These pockets can be completely open at the back (open back mirror) or partially open (semi–open back) (Figure 3.16). For an open back mirror, the core geometry could be generally circular, triangular, or hexagonal, as in the case of sandwich mirrors. Parameters like inscribed circle diameters of the rib pattern, rib thicknesses, and front surface thicknesses affect the stiffness and weight of the mirror. Lightweighting up to 60%, compared to a right circular cylinder, can be achieved (Vukobratovich 1997).

Quilting is another effect that needs to be addressed while discussing lightweight mirror polishing. Quilting is a deformation that takes place due to deflection of the reflecting surface of the face sheet during polishing, which affects image quality. This puts a limit on the polishing pressure

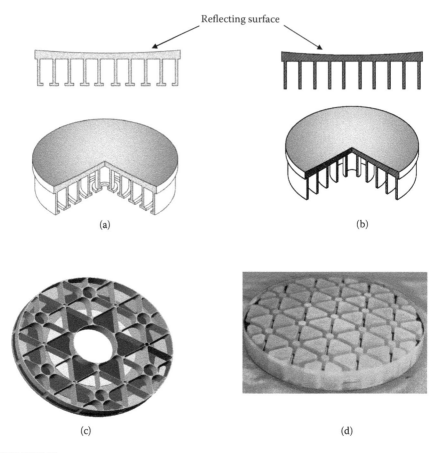

FIGURE 3.16
Cross-sectional view of (a) semi–open back and (b) open back mirror configuration. (c) Typical weight relieving triangular pattern. (d) Photograph of the rear side of the open back mirror. (Courtesy of LEOS/ISRO, Bangalore, India.)

that can be applied. Noncontact polishing such as ion beam sputtering can reduce quilting. In this process, based on the interferometric surface error map, the ion beam dwell time required for removing the material at each location on the optical surface is calculated and the ion beam is delivered accordingly by a specialized machine (Rao et al. 2008). Quilt effect depends on the pattern chosen, inscribed circle, and face sheet thickness. Hexagonal and square patterns are relatively easy to fabricate, but the quilt associated with them is higher. Triangular ribs provide maximum stiffness compared to others at the cost of marginally higher weight.

To produce a rib-stiffened lightweight mirror, material can be removed by machining a solid block through techniques such as computer-controlled milling, abrasive water jet and laser cutting, and/or etching. Alternate methods to produce web-stiffened mirrors include fusing or bonding the core to the front/back plates or casting the mirror with the web stiffeners in place. Realizing the final mirror surface is a very specialized activity, and each mirror has to be supported on a specially designed whiffle tree mount during fabrication to reduce the gravity effect. Each mirror fabrication approach offers its own advantages and disadvantages in terms of fabrication ease, design flexibility, manufacturing time, and cost, and the choice depends on the facility available and the accrued expertise of the team.

3.5.2.2 Optimizing Lightweight Mirror Structure

Optimizing lightweight mirror structure and the support is an area of intense research for decades and the study is still continuing (Schwesinger 1954; Barnes 1969; Genberg and Cormany 1993; Yu Chuan Lin et al. 2011). It is not possible to have a single optimum design to suit all users, as each user has some set of parameters to be optimized to suit his mission. The classical definition of a lightweight mirror is "any mirror that is lighter than a solid mirror of the same diameter and stiffness." What is the optimum lightweight mirror for space use? Ultimately, one is looking for a certain image quality of the telescope of which the lightweight mirror is a part. The optimum lightweight design of a mirror for space optics is more than just minimizing the weight of the mirror without compromising its stiffness or bending strength. We shall define the optimum lightweight mirror as the one that provides the least weight and still has the specified image quality under various mechanical and thermal loads during fabrication, assembly, integration, launch, and in-orbit operation. That is, external mechanical forces due to the manufacturing process, interface effects due to mounting and assembly, static and dynamic loads, gravity release effects, and so on should not produce deformation of the reflecting surface beyond the allowable limit so as to achieve the specified image quality of the telescope. In addition, as the optics size increases it is hard to control temperature gradient on the mirror surface. Therefore, the effect of temperature gradient also needs to be studied and should not degrade the optical performance below the permissible limit within the operating temperature. The deflection of the reflecting surface from the stress-free condition, under the above loads, is the most important critical parameter. Another important parameter is keeping the natural frequency above a certain value, which is determined by coupled analysis with the spacecraft. These studies need to be carried out with the mirror support system (MFD), whose design and location have significant influence on the overall mirror characteristics under various loads. The mirror with its central hole (in the case of a primary mirror) and MFD is a complex structure and is not amenable to conventional closed-form solutions normally applied for classical beam, plate, and shell bending theory. Therefore, optimization is done by the finite element method (FEM). The basic idea of the FEM is to divide a complex structure into a discrete number of smaller "elements." These elements are then modeled mathematically by applying classical equations. Generally, the perturbation of the mirror surface or WFE is expressed by the RMS of the deviation from an ideal surface. Optimization is carried out by varying parameters such as face plate thickness, rib structure, MFD location, and so on so that the WFE is within the specified limits. The WFE of the mirror can also be represented using Zernike polynomials, as discussed in Chapter 2. In addition to optimizing the optical performance, the fundamental frequency of the mirror also needs to be above a certain specified value. The mirror with the best optical performance is not necessarily the lightest or the stiffest in terms of the fundamental frequency. Therefore, one has to arrive at a compromised design such that the telescope meets the desired goal.

3.5.3 Mirror Mounts

It is said "a telescope is as good as the mirrors and the mirrors are as good as their support system." Even if one has fabricated the best mirrors, if the MFDs to the structure are not properly designed it can result in inferior performance of the telescope. The mirror mount should hold the mirror firmly to its structure during the whole operational period against various environmental conditions, without degrading the mirror surface quality achieved in manufacturing. Therefore, the MFD should be designed to isolate the mirror from mechanical (during assembly and launch) and thermally induced forces acting on the telescope structure, so that the effects of these forces on the mirror WFE is within the acceptable limits. The overall stiffness of the mirror and support should ensure that the fundamental frequency should be above a certain specified value. The material should be chosen so that the difference in CTE between the mirror material and the MFD should be as low as possible to reduce the thermal stress at the joint. Another aspect to be considered for mirror mounting design for a space application is the release of gravity effects in orbit.

The mirror mounts are made with a set of flexures, which provides a controlled motion. The simplest is a blade mount, which is schematically shown in Figure 3.17. Here, three blades positioned 120° apart on a mounting ring are attached on the sides of the mirror. It provides radial flexibility to accommodate

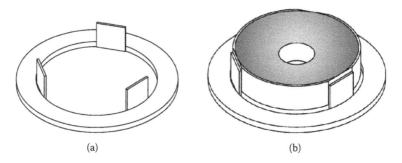

(a)   (b)

FIGURE 3.17
Schematics showing mirror mount using blade flexures: (a) mounting ring with blades and (b) mirror glued to the blade.

differential thermal expansion between the mirror material and supporting structure but does not offer sufficient isolation to the mounting stresses while attaching the mirror support to the telescope structure. To minimize the transfer of mounting stress to the mirror, the planarity of the mounting face of the ring and mounting location on the structure where it is mounted should match with a high degree of tolerance.

3.5.3.1 Bipod Mounts

A bipod consists of two support members arranged in the form of an inverted "V" (Figure 3.18a). To have higher flexibility, local areas are

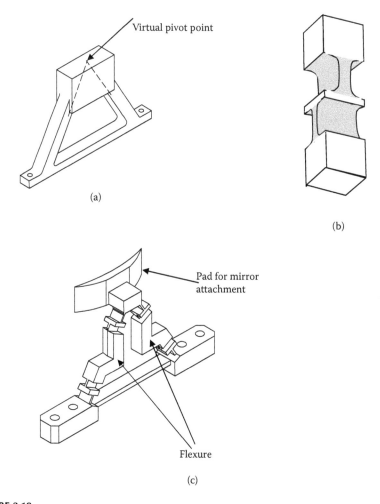

FIGURE 3.18
Schematics showing mirror mounts: (a) basic structure of a bipod, (b) a two-axis flexure, and (c) bipod with flexure and mirror mount.

thinned down considerably. A dual-axis flexure is shown in Figure 3.18b. Such a design has flexibility in two mutually perpendicular axes and hence can isolate misalignment in those two axes. Figure 3.18c shows a scheme of a bipod that has a dual-axis flexure with mirror mounting pad. An advantage of this arrangement is that its "virtual pivot" location can be positioned by adjusting the angle of the two legs in the bipod. This allows locating the virtual pivot point in the mirror's neutral plane even though the bipod is not actually attached to the mirror at that point (Yoder 2006). Three such bipods bonded 120° apart offer six degrees of freedom, thereby providing scope for isolation between the mirror and the telescope structure. The three bipods are connected to a rigid mounting ring, which is attached to the telescope structure. The bipod is usually made out of a monolithic block using electrical discharge machining. The bipods can be either attached to the periphery or at suitably located positions at the back side of the mirror (Figure 3.19).

The mirror mount is usually attached to the mirror by a suitable adhesive. Epoxy-based adhesives, widely used for space applications, usually consist of two parts compound. They are mixed as per the manufacturers' instruction and applied to the interface to be bonded and left to cure either at room temperature or at an elevated temperature to form bond interface. The adhesive for spaceborne optical systems should have low outgassing because outgassing molecules can deposit on cold optical surfaces, which can significantly degrade the light reflectance and spectral characteristics of the

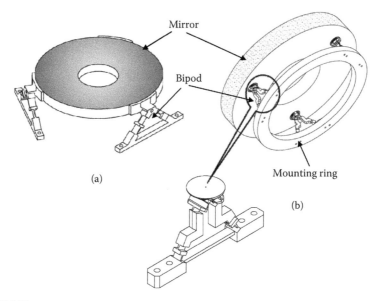

FIGURE 3.19
Schematics showing attaching mirror fixing device to mirror. (a) Bipod flexure attached to the mirror periphery (mounting ring not shown). (b) Bipod attached to the back of the mirror.

optical system. Hence, the outgassing from the adhesive should be within the permissible limit. The choice of the adhesive compound and the process of bonding should ensure that no stress is transferred to the mirror surface. Epoxies shrink while curing and this can transfer stress to the mirror, which can induce deformation of the mirror surface. Another issue is the thermal stress that the epoxy bond can transfer to the mirror when the temperature of the assembly is different from the temperature at which the epoxy was cured. Numerical modeling of these effects is difficult, though attempts have been made by several investigators (Yan et al. 2010). Therefore, it is necessary to understand and evaluate the performance of the epoxy in the bonded condition experimentally. A number of experimental studies are quoted in the literature using different types of adhesives (Patterson et al. 1998; Seo et al. 2007). Kumar and Kumar (2012) have carried out an experimental study on two different adhesives, that is, Epotek-301 and 3M 2216 Gray, particularly on optical surface distortion, lap shear strength, and outgassing property. They have concluded that 3M 2216 Gray with primer 3901 is the favorable choice for optical assemblies for the visible-IR spectrum, exposed to a temperature range of +5° to +40°C and operating at 20°C ± 3°C. It is advisable to carry out experiments on spare optical blanks of the same material as the mirror, to understand the effect of bonding on the mirror surface. The bonding area needs to be optimally chosen. As the bonding area increases the shrinkage effect increases, thereby affecting the surface figure of the mirror. Therefore, to minimize these distortions the contact area of the adhesive with the mirror should be minimized, whereas a small bonding area reduces the strength of the joint. That is, to reduce the thermal distortion the adhesive bonding area should be designed such that it is just large enough to satisfy the stiffness requirement, so as to maintain alignment stability and withstand the launch loads (Mammini et al. 2003).

The design of the bonding pad, which connects the bipod to the mirror should also be designed carefully. The bonding pad is generally designed such that it is thicker at the regions where it connects to the bipods, allowing efficient load transfer between bipod and bonding pad, while the thickness is reduced toward the edge (Figure 3.18c). By thinning the bonding pad to knifelike edges, radial stiffness of the bonding pad is minimized, which facilitates lowering of the stress that is imparted to the mirror during temperature change (Tapos et al. 2005). Long-term dimensional instability of the adhesive needs to be understood as this can affect the alignment of the telescope mirrors (Patterson et al. 1998). Prabhu et al. (2007) have given a general overview of adhesive bonding for optical elements.

3.5.4 Alignment of Mirrors

The mirrors of the telescope have to be integrated in a structure such that the position and orientation of the mirrors are as per the optical design and the whole assembly meets all the environmental specifications and on-orbit operational specifications. The metering structure of the telescope is required to have a high degree of stiffness and dimensional stability over the operating environment. The design should also address the gravity release affecting the performance in orbit. A minimum natural frequency has to be ensured to avoid problems during launch. Except for the "all metal" telescope system, the metering structure is made up of a low thermal expansion material such as invar and carbon-fiber-reinforced plastic. Invar has been used in the IRS PAN and CARTO-1 missions. The Landsat TM telescope structure is constructed using a graphite-epoxy laminate, which has a very low coefficient of thermal expansion and thus eliminates problems due to thermal expansion. However, the graphite-epoxy laminate is hygroscopic and can change dimensions, which needs to be addressed. The Pleadis telescope uses for its structure a carbon/carbon composite, which has a low CTE and is insensitive to moisture.

We shall give a broad outline of the steps involved in aligning a two-mirror telescope system. For a typical two-mirror telescope system, a cylindrical structure is usually chosen with the primary mirror at one end and a spider carrying the secondary at the other end (Figure 3.20). The mechanical parts are fabricated, within the specified fabrication tolerances, to meet the designed values of the placement of the mirrors. To begin with, the components are mounted and aligned to the accuracy possible due to mechanical fabrication. The initial alignment of the primary mirror and secondary mirror in the telescope barrel is done using a theodolite with the primary and secondary rear surfaces taken

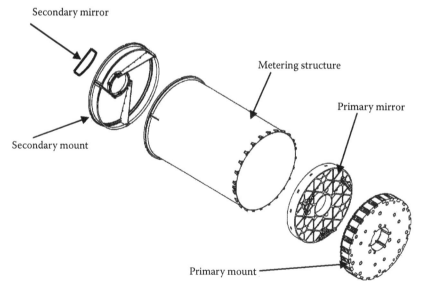

FIGURE 3.20
Schematics showing the components of a two-mirror telescope assembly.

as reference. This will, in general, place the components within about 100 µm, and arc minute accuracy in positioning and tilt respectively. The residual misalignment due to the initial placement produces aberrations of the final image. The next step is to minimize the tilt and decenter errors of the primary and secondary mirrors and maintain the required separation between the mirrors. For this, we have to assess the aberrations and then minimize them to a level such that the overall image quality is as per the set specifications.

Let us recollect how the aberrations are manifested in an imaging optics. A parallel beam incident on an ideal (perfect) imaging optical system emerges as a converging spherical wave front. Any aberration of the optical system produces a wave front that departs from sphericity. Therefore, by measuring the shape of the converging wave front we can estimate the aberrations of the optical system. This is carried out by using an interferometer in an auto-collimating mode. The prime focus of the telescope to be aligned is kept at the focus of the high-quality converging wave front generated by the interferometer. The wave front emerging from the telescope is retro-reflected by a high-quality reference flat (Figure 3.21). The interference between the retro-reflected test beam and the reference beam generated by the interferometer gives rise to interference fringes. The interferogram pattern depends on the nature of aberrations. For a perfect system, we get a null pattern, that is, either fully dark or bright. Figure 3.22 gives the nature of the interference pattern for a few types of misalignment.

If the separation between the two mirrors is not as per design, the emerging wave front would primarily result in defocusing. The interference of this wave front with the reference wave front would give rise to spherical fringes as shown in Figure 3.22a. If the optical centers of the two mirrors do not lie along the optical axis and have decenter error, it would primarily result in coma as shown in Figure 3.22b. If the two mirrors have tilt error, it would primarily result in astigmatism, as shown in Figure 3.22c. Circular fringes would also be obtained if the interferometer focus and the telescope focus do not coincide. It is to be ascertained that the

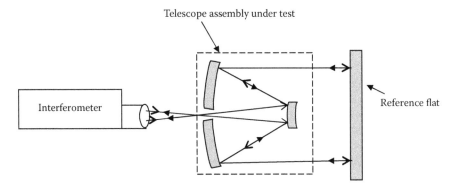

FIGURE 3.21
Optical schematic of the interferometric test setup for telescope alignment.

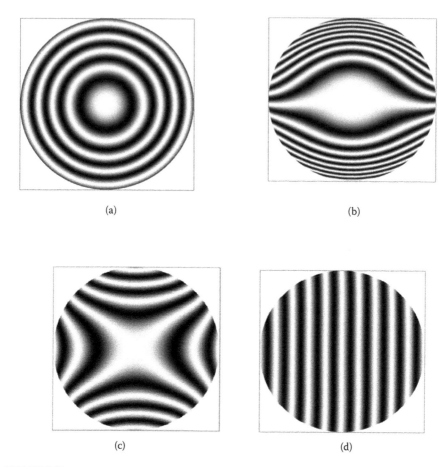

(a)   (b)

(c)   (d)

FIGURE 3.22
Typical interferogram pattern for different types of misalignments: (a) defocus, (b) coma, (c) astigmatism, and (d) tilt.

interferometer focus coincides with the telescope focus. This is ensured by making sure that the interferometer focus is kept at the back focal distance of the telescope. Straight line fringes as shown in Figure 3.22d may be obtained if the two wave fronts are tilted with respect to each other. In reality, misalignments manifest in a combination of tilt, decenter, and defocus, and in an actual system under alignment the pattern can be more complex than what is shown. However, the pattern is not directly used in alignment. Instead, in practice, the interferogram obtained is stored in a computer, which could be part of the interferometer, for further analysis. The computer calculates from the interferogram the optical path difference between an ideal spherical wave front and the wave front from the telescope under test, thereby generating the wave front shape. The wave front shape is expressed in terms of Zernike polynomials from which the type of aberration can be known. The final alignment is carried out by adjusting one or both the mirrors, usually inserting appropriate shims between the mirror mount and the structure. This may need a few iterations until the wave front error is within the specified value. The telescope assembly may be subjected to a few thermal cycles for stabilization and again the alignment stability is verified.

Once the mirrors are assembled and aligned in the structure, we need to integrate the focal plane assembly by placing the detectors at the correct focal plane, to form an imaging system. We shall discuss this aspect in Chapter 6.

References

Ackermann, M. R., J. T. McGraw, and P. C. Zimmer. 2008. Five lens corrector for Cassegrain-form telescopes. *Proceedings of SPIE* 7060: D1–D3.

Ackermann, M. R., J. T. McGraw, and P. C. Zimmer. 2010. An overview of wide field of view optical designs for survey telescopes. http://www.amostech.com/ TechnicalPapers/2010/Systems/Ackermann.pdf (accessed on February 5, 2014).

Hanxiang, B. and S. P. Sadoulet. 2007. Large-format tele-centric lens. *Proceedings of SPIE* 6667: 51–58.

Barnes, W. P. Jr. 1969. Optimal design of cored mirror structures. *Applied Optics.* VIII: 1191–1196.

Bely, Y. P. 2003. *Construction of Large Optical Telescopes.* Springer, New York.

Bergener, D. W., S. M. Pompea, D. F. Shephard, and R. P. Breault. 1985. Stray light rejection performance of SIRTF: A comparison. *Proceedings of SPIE* 0511: 65–72.

Bicknell, W. E., C. J. Digenis, S. E. Forman, and D. E. Lencioni. 1999. EO-1 advanced land imager. *Proceedings of SPIE* 3750: 80–85.

Boies, M. T., J. D. Kinnison, and J. A. Schwartz. 1994. The effect of electron radiation on glass used for space-based optical systems. *Proceedings of SPIE* 2287: 104–113.

Chevrel, M., M. Courtois, and G. Weill. 1981. The SPOT satellite remote sensing mission. *Photogrammetric Engineering and Remote Sensing* 47: 1163–1171.

Czichy, R. H. 1994. Optical design and technologies for space instrumentation. *Proceedings of SPIE* 2210: 420–433.

Epps, H. W. and M. DiVittorio. 2003. Preliminary optical design for a 4.0 m f/2.19 prime focus field corrector with a 2.0 deg. field of view. *Proceedings of SPIE* 4842: 355–365.

ESA. 2004. http://sci.esa.int/science-e/www/object/index.cfm?fobjectid = 34705 (accessed on January 25, 2014).

Fappani, D. and H. Ducollet. 2007. Manufacturing & control of the aspherical mirrors for the telescope of the French satellite Pleiades. *Proceedings of SPIE* 6687: T1–T11.

Faulde, M. and R. N. Wilson. 1973. A three-lens prime focus corrector for parabolic telescope mirrors. *Astronomy and Astrophysics* 26: 11–15.

Figoski, J. W. 1999. The QuickBird telescope: The reality of large, high-quality, commercial space optics. *Proceedings of SPIE* 3779: 22–30.

Firestone, R. F. and Y. Harada. 1979. Evaluation of the effects of solar radiation on glass. *National Aeronautics and Space Administration*, George C. Marshall Space Flight Center, Alabama 35812, Contract No. IAS8-32521, Final Report No. D6139.

Fruit, M., A. Gusarov, and D. Doyle. 2002. Measuring space radiation impact on the characteristics of optical glasses: measurement results and recommendations from testing a selected set of materials. *Proceedings of SPIE* 4823: 132–141.

Gallagher, D., J. Bergstrom, J. Day et al. 2005. Overview of the optical design and performance of the high resolution science imaging experiment (HiRISE). *Proceedings of SPIE* 5874: K1–K10.

Genberg, V. and N. Cormany.1993. Optimum design of lightweight mirrors. *Proceedings of SPIE* 1998: 60–71.

Grabarnik, S., M. Taccola, L. Maresi et al. 2010. Compact multispectral and hyperspectral imagers based on a wide field of view TMA. *International Conference on Space Optics*, Rhodes, Greece. Vol. 4. http://www.rsp-technology.com/ICSO-2010-Surface-roughness%201nm.pdf (accessed on May 20, 2014).

Greenway, P., I. Tosh, and N. Morris. 2004. Development of the TopSat Camera, *Proceedings of the 5th International Conference on Space Optics (ICSO 2004)*, Toulouse, France. ed. B. Warmbein. ESA SP-554 (113).

Hales, W. L. 1992. Optimum Cassegrain baffle systems. *Applied Optics* 31: 5341–5344.

Han Yuan-yuan, Zhang Yu-min, Han Jie-cai, Zhang Jian-han, Yao Wang, and Zhou Yu-feng. 2006. Optimum design and thermal analysis of lightweight silicon carbide mirror. *Proceedings of SPIE* 6148: R1–R5.

JODAS-My Optics. 1999. http://www.myoptics.at/jodas/twomirror.html (accessed on January 25, 2014).

Joseph, G., V. S Iyengar, R. Rattan et al. 1996. Cameras for Indian remote sensing satellite IRS-1C. *Current Science* 70(7): 510–515.

Kirschsteina S., A. Kochb, J. Schöneicha, and F. Döngia. 2005. Metal mirror TMA—Telescopes of the JSS product line. *Proceedings of SPIE* 5962: M1–M10.

Korsch D. 1977. Anastigmatic three-mirror telescope. *Applied Optics* 16: 2074–2077.

Kumar, S. M., C. S. Narayanamurthy, and A. S. Kiran Kumar. 2013. Iterative method of baffle design for modified Ritchey–Chretien telescope. *Applied Optics* 52: 1240–1247.

Kumar, S. M. and A. S. Kiran Kumar. 2012. Adhesives for optical components: An implementation study. *Journal of Optics* 38: 81–88.

Lampton, M. and M. Sholl. 2007. Comparison of on-axis three-mirror-anastigmat telescopes. *Proceedings of SPIE* 6687: 51–58.

Lansing, J. C. Jr. and R. W. Cline. 1975. The four and five band multispectral scanners for Landsat. *Optical Engineering* 14: 312–322.

Lepretre, F. 1994. Lens assemblies for multi-spectral camera, *Proceedings of SPIE* 2210: 587–600.

Mahajan, V. N. 1991. *Aberration Theory Made Simple*. SPIE Press, Bellingham, Washington, DC.

Mammini, P., A. Nordt, B. Holmes, and D. Stubbs. 2003. Sensitivity evaluation of mounting optics using elastomer and bipod flexures. *Proceedings of SPIE* 5176: 26–35.

Marsh, R. G. and H. N. Sissel. 1987. A comparison of wide-angle, unobscured, all-reflecting optical designs. *Proceedings of SPIE* 0818: 168–182.

Myung, C. and R. M. Richard. 1990. Structural and optical properties for typical solid mirror shapes. *Proceedings of SPIE* 1303: 80–88.

Novi, A., G. Basile, O. Citterio et al. 2001. Lightweight SiC foamed mirrors for space applications. *Proceedings of SPIE* 4444: 59–65.

Onaka, T. and A. Salama. 2009. AKARI: Space infrared cooled telescope. *Experimental Astronomy* 27: 9–17.

Paquin, R. A. 1995. Materials for mirror systems: An overview. *Proceedings of SPIE* 2543: 2–11.

Patterson, S. R., V. G. Badami, K. M. Lawton, and H. Tajbakhsh. 1998. The dimensional stability of lightly loaded epoxy joints, UCRL-JC-130589. https://e-reports-ext. llnl.gov/pdf/235105.pdf (accessed on February 7, 2014).

Photonics Handbook. http://www.photonics.com/edu/Handbook.aspx (accessed on January 19, 2014).

Prabhu, K. S., T. L. Schmitz, P. G. Ifju, and J. G. Daly. 2007. A survey of technical literature on adhesive applications for optics, *Proceedings of SPIE* 6665: 71–711.

Rao, H. M. V., R. Venketaswaran, N. Mahale et al. 2008. Ion figuring of light-weighted aspherical mirrors. *Journal of Optics* 37(4): 115–121.

Richard, R. M. and T. M. Valente. 1991. Interference fit equations for lens cell design. *Proceedings of SPIE* 1533: 12–20.

Risse, S., A. Gebhardt, C. Damm et al. 2008. Novel TMA telescope based on ultra precise metal mirrors. *Proceedings of SPIE* 7010: 701016.1–701016.8.

Rodgers, M. J. 2002. Unobscured mirror designs, *Proceedings of SPIE* 4832: 33c60.

Sacek, V. 2006. http://www.telescope-optics.net/dall_kirkham_telescope.htm (accessed on February 7, 2014).

Schwesinger, G. 1954. Optical effect of flexure in vertically mounted precision mirrors. *Journal of the Optical Society of America* 44: 417–424.

Seo Yu Deok, Heayun-Jaung, Kim, Sung-Ke Youn et al. 2007. A study on the adhesive effects on the optical performance of the primary mirror system of a satellite camera. *Journal of the Korean Physical Society* 51: 1901–1908.

Shafer, D. R. 1978. Four-mirror unobscured anastigmatic telescopes with all-spherical surfaces. *Applied Optics* 17: 1072–1074.

Sholl M. J., M. L. Kaplan, and M. L. Lampton. 2008. Three mirror anastigmat survey telescope optimization. *Proceedings of SPIE* 7010: 70103M1–70103M11.

Silny, J. F., E. D. Kim, L. G. Cook, E. M. Moskun, and R. L. Patterson. 2011. Optically fast, wide field-of-view, five-mirror anastigmat (5MA) imagers for remote sensing applications. *Proceedings of SPIE* 8158: 815804.1–815804.13.

Spoto, F., P. Martimort, O. Sy, and P. Laberinti. 2012. Sentinel-2 optical high resolution mission for GMES, Operational services project team, *ESA/ESTEC*. http://www.congrexprojects.com/docs/12c04_doc/4sentinel2_symposium_spoto.pdf?sfvrsn = 2 (accessed on February 7, 2014).

Tapos, F. M., D. J. Edingera, T. R. Hilbya, M. S. Nia, B. C. Holmesb, and D. M. Stubbs. 2005. High bandwidth fast steering mirror, *Proceedings of SPIE* 5877: 587707.1–587707.14.

Thépaut, L., J. Rodolfo, F. Houbre, and R. Mercier-Ythier. 1999. Fabrication and test of high performance wide angle lens assemblies for ocean colour monitor. *Proceedings of SPIE* 3739: 56–62.

Vukobratovich, D. 1997. *Handbook of Opto-Mechanical Engineering*. ed. A. Ahmed. CRC Press, Boca Raton, FL.

Vukobratovich, D. and R. M. Richard. 1988. Flexure mounts for high-resolution optical elements. *Proceedings of SPIE* 959: 18–36.

Wetherell, E. and B. William. 1980. The calculation of image quality. *Applied Optics and Optical Engineering.* Vol. VIII. ed. R. R. Shannon and J. C. Wyant. Academic Press, NY.

Wilson, R. N. 1968. Corrector systems for Cassegrain telescopes. *Applied Optics* 1;7(2): 253–263.

Wynne, C. G. 1968. Ritchy-Cheretin telescopes and extended field systems. *The Astrophysical Journal* 152: 675–693.

Yan, L., YongMing Hu, YingCai Li , YouShan Qu, and JiaoTeng Ding. 2010. Analysis of bonding stress with high strength adhesive between the reflector and the mounts in space camera. *Proceedings of SPIE* 7654: 76541F1–76541F6.

Yoder, P. R. 2006. *Opto-Mechanical Systems Design*, 3rd Edition. CRC Press, Boca Raton, FL.

Yoder, P. R. 2008. *Mounting Optics in Optical Instruments*, 2nd Edition. SPIE Monograph Bellingham, Washington, DC.

Yong Yan, Guang Jin, and Hong-bo Yang. 2007. Design and analysis of large space-borne light-weighted primary mirror and its support system. *Proceedings of SPIE* 6721: V1–V7.

Young, A. T. 1967. Design of Cassegrain light shields. *Applied Optics* 6: 1063–1068.

Yu Chuan Lin, Long-Jeng Lee, Shenq-Tsong Chang, Yu-Cheng Cheng, and Ting-Ming Huang. 2011. Numerical and experimental analysis of light-weighted primary mirror for Cassegrain telescope. *Applied Mechanics and Materials* 52: 59–64.

4

Earth Observation Camera: An Overview

4.1 Introduction

The information collected by Earth observation cameras is meant to identify and map various Earth surface objects. This activity is known as remote
sensing and the instruments used to collect the data are called remote sensors. Remote sensors can be broadly classified as passive sensors and active
sensors. Sensors that sense natural radiation, either emitted or reflected
from the Earth, are called passive sensors. It is also possible to produce electromagnetic radiation of a specific wavelength or band of wavelengths as
a part of the sensor system and the interaction of this radiation with the
target could then be studied by sensing the scattered radiation from the
targets. Such sensors, which produce their own electromagnetic radiation,
are called active sensors. The technology involved in developing sensors is
not the same throughout the electromagnetic spectrum. The technology for
developing microwave sensors is quite different from that of optical-infrared
(OIR) sensors. Therefore, from the standpoint of understanding the design
and realization of the sensors, it is convenient to classify the sensors (both
passive and active) as those operating in OIR region and those operating in
the microwave region. The OIR and microwave sensors can be either imaging
or nonimaging sensors. The imaging sensors give a two-dimensional spatial
distribution of the emitted or reflected intensity of the electromagnetic radiation (as in a photographic camera), while the nonimaging sensors measure
the intensity of radiation, within the field of view (FOV) of the instrument,
and in some cases, as a function of distance along the line of sight of the
instrument (e.g., vertical temperature profiling radiometer [VTPR]). Figure 4.1
gives a possible classification of remote sensors. In this book, we deal with
passive imaging sensors operating in the OIR region.

In this chapter, we shall discuss various performance parameters that
characterize the cameras. Let us try to understand from the viewpoint
of those who use the images from Earth observation cameras, what they
look for in the images. In general, a user is interested in identifying and
characterizing various surface features/objects and measures its physical
properties such as length/distance and height. For this information to be

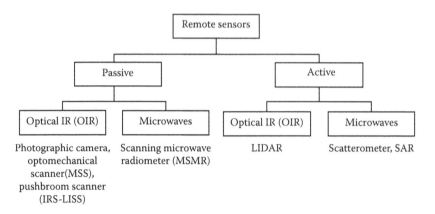

FIGURE 4.1
Classification of remote sensors. (Reproduced with permission from Joseph, G., *Fundamentals of Remote Sensing*, p. 130, Ed. 2, Universities Press, Hyderabad, India, 2005.)

of practical use, the locations of the objects need to be identified with reference to geographic coordinates, that is the end user of the Earth observation data is interested in classification and mapping of the Earth's features. This is achieved by estimating the radiance emitted/scattered from the targets of interest. Therefore, we may say that the image should faithfully represent the geometric and radiometric properties of the ground scene. Such observations have to be made repetitively, depending on the phenomenon under study. It is reasonable to assume that these requirements will depend on the instrument's ability to detect small differences in the emittance/reflectance of the Earth's surface in a number of spectral bands for as small an object as required and as often as necessary. Then the important question is, "What is the optimum set of specifications for a remote sensor?" Unfortunately, there is no unique answer, since the choice of the optimum parameters depends on the theme under study. Even if we identify an ideal set of parameters, the realization of a combination of these ideal parameters (i.e., spatial resolution, number of spectral bands, spectral bandwidth, and signal-to-noise ratio) in a sensor system is a complex problem due to the strong interrelationship these parameters have with one another and the engineering requirements of the sensor (Joseph 2005). We may consider the sensor parameters under four domains: (i) spatial, (ii) spectral, (iii) radiometric, and (iv) temporal.

4.2 Spatial Resolution

The spatial features of relevance in designing an imaging system are the spatial coverage a single frame can provide (primarily specified in terms of the width of the scene) and the spatial resolution. Although the former is straightforward to understand, the term spatial resolution requires elaboration since it is not unambiguously defined and can be interpreted in many ways.

Before we get into spatial resolution in remote sensing, it is useful to understand how the term is used in telescopes. The early use of telescopes was in observing celestial bodies and hence the astronomers wanted to investigate the ability of a telescope to distinguish two close-by stars (point sources) in the FOV. In Chapter 2, we saw that due to diffraction of light, a point source never appears as a point in the image plane. As we saw in Figure 2.11, for a circular aperture, the image has a central bright disc surrounded by a series of fainter rings, with dark minima in between the rings—known as an airy pattern. Thus, diffraction limits the ability of any imaging system to resolve two adjacent objects. The minimum angular separation that can be distinguished is referred to as the angular resolution of the telescope. The generally accepted rule to estimate angular resolution is the Rayleigh criterion, stating that the images of two point objects will be just resolved when the central maximum in the diffraction pattern of one image coincides with the first minimum of the diffraction pattern of the other image (Figure 4.2).

The angular resolution θ is given by

$$\theta = 1.22 \frac{\lambda}{D} \tag{4.1}$$

where λ is the wavelength of the radiation measured, and D the diameter of the telescope aperture (both have to be in the same units, and θ in radians). The θ gives the least angular separation of two point sources so that they are seen as separate when viewed through an optical instrument. When θ is multiplied by the distance of the object from the telescope, we get the resolvable linear distance. This gives the theoretical limit to resolve two objects by an imaging system. It should be remembered when we record an image, other factors like noise introduced by the detecting and recording system further degrades the performance. Though the Rayleigh criterion is not used to define resolution in remote sensing imagery, this formula helps to decide on the minimum diameter of the optics to produce a diffraction-limited image. Let

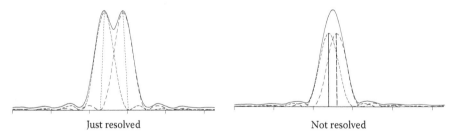

Just resolved   Not resolved

FIGURE 4.2
Resolution limit. Two points are said to be just resolved when the principal maxima of the Airy pattern of one point falls at the first minima of the other point. (Reproduced with permission from Joseph, G., *Fundamentals of Remote Sensing*, p. 131, Ed. 2, Universities Press, Hyderabad, India, 2005.)

us consider the IRS PAN camera, with an entrance aperture of ~22 cm, where the limit of resolution at 0.5-μm wavelength is 2.7×10^{-6} ($1.22 \times .5 \times 10^{-4}/22$) radians. From an altitude of 810 km, the spatial resolution limit is 2 m ($810 \times 10^3 \times 2.7 \times 10^{-6}$), whereas the instrument resolution is 5.6 m. However, if we were to use the same optics for imaging at short wave IR (SWIR) at 2 μm, the diffraction by optics itself limits the resolution to 8 m.

In astronomical applications, the "two point" resolution criteria that we discussed above has a direct practical significance, since in astronomical observations many objects being observed are effectively point sources. However, this is not the case when we deal with the imagery from Earth observation cameras. In remote sensing, we are interested in identifying targets of finite size. In a very general sense, the spatial resolution is a measure of the sensor's ability to image (record) closely spaced objects that are distinguishable as separate objects in the image. Does the spatial resolution mentioned by the sensor manufacturers uniquely represent this capability? When electro-optical sensors using discrete detectors/detector arrays are used for generating imagery (like Landsat MSS, IRS LISS, etc.), the spatial resolution stated denotes the projection of the detector element onto the ground through the optics. Thus, when one says 5.8 m is the spatial resolution of the IRS 1C/D PAN camera, it only means the projection of one CCD element on the ground through the imaging optics from the satellite orbit is 5.8 m, which is the "footprint" of the detector element on the ground depending on the instantaneous field of view (IFOV) of the camera. That is, 5.8×5.8 m is the smallest area from which the radiance is recorded as a separate unit. This footprint is referred to as the instantaneous geometric field of view (IGFOV) (Figure 4.3), that is, the geometric size of the image projected by the detector on the ground through the optical system. It does not guarantee that all foot prints of 5.8-m dimension are distinguishable in IRS1C/D PAN. However, it is also true that one may be able to detect a high-contrast object that is smaller than the IGFOV if its signal amplitude is large enough to significantly affect the gray scale value of that pixel (for example, roads with widths much smaller than 80 m can be seen in Landsat MSS). This is because the instantaneous FOV alone does not adequately define the spatial response of an imaging system. Because understanding this limitation is very important from a users' viewpoint, we shall briefly discuss the constraints using a single number based on IGFOV for spatial resolution. (There is a subtle difference between the term picture element (known as a pixel) and IGFOV though they are used without distinction. "Pixel or picture element is the data sample in the output product to which a radiance value is assigned" (UN A/AC 105/260). That is, it is the smallest unit of a digital image that can be assigned color and intensity. It has dimensions that are not necessarily related to the sensor system parameters since the data can be sampled at different spacing than the detector footprint).

We have seen in Chapter 2 that the imaging optics reduces the contrast, and the amount of reduction of contrast with spatial frequency is given by

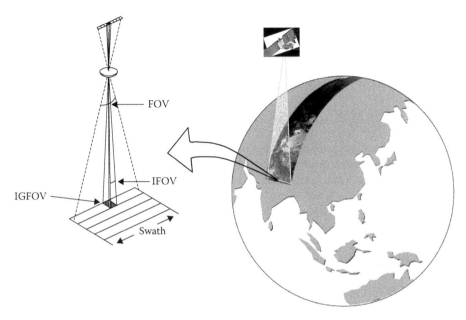

FIGURE 4.3
Schematics showing the concept of FOV, IFOV and IGFOV. (Reproduced with permission from Joseph, G., *Fundamentals of Remote Sensing*, p. 135, Ed. 2, Universities Press, Hyderabad, India, 2005.)

the modulation transfer function (MTF) of the optics. The detector has a finite size and the output of the detector is the average of all the targets' radiance within its footprint on the ground. This spatial averaging is essentially a process of filtering the spatial frequencies available in the scene and it depends on the detector dimension. We can intuitively understand this phenomenon as a detector of larger size averages finer features and hence will have a poor response at higher spatial frequency. This effect can be represented as the MTF of the detector element. MTF of a square detector of size "a × a" can be expressed as

$$\mathrm{MTF}_{\text{detector}}(f) = \frac{\sin(\pi f a)}{\pi f a} \tag{4.2}$$

Here f is represented as line pair per unit dimension. When $f = \frac{1}{a}$, then MTF is zero. This is understandable because at this frequency one line pair is totally covered by the pixel and hence the average value is zero. When $f = \frac{1}{2a}$, then the MTF is 0.63. That is, the detector footprint covers a black or white strip, which represents the IGFOV.

In any electro-optical system, data acquisition is carried out by sampling. The process of spatial sampling further degrades the MTF due to sampling the MTF (Boreman 2001). Other components contributing to the MTF of a camera include the camera electronics, the satellite attitude jitter while in orbit, and even the atmosphere. All these contribute to the overall MTF of the system. As discussed in Section 2.6, the resulting MTF of the system is the product of all the MTFs of its components, that is

$$\mathrm{MTF}_{camera} = \mathrm{MTF}_{optics} \times \mathrm{MTF}_{detector} \times \mathrm{MTF}_{electronics} \times ----- \cdot$$

If we have to distinguish any two adjacent objects, there should be a difference in the radiance from these two objects at the wavelength of observation, which we call contrast. If the MTF brings down the contrast of targets in the recorded image to a level below the detectable threshold of the analyzing system, such objects will not be detected. Therefore, smaller objects of higher contrast may be detected compared to larger objects that have less contrast.

Is it justifiable to specify the sensor quality just in terms of geometric projection without any consideration of the contrast reduction it produces? How does one judge image quality of different sensors with the same IGFOV, but different MTFs? The visual interpretation of imagery depends on the visual contrast threshold, which is the minimum contrast that can be resolved by the analyst. Let us consider target identification when carried out by computers. Let us assume a hypothetical case of data without any noise. In such a situation, the computer can distinguish targets with contrast difference as small as one digital number. However, when noise is present, only those targets can be distinguished whose radiance difference can be measured, which depends on the radiometric resolution of the system NEΔL (Section 4.4). The differential radiance at the image space depends on the contrast in the object space and the MTF of the total system. Thus, the ability to distinguish two targets depends on the object contrast, MTF, and the noise equivalent radiance. (The term noise equivalent modulation (NEM) is not appropriate as modulation by itself cannot set a lower threshold without target contrast.) Thus, a camera with low MTF but high SNR may perform better than one with a higher MTF but lower SNR. Therefore, a better figure of merit (FM) of a camera for spatial resolution is the ratio of MTF at IGFOV to noise equivalent radiance at a reference radiance, say 10% of saturation value, that is, $\left[\dfrac{\mathrm{MTF}}{\mathrm{NE}\Delta\mathrm{L}}\right]$. To make FM independent of units we may use the value of signal to noise ratio (SNR) at the reference radiance instead of Noise Equivalent Radiance. Since system with highest SNR has a better performance the FOM can be rewritten as [MTF × SNR]. The higher the FM values, the better the discriminability between the targets for the same contrast.

Now, let us consider two cameras with MTF curves as in Figure 4.4. Though at IGFOV both have the same MTF, at twice the IGFOV, "A" can discriminate lower contrast objects better than "B." Therefore, the shape of the MTF curve also matters. Although the MTF curve shape is not a convenient way to specify the system MTF, values at twice the IGFOV can be an additional parameter to be specified as part of the system performance parameters. To summarize,

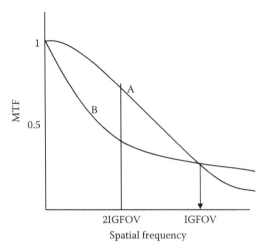

FIGURE 4.4
Schematic showing influence of the shape of the MTF curve on spatial resolution. Though both curves have the same MTF at IGFOV, for larger objects curve A gives a better performance.

apart from the IGFOV, the detectable object size depends on the MTF, the contrast of the targets, and NEΔL.

A concept of effective instantaneous field of view (EIFOV) was introduced by NASA (NASA 1973). EIFOV is defined as the resolution element for which the system MTF is 50%. EIFOV was introduced to compare the performance of different sensors.

Generally, the MTF is considered an indicator of how sharp the edges are in an image. However, the MTF is also a measure of how accurately the actual radiance from a pixel is measured, because a lower MTF indicates contribution from other pixels to the pixel under observation (and vice versa). This could lead to problems in multispectral classification, as the radiance of a measured pixel is dependent on the nature of adjacent pixels. Thus, the same object with different surroundings will have different radiometric response (see Section 4.4.1). This concern was raised by Norwood (1974). Therefore, the question is, "What is radiometrically accurate IFOV?" Joseph (2000), has introduced radiometrically accurate IFOV (RAIFOV) as the resolution for which the MTF is higher than 0.95. Table 4.1 gives these IFOV values for some of the high-resolution cameras.

Another terminology used in relation to spatial characteristics of imagery is the ground sampling distance (GSD). The data can be generated by sampling at certain specified ground distances that could be smaller than IGFOV. GSD is the linear distance between sampling centers on the ground. It is not correct to use GSD as synonym to IGFOV.

All the above explanations point out that it is not possible to define spatial resolution by a single performance parameter. Therefore, to name a single performance parameter to compare different sensors, to establish their capability to resolve smallest object, is an unresolved task. Irrespective of the shortcomings, we shall hereafter refer to spatial resolution as the projection

TABLE 4.1

In-Orbit Measured Values of Various IFOVs for IKONOS, GeoEye, and Cartosat
Cameras

Parameters (m)	IKONOS (1)		WorldView-1 (2)	Cartosat 2A (3)
	PAN	MXL	PAN	PAN
IGFOV	0.82	3.28	0.5	0.8
EIFOV	3.6	12	1.04	2.3
RAIFOV	20	64	12	11

Source: (1) IKONOS Instrument/Product Description, 2009; (2) GeoEye, Instrument/Product
Description, 2009 (1&2credits Digital Globe); (3) Kumar Senthil, M., 2014, Personal communication; and Srikanth et al., *J. Spacecraft Tech.*, 20(1), 56–64, 2010.

of the detector element by the optics onto the ground. Based on this definition, the spatial resolution of Landsat MSS is 79 m, SPOT (multispectral
channel) 20 m, and IRS LISS-II 36 m.

Before concluding this section, we shall briefly discuss the classical John's
criteria for detection, identification, and recognition (DIR) of military objects.
The criteria was established in 1958 by John Johnson of the Army Night
Vision laboratory in Fort Belvoir, Virginia, by relating visual discrimination
levels to the Air Force tri-bar target frequencies using minimum dimension
of the target as reference (Minor 2002). He established minimum line pairs
per dimension to achieve a 50% probability of success for various tasks. To
detect whether a target is present in the scene, there should be a minimum of
1-line pair, whereas for recognition, that is, to which class the object belongs
(i.e., the target is tank and truck) requires 3-line pairs. For identification, the
type within a class (i.e., type of tank) to which an object belongs requires
6-line pairs. This is a very useful guideline for military target surveillance.

Figure A.2 shows how the discriminability of features gets better as the
spatial resolution increases.

4.3 Spectral Resolution

In multispectral remote sensing, the variation in reflected/emitted spectral
radiation is used to distinguish various features. However, it is not essential in
most cases to get continuous spectral information. (In hyper spectral imaging,
we measure continuous spectra, discussed in Chapter 8.) In multispectral imaging cameras, we sample the reflected/emitted spectrum by making measurements at a few selected wavelengths. There are three aspects to be considered in
the spectral domain: (1) location of the central wavelength, (2) the bandwidth,
and (3) the total number of bands. The wavelength region of observation, usually called spectral bands, is defined in terms of a "central" wavelength (λ_c) and
a "bandwidth" ($\Delta\lambda$). The bandwidth is defined by lower (λ_1) and upper ((λ_2)
cutoff wavelengths. The spectral resolution $\Delta\lambda$ is given by ($\lambda_2 - \lambda_1$).

The spectral resolution describes the wavelength interval at which the observation is made. The smaller the $\Delta\lambda$, the higher is the spectral resolution. Though the definition looks simple, the exact location of λ_1 and λ_2 in a practical system is difficult. When one says the spectral bandwidth of IRS LISS, Band 1 is 0.45–0.52 µm, what does it really mean? In an ideal system, the response should be 1 between the wavelengths 0.45 and 0.52 µm, and 0 for wavelengths outside this range (Figure 4.5a). Such a rectangular or box-car-shaped filter is not practically feasible and a realistic filter has finite rise and fall characteristics in the spectral response curve. Therefore, there should be spectral bandwidth normalization to convert the spectral response to an equivalent box-car-shaped response with well-defined wavelength limits and a constant pass band response. One method usually adopted is to define the bandwidth in terms of 50% of the peak value on either side (full width at half maximum— FWHM) (Figure 4.5b). This definition holds best if the spectral response of the system is Gaussian or close to that. However, in a practical filter, the pass band response has a number of "ringings" (Figure 4.5c) and assigning the peak value is not straightforward. Also, this method may not be appropriate when the response is skewed. Palmer (1984) suggested a technique called the "moments" method to compute λ_c and $\Delta\lambda$, which avoids the above problem of identifying the peak value. The analysis is based on determining the first and second moments of the spectral response curve R (λ) (Palmer 1980). Following Palmer (1984), the central wavelength λ_c and $\Delta\lambda$ are given as

$$\lambda_c = \frac{\int_{\lambda_{min}}^{\lambda_{max}} \lambda R(\lambda)d\lambda}{\int_{\lambda_{min}}^{\lambda_{max}} R(\lambda)d\lambda} \tag{4.3}$$

$$\lambda_1 = \lambda_c - \sqrt{3}\sigma \text{ and } \lambda_2 = \lambda_c + \sqrt{3}\sigma \tag{4.4}$$

$$\Delta\lambda = 2\sqrt{3}\sigma \tag{4.5}$$

where σ is given by

$$\sigma^2 = \frac{\int_{\lambda_{min}}^{\lambda_{max}} \lambda^2 R(\lambda)d\lambda}{\int_{\lambda_{min}}^{\lambda_{max}} R(\lambda)d\lambda} - \lambda_c^2$$

λ_{min} and λ_{max} are the minimum and maximum wavelengths beyond which the spectral response is zero. The merit of this method is that the values are not dependent on the spectral response shapes. Pandya et al. (2013) have used this technique to compare the spectral characteristics of sensors on board the Resourcesat-1 and Resourcesat-2 satellites. Table 4.2 gives a comparison of the values for Resourcesat-2 LISS3 bands using FWHM and the moment

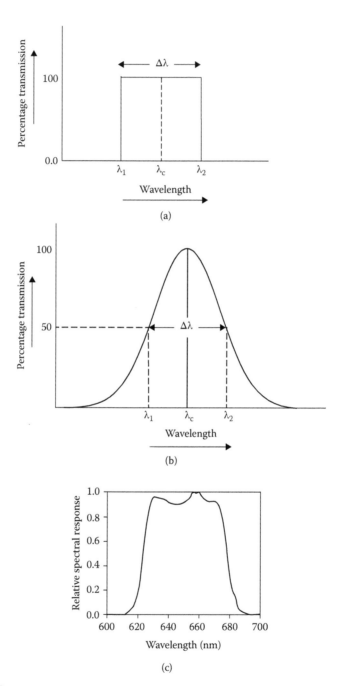

FIGURE 4.5

Spectral bandwidth definitions. (a) An ideal response, wherein response is zero below λ_1 and above λ_2. (b) More close to a practical filter, the bandwidth $\Delta\lambda$ is the full width at half maxima (FWHM). (c) Spectral response of a practical filter. The peak response is not well defined. (Reproduced with permission from Joseph, G., *Fundamentals of Remote Sensing*, p. 139, Ed. 2, Universities Press, Hyderabad, India, 2005.)

TABLE 4.2

Comparison of Bandwidth (Δλ) and Central Wavelength (λc) for Resourcesat-2
LISS-III Using FWHM and Moment Method

Bands	FWHM (μm)		Moment Method (μm)	
	λc	Δλ	λc	Δλ
B2	559	66	561.5	69
B3	651.0	56	651.5	58.0
B4	812.0	85	811.8	95.1
B5	1620.0	132	1619.0	149.3

Source: Pandya, M. R. 2013, personal communication.

method. As seen from the table, the central wavelengths by the two methods are very close, but the bandwidth by the moment method is lower than the conventional FWHM method by about 5% to 13%. It is suggested the moment method be adopted by all the sensor manufacturers so that intercomparison is meaningful.

Another spectral parameter that is of concern but often not seriously considered is the out-of-band response. This is particularly important when a narrow bandwidth is used for spectral selection as in the case of ocean color monitoring sensors such as SeaWifs and OCM. In the case of SeaWifs at the data product level, the specification is less than 5% of the within-the-band value. Within band in this case is defined as the wavelength interval between the lower and upper 1% response (Barnes et al. 1999).

At the instrument level, the overall spectral response depends on the optics response, filter response, and the detector response. At the data product level, the spectral content also depends on the input source spectral response shape, which is difficult to assess in remote sensing as it depends on the target characteristics. The band shape of the interference filter chosen and its out-of-band rejection property should be consistent with the overall camera performance requirements.

The selection of bandwidth is a trade-off between the energy to be collected and the spectral shape of the feature to be observed. If one is interested in observing some characteristic emission lines, then obviously the bandwidth has to be much smaller than the line width so as to sample a few points in the line spectra. For land observations in the visible and near IR (VNIR), a few tens of nanometer is usually used, while for ocean observation 10–20 nm is used.

Other important parameters in the spectral domain are the number of spectral bands and the central wavelength; both are given to the camera designer. The most important criterion for the selection of band location is that the spectral bands should be away from the absorption bands of atmospheric constituents, i.e., should be within the atmospheric window. The Landsat MSS bands were chosen rather arbitrarily. However, field studies have shown that certain spectral bands are best suited for specific themes. The thematic mapper bands are selected based on such investigations. The bands selected should be uncorrelated to the extent possible, because correlated bands give redundant information, which does not help in improving the discriminability. There are a number of statistical methods to evaluate correlation between the bands (Kondratyev and Pokrovsky 1979). Unfortunately, the outcome of such studies is scene dependent and hence a certain amount of generalization is required when one designs a camera for global coverage.

From the classification accuracy point of view, the increase from one to two bands produces the largest improvement in accuracy. However, after about four bands, the accuracy increases very slowly or may even flatten out. Because the data rate increases directly with the number of bands, selection of an optimum number of bands is essential. What is important is to have a minimum set of optimal bands. The study of Sharma et al. (1995) shows that the addition of a middle IR band with any other band combination of TM gives an improved separation of classes in agricultural classification.

4.3.1 Interference Filter

Though in some of the Earth observation cameras, spectral selection has been carried out by using conventional devices like prisms and gratings, most of the modern multispectral imaging systems use interference filters. So, in this section, we shall briefly discuss the functioning of interference filters. Interference filters consist of very thin multiple layers of alternating high and low refractive index material coated on a suitable substrate. Usually, three or five layers are grouped into units called "cavities" and they are separated by a thick spacer. The wavelength selection is by virtue of the interference effects that take place between the incident and the reflected waves at the thin-film interface. By appropriate choice of thicknesses, the range of wavelengths that the filter transmits (i.e., the passband) can be adjusted. The number of cavities in a filter decides the overall shape of the filter. As more cavities are included, the slope of the spectral response on the lower and upper wavelength bands of the response curve become steeper and the peak becomes increasingly flat (Schott Catalogue). The IRS cameras use four cavity filters. The filter characteristics are specified for normal incidence. Because the path length of the light within the filter changes with the incidence angle, the filter characteristics change if it is tilted or used in a strongly convergent or divergent beam. This is of concern when cameras of wide FOV have to use interference filters for spectral selection. The peak transmission will be shifted to a shorter wavelength dependent on the "effective refractive index" of the filter (n_e) in accordance with the following formula:

$$\lambda = \lambda_0 \left[1 - \left(\frac{n_0}{n_e} \right)^2 \sin^2 \theta \right]^{\frac{1}{2}} \tag{4.6}$$

where λ_0 is wavelength at normal incidence, λ is shifted wavelength at angle of incidence θ, n_e is effective refractive index of the filter, and n_0 is refractive index of external medium (1 for air).

Filters are also sensitive to temperature change; the amount of shift depends on the design.

4.4   Radiometric Resolution

The sensors respond only to those radiance values lying between the lower and upper radiance settings. The lower level is usually set to zero radiance. The upper level is usually referred to as the *saturation radiance* (SR) since for any radiance input beyond SR, the output of the sensor remains constant. Thus, the maximum radiance we can measure is SR. The SR setting depends on the mission objective. For example, for cloud/snow radiance measurement, the value is kept at or a little above 100% reflectance of solar irradiance, whereas for the ocean color measurement ~10% reflectance is adequate, as ocean-leaving radiance itself is only a few percent of solar irradiance. In electro-optical sensors, the output is usually digitized, to produce discrete levels. The digitization is referred as quantization and is expressed as "n" binary bits. Thus a 7-bit digitization implies 2^7 or 128 discrete levels (0–127).

Radiometric resolution is a measure of the capability of the sensor to differentiate the smallest change in the spectral reflectance/emittance between various targets. This is represented as the noise equivalent reflectance (or temperature) change—NE$\Delta\rho$ (or NEΔT). This is defined as the change in reflectance (temperature) that gives a signal output of the sensor equal to the root mean square (rms) noise at that signal level. Instead of NE$\Delta\rho$, it is more appropriate to use noise equivalent differential radiance (NEΔL), which we can measure at the instrument level. This can be defined as the radiance change at the input of the sensor, which produces an output signal change equivalent to rms noise at that signal level. It depends on a number of parameters such as the signal to noise ratio (S/N), the saturation radiance setting, and the number of quantization bits. SPOT HRV has 8-bit quantization, whereas the IRS LISS1/2 cameras have 7-bit quantization. If the S/N and saturation setting are properly chosen, in principle, for a specific radiance, both can have the same NEΔL. Current systems are being designed with 11 or more bit digitization. Such systems, unless they have a corresponding S/N, do not imply a better radiometric resolution. We shall illustrate this by comparing two hypothetical sensors. Let us consider two sensors with a linear transfer function, i.e., output voltage is linear with respect to input radiance, with the characteristics given in Table 4.3. Sensor-2 can sample at a smaller radiometric interval compared to sensor-1 and no measurement can be made better than that given

TABLE 4.3

Comparison of Design Parameters for Two Sensors and the Trade-Offs in Instrument Performance

Parameters	Sensor-1	Sensor-2
(a) Saturation radiance (mw/cm²/sr/μm)	20	35
(b) No. of bits (levels)	7(128)	8(256)
(c) Radiometric sampling interval (a/b) (mw/cm²/sr/μm)	0.16	0.14
(d) System S/N at 15mw/cm²/sr/μm	90	75
(e) NEΔL (15/d) (mw/cm²/sr/μm)	0.17	0.20

Source: Joseph George, *ISPRS J. Photogramm.*, 55(1), 9–12, 2000. Reproduced with permission.

by "c" in Table 4.3. However, sensor-1 gives a better performance in terms of the noise equivalent radiance at the reference radiance. Thus, the quantization alone does not necessarily give an idea of the radiometric resolution capability of an instrument. Nevertheless, there is an advantage to having a higher number of quantization levels. A camera in orbit has to collect imagery across a wide range of illumination levels depending on the latitude and the target reflectance—more than 90% for snow to a few percent for water bodies. The dynamic range required to cover a wide range of radiance received by the camera requires a nonlinear gain or capability to change gain in flight anticipating expected radiance for a latitude and season. If we have a higher number of bits, it increases the dynamic range, so that the measurement of objects with radiance varying from ocean to snow can be performed without any gain change.

4.4.1 Radiometric Quality

Radiometric quality of the image depends primarily on the radiometric resolution, the calibration accuracy, and the MTF. Resolution, in general, is the minimum difference between two discrete values that can be distinguished by a measuring device. However, high resolution does not necessarily imply high accuracy. Accuracy is a measure of how close the measurement is to the true value.

The radiometric accuracy could be of two kinds.

(1) Absolute accuracy: An electro-optical sensor output, using suitable calibration, can be represented as, say, mw/cm²/sr/μm. How close the sensor data is to a primary standard of radiance is the measure of absolute accuracy.

(2) Relative accuracy: This refers to the relative accuracy among the bands with respect to a primary standard. For example, for a four-band multispectral camera, though each band does not represent accurately the radiance value, if their ratio with reference to one band is the same as the true value ratio, then there is no relative error.

In computer classification of the features from the remotely sensed data (as in the maximum likelihood classifier), normally one looks at the statistical difference in the reflected value measured by the instruments. In this context, the absolute value of the reflectance (or radiance) is not important because one compares the relative reflectance values between the pixels. However, when one is interested in using two-date data for classification or combining information from two different sensors to find out how the reflectance changed with time, it is necessary to have the absolute radiance value.

Radiometric errors are also introduced due to the MTF of the camera system. As mentioned earlier, the MTF essentially shows the contrast reduction from object space to the image plane. Because of this, the radiance measured by the instrument does not represent the actual reflectance of the pixel as the signals from the adjacent pixels spill over. This could lead to problems in multispectral classification, because the radiance of a pixel measured is dependent on the nature of the adjacent pixels. This is shown in Figure 4.6.

Another contribution to the radiometric error is due to atmosphere, because the atmosphere corrupts the actual reflectance reaching the sensor. The scattering in the atmosphere adds an additional amount of radiance in the observation geometry-path radiance, which in turn reduces the contrast and also adds to the actual reflectance value. The atmospheric effect of nearby pixels on the radiance of a given pixel (adjacency effect) also affects the spatial resolution and introduces radiometric error (Kaufman 1984). The radiometric

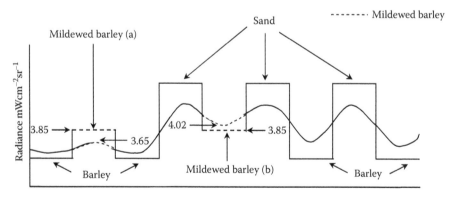

FIGURE 4.6

Example showing the MTF effect on radiometric accuracy. The "square wave" pattern gives the actual radiance from three targets—barley, mildewed barley (MB), and sand. There are two fields of mildewed barley—one among barley (a) and the other among sand (b). The "sine curve" gives the radiance when the scene radiation is measured through a radiometer. The numbers represent the radiance values. The MTF of the radiometer modifies the radiance value. Mildewed barley (a) and (b) have the same radiance (3.85). But due to the radiometer MTF, the MB at "a" shows a lower value (3.6), whereas MB at "b" shows a higher value (4.02). In the case of "b" the higher radiance from the sand spills over to MB, whereas the reverse happens in case "a." Thus, the redistribution of the radiance due to the MTF can give an erroneous radiance value compared to the actual field value. (From Norwood, V. T., *Proc. of SPIE* 51, 37–42, 1974. Reproduced with permission from SPIE.)

errors generally lead to poor classification accuracy. These effects can be corrected to a certain extent using appropriate models.

4.5   Temporal Resolution

One of the main advantages of satellite remote sensing is its ability to repeatedly observe a scene at regular intervals. Temporal resolution refers to the temporal frequency with which a given scene can be imaged, usually expressed in days. The IRS-1C LISS-III camera has a temporal resolution of 24 days. That is any part of the globe (except around the pole) can be imaged every 24 days. This is also called repetivity. Repetivity depends on the orbit characteristics and the swath. The larger the swath, the higher will be the temporal resolution. Thus, SeaWiFS with a swath of about 3000 km has a temporal resolution of 1 day. Of course, the highest temporal resolution is possible by geosynchronous observation systems such as Meteosat, Insat, and VHRR, with a temporal resolution ranging from a few minutes to 30 minutes, depending on the mode of operation. Higher temporal resolution enables monitoring of rapid changes, such as forest fires, floods, etc. and also improves the probability of obtaining cloud-free imagery over areas that experience frequent cloud cover.

Here, the repetivity of 24 days of IRS-1C means that the subsatellite track repeats (except for slight orbit perturbations) every 24 days. Therefore, the images taken every 24 days have the same instrument view angle for any location, which is important so that the Bidirectional Reflectance Distribution Function (BRDF) differences do not influence the data. With the launch of SPOT, we have, for the first time, the capability of imaging an area by tilting the view direction of the camera across track. This mode increases the frequency of observation of a specific site and the term "revisit capability" is added to address this capability. This novel concept is an excellent idea to image a specific area at shorter intervals than the temporal resolution of 26 days of SPOT by across-track pointing of the sensor. However, the "revisit capability" should not be misconstrued as temporal resolution (also referred to as repeat cycle). The revisit of a location is carried out at the cost of not acquiring data over some other location. Therefore, if revisit capability is exercised, it is not possible to have systematic coverage of the whole globe within a specified time duration, as is possible with Landsat TM/IRS LISS cameras. This subtle difference should be understood.

It should be borne in mind that when the engineering design of the sensor is carried out, there are trade-offs between the four resolutions. For example, if we want a high spatial resolution, the IFOV has to be reduced, which reduces energy collected by the sensor, thereby producing a poor S/N leading to a poor radiometric resolution. On the other hand, keeping the spatial resolution the same, one can improve the radiometric resolution by increasing the spectral bandwidth (thereby collecting more energy), giving, however, poor spectral resolution. Again, higher spatial resolution and increased swath results in increased data rate. Thus, it is very essential to have a judicial choice of sensor parameters, to meet the data utilization requirement.

4.6 Performance Specification

The best use of any instrument depends on how well the user understands the performance characteristics of the instrument. Many of the Earth observation camera parameters are not unambiguously defined and much worse they are interpreted in many ways. There is a need to specify the optimal parameters of a camera to understand its potential and for intercomparison with other sensors. Considering these aspects, the author, who was president of Commission-I (1996–2000) of the International Society for Photogrammetry and Remote Sensing (ISPRS) drafted a white paper—"How well do we understand Earth observation electro-optical sensor parameters?"—outlining the confusion among various terms used to specify the sensor parameters (Joseph 2000). The paper also suggested definitions for several of these terms. Dr. Stan Morain, after taking over the presidency of Commission-I for the period 2000–2004 conducted an *International Workshop on Radiometric and Geometric Calibration* in December 2003 in which over 80 experts from seven countries participated (Morain and Budge 2004). Though there was a consensus that commercial satellite data providers and space agencies should present data about sensors in a way that would ensure one's ability to intercompare sensors and products, a concrete outcome is still awaited. We present below a set of instrument parameters that should be specified for every Earth observation camera.

1. Spatial domain
   a. IFOV/IGFOV
   b. EIFOV
   c. RAIFOV
   d. FOV/Swath
   e. MTF at IFOV
   f. MTF at twice IFOV
2. Spectral domain
   a. Central wavelength
   b. Bandwidth (using the moment method)
   c. Out-of-band contribution

3. Radiance domain
   a. Saturation radiance (SR)
   b. S/N (i) at 90% SR (ii) at 10% SR
   c. Number of digitization bits

Though NEΔE can be derived from S/N, it is better if it can be explicitly expressed. These measurements are normally carried out in a laboratory. It is also necessary to have a broad agreement among the manufacturers and agencies who own the system regarding the procedures to be adopted to measure each of the above parameters. This will make the comparison between various sensors more meaningful. Some of the above parameters are carried out after launch as part of the on-board characterization of the camera. We shall discuss them in a Section 4.8.

4.7 Imaging Modes

In general, imaging can be carried out in three different ways (Joseph 2005).

(1) Frame by frame: Here, a snapshot is taken at one instant covering a certain area on the surface depending on the sensor characteristics and platform height. A typical example is the conventional photographic camera. The imaging carried out by an area array CCD or other types of area detectors also produce an image in this mode of operation. (At times it is referred to as the *stairing mode*.) Successive frames image a strip of terrain depending on the camera orientation. Generally, the successive frames are taken with certain overlap (Figure 4.7a).

(2) Pixel by pixel: Here in the basic configuration the sensor collects the radiation from one IGFOV at a time. Generally, a scan mirror directs the sensor to the next pixel in the cross-track direction, and by the scan mirror motion, one cross-track line of width equal to one IGFOV is imaged. Successive scan lines are produced by the motion of the platform. This is the way optomechanical scanners image. This mode of imaging is also referred to as *whiskbroom scanning* (Figure 4.7b).

(3) Line by line: Here the sensor collects radiation from one "line" in the cross-track direction at one instant using a linear array of detectors. Successive lines are generated by platform motion. This is also called *pushbroom scanning* (Figure 4.7c).

We shall discuss details of realizing cameras operating in modes (2) and (3) in Chapters 5 and 6, respectively.

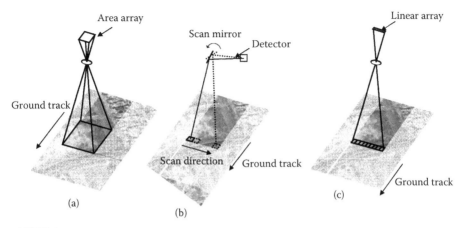

FIGURE 4.7
Schematic showing various modes of image generation from space. (a) Area imaging. (b) Imaging pixel by pixel—whiskbroom scanning. (c) Line by line scanning—pushbroom scanning. (Reproduced with permission from Joseph, G., *Fundamentals of Remote Sensing*, p. 153, Ed. 2, Universities Press, Hyderabad, India, 2005.)

4.8 On-Orbit Performance Evaluation

The raw data received at the ground station has a number of errors/deviations produced due to the sensor, platform, intervening atmosphere and the data transmission and reception system. Therefore, one has to "restore" the distorted image data to a more faithful representation of the original scene. The user is interested in knowing the location of points in the image in geographic coordinates, i.e., georeferencing. The function of georeferencing is to transform the received image coordinates to a specific map projection. For the end users, the image product given to them should faithfully represent the geometric and radiometric properties of the ground scene. Assessing how best the imagery meets the specified goal is a major activity. The quality of the data products generated depends on the sensor performance and the various models used for correction, georeferencing, etc. In this section, we shall discuss aspects related to monitoring the in-orbit performance of a sensor.

Before a camera is put on board a satellite, extensive prelaunch performance validation is carried out. Though cameras are well characterized in the laboratory, their performance can vary in orbit due to various reasons such as misalignment due to launch load, gradual component degradation due to on-orbit environment, etc. Therefore, it is essential to periodically monitor the in-orbit camera performance. To monitor the radiometric sensitivity, the cameras usually carry an in-flight calibration scheme as part of the sensor system. The onboard calibration system essentially monitors the stability of the radiometric response of the sensor. There are different designs to accomplish this task. The basic concept for the solar reflective channel is to bring to the focal plane radiation from a light source or diffused solar radiation. However, the basic design varies from instrument to instrument. The IRS LISS cameras use light emitting diodes (LEDs) to illuminate the focal plane assembly. The calibration cycle is performed over ocean during night. By varying the LED current and using the on–off combination of the LEDs, 12 nonzero intensity levels were generated for LISS1/2 cameras. Though such a system does not allow monitoring of the degradation effect of the optics, it is found to be good enough to monitor the stability of the detector response. Landsat 8 uses two solar diffusers (one is normally used, while the other, pristine, is used only occasionally as a reference), which when positioned reflect light diffusely into the instruments' aperture to provide a full system full aperture calibration. A shutter, when closed, provides a dark reference. In addition, two lamp assemblies, operated in redundant configuration, illuminate the full telescope system (Markham et al. 2008). Though onboard calibration systems have been used in almost all Earth observation cameras, the stability of the calibration system itself is at times doubtful, thus defeating the whole purpose of using them. An alternative approach to monitor the post-launch radiometric stability is to use external stable calibration sources, whose radiances are known (Slater et al. 1996). These sources could be bright targets such as snow, ice, or desert; dark targets over ocean; or celestial objects like the moon.

In situ measurement of natural or artificial sites on the surface of the Earth along with models is widely adopted to provide an independent method to establish the performance of the Earth observation cameras. This methodology is known as vicarious calibration. The number of concurrently operating Earth observation cameras has increased since the launch of Landsat 1. To fruitfully use the data from multiple satellites, it is necessary to have a common calibration methodology to establish each satellite's radiometric accuracy and sensitivity. The goal is to generate the camera radiometric transfer function, which gives the relation between the camera output digital number (DN) and the corresponding radiance at the input of the camera in physical units.

The first step is to identify natural Earth surfaces that are homogeneous over space and stable over time. The reflected radiance from the test site is measured at the time of overpass of the sensor under test. Along with surface measurements, various atmospheric parameters such as water vapor, ozone, and optical depth are also concurrently measured/derived. The surface reflectance data and the atmospheric information are used in a radiative transfer model to calculate the top of the atmospheric radiance value at the sensor.

The selection of the test site is very important. It should have high reflectance (>0.3) to reduce the impact of atmospheric errors. In addition, the site should have near-Lambertian reflectance, and a spatial uniformity so that a large area can be characterized from only a few measurements, to minimize solar angle change during measurements. The site should have minimum aerosol loading and cloud cover. The three most frequently used sites in the United States are the White Sands National Monument in New Mexico, the Railroad Valley Playa in Nevada, and the Ivanpah Playa in California (Thome et al. 2008).

However, from one natural test site we can get only one radiance value. It is possible to produce reflectance panels with varying reflectance values to cover the entire dynamic range. The panels should be placed on low reflectance background locations to avoid adjacency effects. The size of each panel should be much larger than RAIFOV so that the measured value is not influenced by MTF. Because of size constraints, this is practically possible only for high spatial resolution systems. Another approach is to use the Moon as a spectral radiance standard (Barnes et al. 1999; Kieffer and Wildey 1996; Kieffer et al. 2003).

Another sensor performance to be assessed is the spatial response, which is reflected through the MTF. Though the camera MTF is well characterized in the laboratory, in orbit, the sensor faces situations different from the laboratory. For example, the attitude jitter of the satellite and the consequent image motion (satellite motion during the integration time) experienced by the camera can reduce the MTF compared to a laboratory measurement. As we have discussed in Chapter 2, MTF is the modulus of the system optical transfer function (OTF), which is the Fourier transform of the system point-spread function (PSF). Therefore, the MTF of an imaging system can be determined, at least in principle, if we can have the response of the system to a point source. However, point sources are hard to find in nature, though artificial sources are tried without much success (Leger et al. 1994). It is also possible to obtain the MTF from the line-spread function (LSF) by imaging an ideal line object (e.g., infinitesimally thin structure in a dark background) giving a LSF. The MTF in one dimension is the magnitude of the Fourier transform of the LSF. Yet another method is to image an edge—that is an object that reflects on one part and is perfectly black on the other—a target with a straight edge with sharp discontinuity, that is, a step function. The image of such a target gives an edge-spread function (ESF). The derivative of the ESF is the LSF, the Fourier transform of which yields the one-dimensional MTF. Thus, the targets that can be used for MTF evaluation are points, lines, and edges (Figure 4.8).

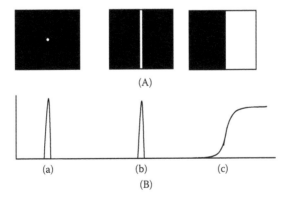

FIGURE 4.8
(A) Types of targets for MTF measurement. (B) The corresponding response of a line scanner: (a) PSF, (b) LSF, and (c) ESF.

Though the procedure we have given above seems to be straightforward, there are a number of practical limitations. In the first place, finding an ideal (or a near-ideal) target in natural scenes is not a trivial task. Bridges, roads, and coastlines are some of the natural objects identified as targets of opportunity for MTF evaluation. Whether they form an edge/line for a sensor depends on the sensor resolution. A target can be considered a line source if the target and background are uniform (spectrally and spatially) and the line width is 0.2 of a pixel or less (Rojas 2002). There should be adequate contrast between the target and the background so that subpixel data can be extracted. The San Mateo-Hayward Bridge near San Francisco was used as a line source for evaluating TM MTF (Rauchmiller and Schowengerdt 1988). Artificial targets also have been used to generate PSF.

The next issue is how to extract the PSF from the image. The response of the camera to point/line/edge depends on where on the array the target falls. If the target is aligned so that most image irradiance falls completely on one single-detector element, then the signal produced is spatially compact. If the target falls on two adjacent columns, the flux from the source is split into a larger number of detector elements giving a broader response (Figure 4.9). That is, although identical sources are imaged in both cases, significant differences are seen in the sampled images, because of the different spatial phase of each source relative to the sampling grid. If we compute the MTF for such a sampled system, the Fourier transform will depend on the alignment of the target with respect to the detector. That is, the image appears degraded if the sampling grid is moved across the scene. We can conceptualize this phenomenon as a result of GSD being too large compared to the image of a point; in other words, the PSF is undersampled. This effect of sample-scene phasing, that is, the uncertainty in the location of the scene with respect to the sampling grid of the system, is seen in all digital imaging systems (Rauchmiller and Schowengerdt 1986). Such a system is called spatially variant. There are different techniques used to overcome this issue. The basic concept is to have a target with different phasing with respect to the sampling grid. Owing to this target phasing, each imaged point/edge source exhibits a different amount of intensity distribution in the digital image. That is, there are several images with different subpositions on the sampling grid to rebuild an oversampled

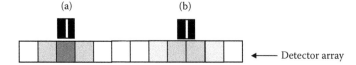

FIGURE 4.9

Schematic showing the response of a linear array for a line target. The target (a) falls at the center of a pixel, so that most image irradiance falls completely on one single-detector element. The target (b) falls at the border of two adjacent detector elements, and hence most of the irradiance is split into two detectors, with further spillover to adjacent pixels thus producing a broader PSF. The intensity levels are represented in grey tones.

PSF. The target pixels may then be suitably recombined according to their known relative positions to generate the system PSF.

Rauchmiller and Schowengerdt (1986) used an array of 16 large black squares set in four rows by four columns against a white background (at the White Sands Missile Range in New Mexico) to evaluate the MTF of Landsat TM. The design of the target creates four-pixel shifts of *point sources* throughout the 30-m IFOV of the TM thereby facilitating sample-scene phasing to effectively resample the PSF at a 7.5-m sample interval. The target pixels are then recombined according to their known relative positions to form a single-system PSF.

Though there are several on-orbit MTF estimation methods using lines and points, there are no ideal point or line targets to perform on-orbit MTF estimation. An alternative is to remove the influence of point size or line width to use a knife-edge as a target to estimate on-orbit MTF. The theoretical foundation of the knife-edge method is rigorous, and its processing method is straightforward (Li et al. 2010). In addition, a knife-edge target is easy to deploy or select and has a relatively high S/N and hence it is widely and successfully used for on-orbit MTF estimation of several satellite electro-optical sensors (Anuta et al. 1984; Helder et al. 2006; Kohm 2004; Leger et al. 2004; Nelson and Barry 2001). To avoid undersampling of the ESF, we need to have the target edge located with different phasing with respect to the sampling grid. Therefore, the edge is oriented at a skew angle to the subsatellite track, that is, the sensor array. This provides data sampled at slightly different sample-scene phases relative to the target. The selection of the target is important. The width of the dark and bright regions should be much more than the expected system PSF projection on the ground. The length of the target must ensure the number of sample points for ESF curve-fitting is large enough to suppress the effect of random noise. The angle between the target and sensor scan line should not be too large so that a reasonable number of *phased* samples are obtained to construct the ESF. Figure 4.10 shows how the edge phasing takes place with each scan line. The next task is to determine the edge location of each image profile across the knife-edge to subpixel level by suitable interpolation algorithm (Tabatabai and Mitchell 1984; Mazumdar et al. 1985; Mikhail et al. 1984). For various reasons, the subpixel edge locations of all image profiles do not lie along a straight line. A least-square-fit is carried out to bring the subpixel edges to a straight line and then they are aligned to a common coordinate according to their pixel location. To reduce the influence of random noise, the oversampled data points are fitted to a curve producing the ESF. The ESF is then differentiated to obtain the LSF, and the Fourier-transformed LSF gives the corresponding MTF (Li et al. 2010; Robinet and Léger 2010).

Another MTF estimation technique used is a two-image analysis in which a high-resolution image is used as reference. Rojas et al. (2002) derived MODIS on-orbit MTF using ETM as reference data. In this case, as the two imageries were from different satellites, image registration is a major effort.

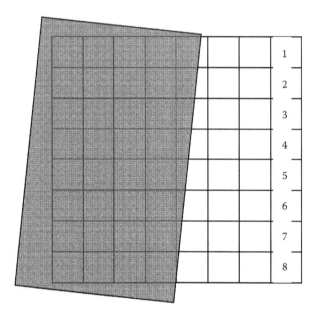

FIGURE 4.10
Principle of oversampling an edge. Eight successive scan lines are shown. As the edge is scanned from 1 to 8, during each scan the edge covers a different proportion of the pixel. Subpixel detection gives the location of the edge during each scan.

However, now multiresolution data is available from the same satellite as in IKONOS, IRS-Resourcesat, etc., this technique can be applied in a better way. Resourcesat has an added advantage as both the high resolution (LISS-4) and low resolution (LISS-3) cameras have similar spectral response. A two-image comparison method using LISS-3 and LISS-4 was carried out by Raghavender et al. (2013) to derive the MTF of the coarser resolution LISS-3 camera. In this study, they derived the MTF of the LISS-3 sensor in three bands, by replacing the real scene with corresponding LISS-4 images.

References

Anuta, P. E., L. A. Bartolucci, M. E. Dean et al. 1984. Landsat-4 MSS and thematic mapper data quality and information content analysis. *IEEE Transactions on Geosciences and Remote Sensing* 22: 222–236.

Barnes, R. A., R. E. Eplee, F. S. Patt, C. R. McClain. 1999. Changes in the radiometric sensitivity of SeaWiFS determined from lunar and solar-based measurements, *Applied Optics* 38(21): 4649–4664.

Boreman Glenn D. 2001. *Modulation Transfer Function in Optical and Electro-optical Systems*. SPIE Press, Bellingham, Washington, DC.

GeoEye-1. http://gmesdata.esa.int/geonetwork_gsc/srv/en/resources.get?id=375& fname=GeoEye-1_Product_guide_v2.pdf&access=private. (accessed on January 25, 2014).

Helder, D., J. Choi, C. Anderson. 2006. On-orbit Modulation Transfer Function (MTF) Measurements for IKONOS and Quickbird, *Civil Commercial Imagery Evaluation Workshop,* Reston, VA. http://calval.cr.usgs.gov/wordpress/wp-content/uploads/ JACIE_files/JACIE06/Files/28Helder.pdf (accessed on February 8, 2014).

IKONOS Instrument/Product Description. 2009. http://gmesdata.esa.int/ geonetwork_gsc/srv/en/resources.get?id=376&fname=IKONOS_Product_ guide_v2.pdf&access=private (accessed on January 25, 2014).

Joseph, G. 2000. How well do we understand Earth observation electro-optical sensor parameters? *ISPRS Journal of Photogrammetry and Remote Sensing* 55(1): 9–12.

Joseph, G. 2005. *Fundamentals of Remote Sensing.* Universities Press, Hyderabad, India.

Kaufman, Y. J. 1984. Atmospheric effect on spatial resolution of surface imagery. *Applied Optics* 23(19): 3400–3408.

Kieffer, H. H., T. C. Stone, R. A. Barnes et al. 2003. On-orbit radiometric calibration over time and between spacecraft using the moon. *Proceedings of SPIE* 4881: 287–298.

Kieffer, H. H. and R. L. Wildey. 1996. Establishing the moon as a spectral radiance standard. *Journal of Atmospheric and Oceanic Technology* 13(2): 360–375.

Kohm, K. 2004. Modulation transfer function measurement method and results for the ORBVIEW-3 High Resolution Imaging Satellite. http://www.cartesia.org/ geodoc/isprs2004/comm1/papers/2.pdf (accessed on February 7, 2014).

Kondratyev K. Ya, O. M. Pokrovsky. 1979. A factor analysis approach to optimal selection of spectral intervals for multipurpose experiments in remote sensing of the environment and earth resources. *Remote Sens Environment* 8(1): 3–10.

Leger, D., J. Duffaut, F. Robinet. 1994. MTF measurement using spotlight. *Geoscience and Remote Sensing Symposium IGARSS* 94: 2010–2012.

Leger, D., F. Viallefont, P. Deliot, C. Valorge. 2004. On-orbit MTF assessment of satellite cameras. In *Post-Launch Calibration of Satellite Sensors,* edited by S. A. Morain and A. M. Budge. Taylor and Francis, London, United Kingdom.

Li, X., X. Jiang, C. Zhou, C. Gao, and X. Xi. 2010. An analysis of the knife-edge method for on-orbit MTF estimation of optical sensors. *International Journal of Remote Sensing* 31: 4995–5011.

Markham, B. L., P.W. Dabney, J. C. Storey et al. 2008. Landsat Data Continuity Mission Calibration and Validation. http://www.asprs.org/a/publications/ proceedings/pecora17/0023.pdf (accessed on February 7, 2014).

Mazumdar, M., B. K. Sinha, and C. C. Li. 1985. Comparison of several estimates of edge point in noisy digital data across a step edge, 19–23. In Proc. Conf. on Computer Vision and Pattern Recognition, IEEE Computer Society.

Mikhail, E. M., M. L. Akey, and O. R. Mitchell. 1984. Detection and subpixel location of photogrammetric targets in digital images. *Photogrammetria* 39(3): 63–83.

Minor John, L. 2002. Flight test and evaluation of electro-optical sensor systems, *SFTE 33rd Annual International Symposium 19-22 August 2002, Baltimore, MD.* http:// www.americaneagleaerospace.com/documents/Electro-Optics/Flight%20 Test%20%26%20Evaluation%20of%20Electro-Optical%20Sensor%20Systems. pdf (accessed on February 8, 2014).

Morain, S. A. and M. B. Amelia. (Ed). 2004. *Post-Launch Calibration of Satellite Sensors: Proceedings of the International Workshop on Radiometric and Geometric Calibration,* 2–5 December 2003, Gulfport, MS. Taylor and Francis, London, United Kingdom.

Nelson, N. R. and P. S. Barry. 2001. Measurement of Hyperion MTF from on-orbit scenes. *IEEE Transactions on Geosciences and Remote Sensing* 7: 2967–2969.

NASA. 1973. Special Publication #335, Advanced scanners and imaging systems for Earth observation. Working Group Report.

Norwood, V. T. 1974. Balance between resolution and signal-to-noise ratio in scanner design for earth resources systems. *Proc. SPIE* 51: 37–42.

Palmer, J. M. 1980. Radiometric bandwidth normalization using root mean square methods. *Proc. SPIE* 256: 99–105.

Palmer, J. M. 1984. Effective bandwidths for LANDSAT-4 and LANDSAT-D' multispectral scanner and thematic mapper subsystems. *IEEE Transactions on Geoscience and Remote Sensing* 22(3): 336, 338.

Pandya, M. R., K. R. Murali, A. S. Kirankumar. 2013. Quantification and comparison of spectral characteristics of sensors on board Resourcesat-1 and Resourcesat-2 satellites. *Remote Sensing Letters* 4(3): 306–314.

Raghavender, N., C. V. Rao, A. Senthil Kumar. 2013. NRSC Technical note, NRSC-SDAPSA-G&SPG-May-2013-TR-524.

Rauchmiller, R. F. Jr., and R. A. Schowengerdt. 1988. Measurement of the Landsat Thematic Mapper MTF using an array of point sources. *J. Optical Engineering* 27(4): 334–343.

Rauchmiller, R. F., and R. A. Schowengerdt. 1986. Measurement of the Landsat Thematic Mapper MTF using a 2-dimensional target array. *Proc. SPIE* 0697:105–114.

Robinet, F. V., D. Léger. 2010. Improvement of the edge method for on-orbit MTF measurement *Optics Express* 18(4): 3531–3545.

Rojas, Francisco. 2002. Modulation transfer function analysis of the moderate resolution imaging spectroradiometer (MODIS) on the Terra satellite, PhD thesis, Department of Electrical and Computer Engineering, The University of Arizona. http://arizona.openrepository.com/arizona/bitstream/10150/280247/1/azu_td_3073306_sip1_m.pdf (accessed on February 8, 2014).

Schowengerdt, R. A. 2001. Measurement of the sensor spatial response for remote sensing systems. *Proc of SPIE* 4388: 65–71.

Sharma, S. A., H. P. Bhatt, and Ajai. 1995. Oilseed crop discrimination, selection of optimum bands and role of middle IR photogrammetry and remote sensing. *Photogrammetry and Remote Sensing* 50(5): 25–30.

Schott Catalog: http://www.schott.com/advanced_optics/english/download/schott_interference_filters_propert_2013_eng.pdf (accessed on February 8, 2014).

Slater, P. N., S. F. Biggar, K. J. Thome, D. I. Gellman, and P. R. Spyak. 1996. Vicarious radiometric calibrations of EOS sensors. *Journal of Atmospheric and Oceanic Technology* 13:349–359.

Srikanth, M., V. KesavaRaju, M. SenthilKumar, and A. S. KiranKumar. 2010. Stellar Imaging Operations in Cartosat- 2A. *Journal of Spacecraft Technology* 20(1): 56–64.

Tabatabai, A. J. and O. R. Mitchell. 1984. Edge location to subpixel values in digital imagery. *IEEE Transactions on Pattern Analysis and Machine Intelligence* 6(2): 188–201.

Thome, K., J. McCorkel, J. Czapla-Myers. 2008. Inflight intersensor radiometric calibration using the reflectance-based method for Landsat-type sensors. *Pecora 17.* Available at http://www.asprs.org/a/publications/proceedings/pecora17/0040.pdf (accessed on February 8, 2014).

UN A/AC 105/260, Committee on the peaceful uses of outer space. Report on effective resolution element and related concepts, http://www.fas.org/irp/imint/docs/resunga.htm (accessed on February 8, 2014).

5

Optomechanical Scanners

5.1 Introduction

Civilian Earth observation started with the launch of TIROS-1, carrying an electron beam imaging (TV camera) system. The need to cover larger spectral regions soon led to the replacement of electron beam imagers with systems using discrete detectors. In this arrangement, radiance from the scene is directed to the optics with the help of a mechanical scanner and, hence, such systems are known as optomechanical scanners. Because of the mode of scanning, they are also known as *whiskbroom* scanners.

5.2 Principle of Operation

Consider a collecting optics—say, a telescope—at whose prime focus along the optical axis a detector is placed. A plane mirror is kept inclined at 45° to the optical axis of the telescope, as shown schematically in Figure 5.1. The radiation emitted or reflected from the scene is intercepted by the plane mirror, which diverts the radiation to the telescope. The telescope focuses the radiation onto a detector kept at the focal plane. The detector receives radiation from an area on the ground corresponding to the projection of the detector by the telescope onto the ground, which is determined by the detector size, focal length of the optics, and height of the optical system above the ground. The ground parcel so seen is called the instantaneous geometric field of view (IGFOV). If the plane mirror (hereafter, we refer to it as the scan mirror) is rotated with the axis of rotation along the optical axis, the detector starts looking at adjacent regions on the ground lying in the scan direction. Thus, the detector collects radiance from one line on IGFOV by IGFOV basis. Consider such an instrument mounted on a moving platform, like an aircraft or a spacecraft, such that the optical axis is along the velocity vector of the vehicle. The rotation of the scan mirror collects information from a strip on the ground at right angles (close to) to the direction of motion of the platform,

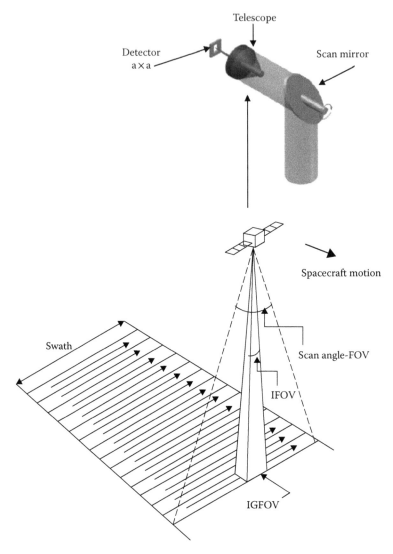

FIGURE 5.1
Schematic showing the operation of an optomechanical scanner.

generating one line of data in the cross-track direction. The vehicle motion produces successive scan lines. If the mirror scan frequency is adjusted such that by the time the platform moves through one IGFOV, the scan mirror is set to the start of the next scan line, then successive and contiguous scan lines can be produced. To produce a contiguous image, the scan frequency has to be correctly adjusted, depending on the velocity of the platform and IGFOV dimension. To summarize, in the cross-track direction information is collected from each IGFOV by the motion of the scan mirror, thereby

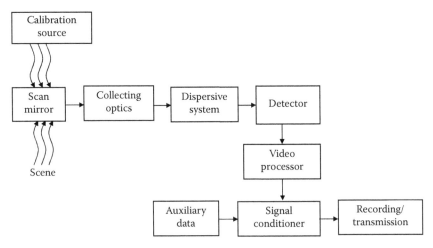

FIGURE 5.2

Functional block diagram of an MSS. (Reproduced from Joseph, G., Fundamentals of Remote Sensing, Universities Press (India) Pvt Ltd, Hyderabad, 2005. With permission.)

producing one line of the image, and in the along-track direction successive lines of image in contiguous fashion are produced by the platform motion. This mode of scanning is referred to as whiskbroom scanning.

To produce multispectral imagery, the energy collected by the telescope is directed to a spectral dispersing system for further processing. Such systems that can generate imagery simultaneously in more than one spectral band are called multispectral scanners (MSS). The instrument usually carries a calibration system to monitor the radiometric performance and stability of the sensor. Figure 5.2 gives the functional block diagram of a MSS. Thus, the MSS has got a scan mirror, collecting optics, dispersive system (which essentially spreads the incoming radiation into different spectral bands), and a set of detectors appropriate for the wavelength regions to be detected. The outputs of the detectors go through electronic processing circuits. The data from the scene along with other data like attitude of the platform, temperature of various subsystems, and so on are formatted, and the combined information is either recorded on a suitable storage medium, as is usually the case with aircraft sensors, or transmitted for spacecraft sensors.

5.3 Scanning Systems

In all electro-optical imaging systems, the final image is recorded after some scanning scheme. In TV cameras, the image formed on the faceplate is scanned using an electron beam: an electronic image plane scanner. In an optomechanical imager, the scanning can be carried out either in the object space or in the image space. In image space scanning, the scan mirror is kept after the collecting optics, near the focal plane (Figure 5.3a) and the mirror directs each point in the focal plane to the detector. Obviously, such a system requires the collecting optics to be corrected for the total field of view (FOV), which is quite demanding, especially if a reflective system has to be used. However, it requires a relatively small size scan mirror. Though image space scanning has been used in some of the early optomechanical multispectral scanners, as in the Skylab Multispectral Scanner, S-192 (Abel and Raynolds 1974), it is not generally used in current systems because the collecting optics has to be corrected for the total FOV.

In object space scanning, the rays from the scene fall onto the scan mirror, which reflects the radiation to the collecting telescope (Figure 5.3b). Here, for the rays reaching the detector the direction of rays at the collecting optics remains the same irrespective of the scan mirror position. Thus, when object space scanning is used the collecting optics needs to be corrected only for a small FOV around the optical axis. The extent of field correction depends on IFOV and the distribution of detectors in the focal plane. Usually, the focal plane of whiskbroom scanners consists of a number of detector elements

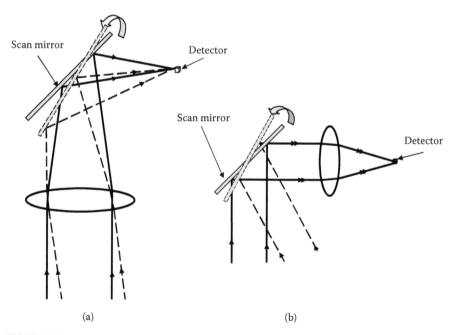

(a)   (b)

FIGURE 5.3
Schematic of object plane and image plane scanning. (a) Image plane scanning. Here, the scan mirror is located between the imaging optics and the image plane. (b) Object plane scanning. Here, the scan mirror is placed before the entrance of the imaging optics. (Reproduced from Joseph, G., *Fundamentals of Remote Sensing*, Universities Press (India) Pvt Ltd, 2005. With permission.)

along track to improve the signal to noise ratio (SNR) by reducing scanning frequency or across track for additional spectral bands.

Different techniques are used for realizing object plane scanning. In the simplest case, a plane mirror is kept at 45° to the optical axis and cross-track scanning is accomplished by continuously rotating the mirror around the optical axis. In this configuration, all the radiation that is falling at 45° to the scan mirror normal and at right angles to the axis of rotation is reflected along the rotational axis (Figure 5.4a). This type of scanning has been extensively used in meteorological payloads like VHRR/AVHRR on board satellites such as NIMBUS, NOAA, and so on. The AVHRR scan mirror is rotated at 360 rpm using an 80 pole hysteresis synchronous motor. This scheme of scanning has poor scan efficiency, that is, the amount of time taken to observe the scene during one scan (in this case one rotation). As we will discuss in Section 5.7, scan efficiency is an important parameter for collecting the maximum signal. From a satellite altitude of 1000 km, even for horizon to horizon data collection the total angle to be covered is less than about 120°. Therefore, the time for which the scan mirror views the Earth's surface (scan efficiency) with such a scanner is less than 30%. During the rest of the period, the scan mirror views either deep space or the instrument body. However, a part of this time (while viewing deep space/instrument body) is used for calibration purposes. The scan efficiency can be increased by having the scan mirror in the form of a prism with multiple faces (Figure 5.4b). Thus, if there are n faces then n scan lines can be generated in one revolution with each face covering an angular swath corresponding to $(360/n)°$. However, such a scheme is rarely used in spaceborne imaging cameras.

Another mode of scanning used to increase the scan efficiency is to oscillate the mirror to and fro to cover the required swath. The oscillation will be carried out such that during the active period (when data is being collected) the mirror sweeps across track with uniform angular velocity and at the end of the scan it returns back as quickly as possible (Figure 5.5). Such systems in practice can provide a scan efficiency of about 45% as in the case of Landsat MSS. Accurate knowledge of the angular position of the mirror with respect to a known reference of the instrument is essential to

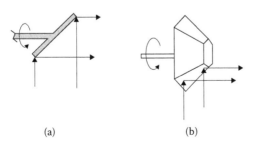

(a)   (b)

FIGURE 5.4
Schematic representation of different object plane scan mechanisms. (a) Single scan mirror. (b) Multiface prism scanner.

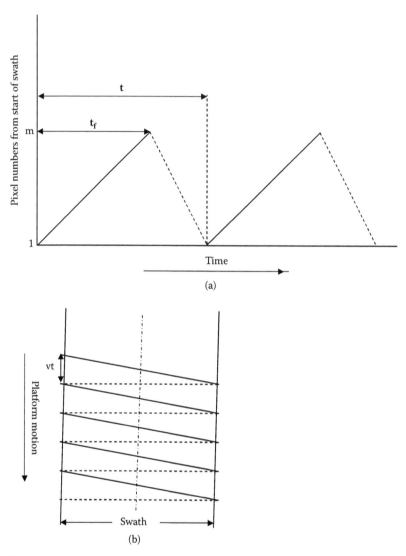

FIGURE 5.5

(a) Schematics showing the scan mirror motion. The scan mirror moves from pixel 1 to m (which defines the swath). After viewing pixel m of the first scan line, the scan mirror comes back to pixel number 1 of the next line. The term *t* is the time taken for one complete cycle, that is, the forward and retrace. (b) Projection of scan lines on ground. Due to orbital motion the scan lines are not at right angles to the sub-satellite track. The dotted lines are the return path of the scan mirror. The central line is the subsatellite track. The term *v* is the subsatellite velocity.

reconstruct the geometry of the imagery. The MSS scan mechanism consists of a 2.5-mm thick elliptical beryllium mirror with a major axis of 35.5 mm suspended on flexure pivots (Lansing and Cline 1975). A scan position monitoring device, consisting of a galliumarsenide (GaAs) diode source, which experiences multiple reflections from the scan mirror, is used to generate angular position information with accuracy better than ±10 μrad.

Further improvement in scan efficiency can be achieved by using both forward and reverse motions of the mirror to collect data, as used in Landsat Thematic Mapper (TM). This method of bidirectional scanning increases scan efficiency to better than 80%. The Landsat TM scan mechanism is the most complex system used in any optomechanical scanner, and we shall discuss its operation in Section 5.8.

The scanning scheme that we hitherto discussed is for imaging from a low Earth orbiting platform, where there is relative motion between the satellite and the Earth. Imaging from the geostationary orbit started in 1966 with the launch of Application Technology Satellite-1 (ATS-1) of the United States, primarily to get information important to understanding meteorological phenomena. A satellite at the geostationary platform appears stationary with respect to the Earth and hence cannot produce successive scan lines as we have seen earlier. Therefore, the camera should have a provision to scan in two orthogonal directions to generate two-dimensional images. The geostationary satellite attitude is fixed in space by spinning the satellite around one axis, thereby keeping that axis relatively fixed in the inertial space, or the spacecraft may be designed to have all three axes fixed in space. The scan mechanism design differs in both the cases.

All the early Earth observation systems from geostationary orbit, like GOES (United States), METEOSAT (European Space Agency), GMS (Japan), and so on, were spin stabilized, with the spin axis in the north–south direction oriented parallel to the Earth's axis. The spin of the satellite (about 100 rpm) provides a highly linear scan in the east–west direction. The camera should have a provision to scan in the north–south direction. The latitude scan can be achieved through different schemes. In one of the configurations, the optical axis of the imager is aligned at right angles to the spin axis such that the primary mirror of the telescope directly views the Earth. After every spin, the imaging telescope is tilted in the south–north direction to generate a latitude scan. The tilt angle depends on the width of the strip covered in the north–south direction during the spin so as to generate contiguous scan lines. After completing the south–north scan in 20–25 minutes, depending on the design, the telescope is brought to its starting position at a faster rate. The retrace can cause some disturbance to the spacecraft and once the spacecraft has stabilized, the next image collection sequence starts. Typically, from a spinning satellite the camera provides a full Earth disk image in every 30 minutes.

This concept was first realized in the spin scan camera flown on ATS-1 in 1966. Here, the entire camera was tilted stepwise on flexural pivots so as to scan in the south–north direction (Suomi and Krauss 1978). A modified scheme is used in the Visible and Infrared Imager (MVIRI) instrument on the first-generation METEOSAT. Here, the telescope mirror assembly is tilted to keep the focal plane assembly fixed. The optical beam from the telescope is taken out through a fold mirror so that the detectors remain fixed on the spacecraft (Figure 5.6a). Here, the telescope mirror assembly is mounted on two flexible plate pivots and the scanning mechanism is a high precision jack screw driven by a stepper motor via a gearbox (Hollier 1991). After every east–west scan, the spin clock delivers a signal to the scanning motor electronics to rotate the telescope through an angle of 1.25×10^{-4} rad, corresponding to one image strip covered in the north–south direction during the east–west scan. The other mode for the latitude scan is to align the optical axis along the spin axis. The radiation from the Earth is directed to the collecting optics via a scan mirror kept at 45° to the optical axis (Figure 5.6b). This scheme has been adopted in the early GOES series and Japanese GMS. The GOES satellite spin provides a west to east scan motion while latitudinal scan is accomplished by sequentially tilting the scanning mirror north to south at the completion of each spin.

From a geostationary orbit, the Earth subtends about 17.4°. Hence, the scan efficiency (the time for which the camera views the Earth compared to the spin period) of a spin scan system is less than 5%. Therefore, a spinning satellite is not suitable when one wants to generate high spatial resolution imageries from a geostationary altitude. The next logical step is to have a three-axis stabilized platform. In such a configuration the three axes (roll, pitch, and yaw) of the satellite are maintained stable with respect to the Earth and hence the camera "stares" at the Earth. Therefore, the camera should have provision for scanning in the longitude and latitude directions. The first three-axis stabilized imaging system from a geostationary orbit was the ATS-6 of the United States launched in 1974. The operational use of a three-axis stabilized imaging system from geostationary orbit started with the launch of India's multipurpose satellite INSAT in 1982. We shall briefly present the scanning system used for INSAT VHRR. The Earth disk is imaged by scanning in the in the east–west/west–east direction (fast scan) and in the south–north direction (slow scan). This is accomplished by mounting a beryllium scan mirror on a two-axis gimbaled scan mechanism. Fast and slow scan motions are realized by separate brushless direct current (DC) motors with an inductosyn position encoder and controlled by independent servo systems (Krishna and Kannan 1966). During the 1-second east–west scan, the telescope collects data for one "effective" IFOV strip. To increase the scan efficiency, the data are collected during both east to west and west to east scans. To achieve a bidirectional scan, at the end of the fast scan in one direction, the direction of scan is reversed and the scan mirror is stepped in the north–south direction through one "effective IFOV" strip. To avoid any residual disturbance during imaging due to the velocity reversal of the mirror, the scan rate is reduced in steps and increased in similar steps in the reverse direction. The fast scan also covers 21.4°, allowing adequate time for the mirror to be stable during the imaging period. The scan mirror moves through ±5.35° in the fast scan direction and ±10° in the slow scan direction to generate 21.4° × 20° image frame (Joseph et al. 1994). To observe an event frequently, the mirror can be positioned anywhere in the north–south direction in steps of 0.5° and image a strip covering 4.5° in the north–south direction in approximately 7 minutes. This sector scan mode is particularly suited to follow severe weather conditions.

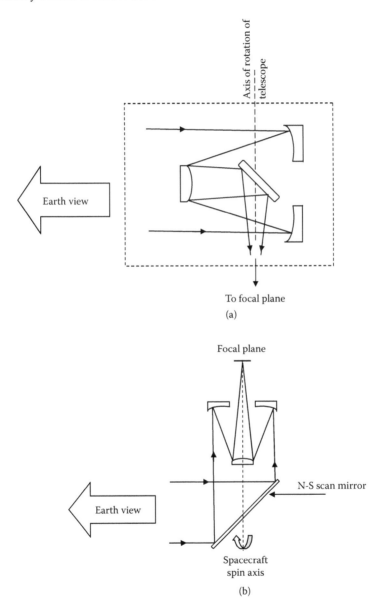

FIGURE 5.6
Schematics of scanning by a spinning satellite from geostationary orbit. (a) METEOSAT MVIRI camera scanning scheme. The spin axis of the satellite is normal to the plane of paper. (b) GOES/GMS camera scanning scheme. The optical axis is aligned to the spin axis of the satellite. The radiation from the Earth is directed to the telescope through a scan mirror.

Realizing a scanning system meeting all the optical and mechanical requirements is a challenging task, which requires a team with expertise in various disciplines such as optics, mechanical and electrical engineering, and manufacturing. The weight of the mirror has to be minimized to have a low inertia and at the same time be stiff enough to take the mechanical loads. The most frequently used material for scan mirrors is beryllium because of its low density and high Young's modulus. It is 45% lighter than aluminum and approximately five times as stiff and this makes it a low inertia component for scanning applications. The beryllium blank is lightweighted by scooping material from the back (i.e., egg crate core) as per the computer design to minimize weight and achieve the required stiffness. We discussed in Section 3.4.1 the precautions and processes to be followed if beryllium is to be used as an optical component. Another material being considered for scan mirrors is carbon/silicon carbide (C/SiC) material, which has light weight, high stiffness, low coefficient of thermal expansion, and good thermal conductivity (Harnisch et al. 1994, 1998).

When using a rotating scan mirror as in the case of AVHRR, dynamic balancing of the mirror is an important step. The mirror optical quality can also be degraded by dynamic loads. The mirror dynamic deformation can be calculated using the finite element analysis method, and the imaging quality of the scan mirror can be estimated as a guideline for mechanical design of the scan mirror (Feng et al. 1994).

The linearity of mirror motion (angle vs. time) is an important performance requirement, and meeting the linearity requirement is a challenge. The TM design requirement is that the total angular vibration of the scan mirror should be less than 2 μrad peak to peak during the data collection period. Even with a perfect electrical design, the scan mirror motion can induce vibrations in the structure and the linearity can be influenced by structural vibration. TM scan mirror assembly is one of the most complex designs, and the details are given by Starkus (1984).

5.3.1 Scan Geometry and Distortion

When we talk about scan linearity, it means that the mirror sweeps a constant angle per unit time. The image data are acquired at regular time intervals and hence at constant angular intervals. However, this does not mean that the projection of the detector on the ground, that is, IGFOV size, remains the same across the swath. This is because the distance of the telescope from the imaging section on the ground varies across the track. Moreover, if swath coverage is very large curvature of the Earth's surface also contributes to distortion. For an instrument with IFOV β rad on a platform of height h km, the nadir pixel size is βh km. When the scan mirror is at an angle θ from nadir, from simple trigonometry it can be shown that in the direction of spacecraft travel (along track) the resolution varies as $\beta h \sec \theta$ and across track as $\beta h \sec^2 \theta$ (assuming a flat Earth surface) (Figure 5.7). To get around this issue, we should have a scanning scheme such that the scanned strip on the

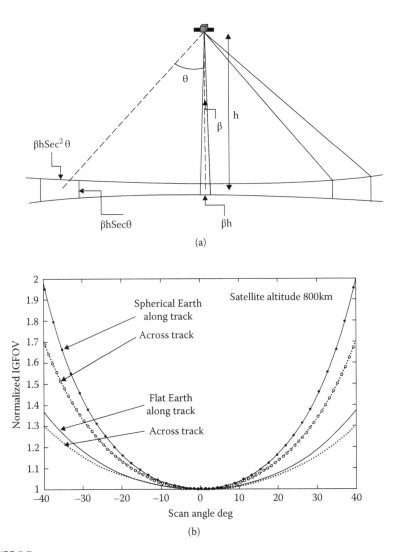

FIGURE 5.7

(a) Effect of scan geometry on ground resolution for an optomechanical scanner. h, satellite height; β, IFOV; and θ, off-nadir view angle. Curvature of the Earth not considered. (b) Plot showing degradation of resolution with scan angle, when curvature of the Earth is also taken into account. In comparison with flat Earth values, the degradation increases faster when Earth's curvature is taken into account. (From Kurian, M., 2014, personal communication.)

ground makes equal distance to the scanner. This can be achieved by conical scanning in which the scanned strip is annular (Figure 5.8). This method of scanning has been carried out in Multispectral Scanner S192 flown on board Skylab and in the Along Track Scanning Radiometer carried in the ERS1/2 satellites.

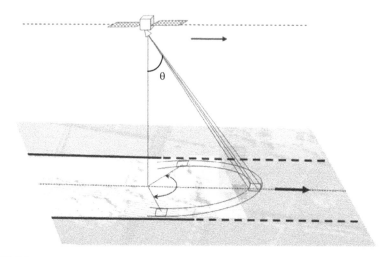

FIGURE 5.8
Schematics showing geometry of conical scanning. The look angle θ remains constant across the swath, maintaining constant instantaneous geometric field of view.

5.4   Collecting Optics

All the current optomechanical scanning systems implement object space scanning. As discussed earlier, in the cameras employing object space scanning the image is collected close to the optical axis and hence the telescopes need to be corrected only for a small field to cover the IFOV (or n times IFOV, if n detectors are used per spectral band) in the along-track direction. The across-track field requirement depends on the arrangement of the detectors in the focal plane. In practice, the optics for optomechanical scanners needs to be corrected for less than 0.1°. In addition, the camera has to cover a broad wavelength region. Therefore, in Earth observation cameras two mirror systems are normally used for optomechanical scanning imagers. Landsat MSS/TM, INSAT VHRR, GOES, and a number of other missions have an RC telescope as the collecting optics. They are discussed in detail in Chapter 3. There are a few cameras like INSAT-1 VHRR, TIROS-N AVHRR, and Heat Capacity Mapping Mission Radiometer that have an afocal Mersenne telescope as the collecting optics (Figure 5.9). Here, the secondary focus coincides with the focal point of the primary. When the primary and secondary mirrors are well aligned (both optical axes collinear), a collimated beam parallel to the optical axis incident on the primary emerges parallel but with a reduced cross section. The advantage of this configuration is that beam splitters used to separate the focal plane for accommodating different bands can be placed in the beam at an angle without introducing any aberrations. However, a small misalignment between the primary and the secondary mirror can affect output beam collinearity.

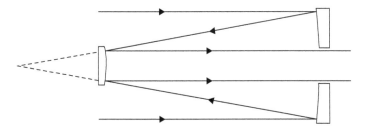

FIGURE 5.9
Schematics of the Mersenne afocal reflective telescope. The focus of primary and secondary coincides, thereby producing a parallel output beam for a point source.

5.5 Dispersive System and Focal Plane Layout

The radiance reaching the focal plane through the collecting optics contains all the spectral components emanating from IGFOV, subject to the spectral response of the scan mirror and collecting optics. The next task is to measure the radiance at the focal plane over a specific portion of the electromagnetic spectrum of our interest. That is, we have to separate the spectra received into the wavelength region of our interest for further measurement using appropriate detectors. When one is interested in recording information of continuous spectra as in a spectrometer, one uses prisms or gratings as dispersive elements. Such schemes have been used in some of the earlier space-borne sensors such as Skylab S192, and in aircraft scanners. We shall discuss their use in Chapter 8 when we deal with hyperspectral imagers. When we want to sample the spectrum in a few specific wavelength regions, spectral selection is more efficiently carried out using interference filters.

To accommodate different types of detectors, the primary focal plane is optically "split" into several focal planes, thereby creating physical space to house the detectors. That is, the incoming radiation collected by the telescope has to be divided into multiple streams. To explain the concept, let us first consider how the focal plane can be split for a two-band camera. Figure 5.10 gives the scheme to accommodate two bands using a single collecting optics. Here, the incoming beam A from the collecting optics is split using a beam splitter to B and C. Thus, the beam splitter facilitates partitioning of the beam in two directions to help the placement of detectors. This concept can be extended to accommodate more bands by placing more beam splitters in the optical path. Thus, the spectral dispersion is done by a set of beam splitters at appropriate locations and final spectral selection is carried out by interference filters placed close to the detector system. However, if the beam splitter is put in a converging beam its effect on image quality should be taken into account. Beam splitters are generally polarization sensitive, which also needs to be taken into account.

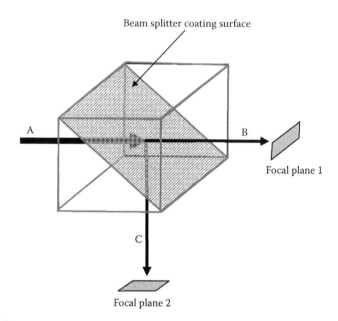

Focal plane 2

FIGURE 5.10
Schematics showing how by using a beam splitter the focal plane of a single collecting optics can be separated to accommodate two detectors for spectral selection. Here, two right-angled prisms are cemented together at the hypotenuse, which is coated with beam splitter coatings. The incoming beam A is split into two beams B and C. If the coating is semitransparent, the input radiance is divided between B and C of approximately equal value. A dichroic coating can optimize the transmitted and reflected radiation of a particular wavelength region. A cube is preferred to a plate, so as to keep the optical path length the same for the two beams. If the beam splitter is in the converging beam, the telescope optical design should take into account this additional optical path.

A simple beam splitter is a semisilvered plate kept at 45°, which splits the beam with approximately equal intensity in two directions. However, such an arrangement reduces the irradiance at the detector. To overcome this problem, dichroic beam splitters are used. Dichroic beam splitters reflect most of the radiation in one specified spectral region while optimally transmitting in another spectral region (Figure 5.11). Dichroics can be designed with the transition region at different wavelengths, with reflectivity typically reaching values greater than 95% and transmission typically exceeding 85%. The dichroic should be chosen such that the wavelength of interest does not fall in the transition region. This scheme of using a beam splitter and interference filter has been used in most of the optomechanical scanners, such as NOAA AVHRR, INSAT VHRR, and radiometers in the GOES series of satellites, to name a few. Figure 5.12 gives the focal plane arrangement using multiple dichroic beam splitters for the GOES N imager.

If we have a number of detectors on the same substrate overlaid with appropriate filters, one can avoid the beam splitters to separate the focal plane. Thus, the focal plane will have an array of detectors with spectral

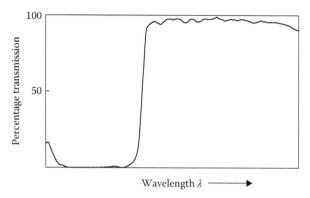

FIGURE 5.11
Schematics showing spectral response of a dichroic beam splitter. Where the transmission is minimum, that wavelength region will be reflected.

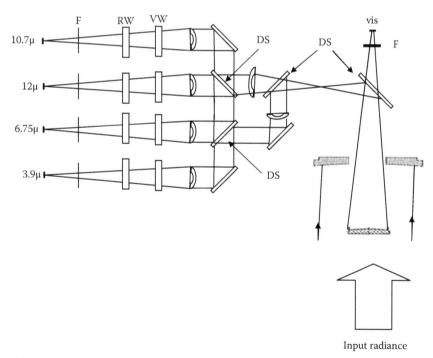

FIGURE 5.12
Schematics showing focal plane separation for accommodating detectors using dichroic splitters for GOES-N imager. RW, radiation window; VW, vacuum window; DS, dichroic splitter; and F, filter. (Adapted from NASA, http://goes.gsfc.nasa.gov/text/GOES-N_Databook_RevC/Section03.pdf, accessed on May 14, 2014, 2009.)

selection filters. A similar concept in a different way was implemented in Landsat MSS. During the development phase of MSS, integrated detector arrays had not yet been developed for space use, so a matrix of fiber optic elements was used at the focal plane to separate the four spectral bands. Each band had 6 detectors, and hence 24 fibers were arranged in a 4 × 6 array (corresponding to the 4 spectral bands and the 6 lines scanned per band), forming the focal plane assembly whose square end forms a field stop. The radiation falling on the fiber array was carried by a fiber optic pigtail to the detector filter assembly. As the technology advanced, as in the case of Landsat-4 TM, the focal plane had a monolithic detector array overlaid with filters. Both Landsat MSS and TM used relay optics following the prime focus to create a separate focal plane to accommodate detectors that required cooling, referred to as the cold focal plane. However, it should be noted that in this scheme there is no inherent band-to-band registration at the instrument level. Figure 5.13 shows how the Landsat MSS bands are separated in the across-track direction. As this is a fixed bias, it can be the correction procedure to produce band to band registered data products. However, there could be other time-varying components such as scan mirror performance

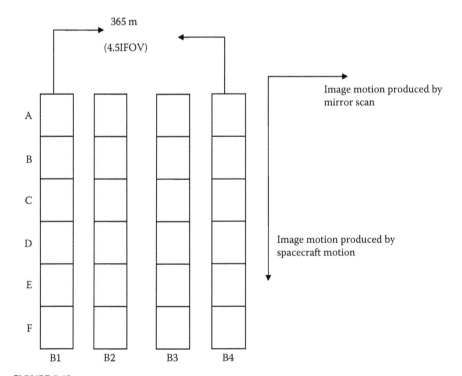

FIGURE 5.13
Projection of Landsat MSS detectors onto the ground. Each band (B1–B4) has six detectors (A–F). B1 and B4 are separated by about 4.5 instantaneous field of view.

changes and drift/jitter of the spacecraft that require modeling and more complex correction procedure. This calls for accurate knowledge of the scan mirror angular position and platform dynamics.

5.6 Detectors

To measure the radiance received at the focal planes, it has to be converted into an electrical signal—voltage or current. This is accomplished by electro-optical detectors. The detectors can be broadly classified as follows:

1. Thermal detectors
2. Photon detectors

The photon detectors may be further subdivided into photoemissive, photoconductive, and photovoltaic detectors. Before we proceed further, we discuss the performance parameters of these detectors.

5.6.1 Detector Figure of Merit

To compare different detectors, a number of figures of merit have been evolved over many years. The camera designer chooses those detectors whose figures of merit are best suited for the end use. We shall briefly describe the various figures of merit in this section.

Responsivity: Responsivity, R, is essentially a measure of how effective (sensitive) a detector is in converting the radiant power to an electrical signal—current or voltage. For a detector that generates current output, the responsivity is the ratio of the root mean square (rms) signal current (in amperes) to the rms incident radiant power, that is, amperes/watts. For a detector that generates output voltage, it is given as volts/watts. As in general the responsivity is not constant over the wavelength region, R is specified for a particular wavelength, R_λ. If I_s is the rms current generated for a spectral radiant rms power of Φ_λ, then

$$R_\lambda = \frac{I_s}{\phi_\lambda \, d\lambda} = \frac{I_s}{E_\lambda A_d \, d\lambda} \tag{5.1}$$

where E_λ is the spectral irradiance (W/cm^2/μm) and A_d is detector area in square centimeters. It should be noted that responsivity is dependent on the operating condition, such as detector bias (wherever it applies), temperature, chopping frequency, and so on. Knowledge of responsivity allows the user to determine how much detector signal will be available to decide on the preamplifier configuration.

Quantum efficiency (QE): In photon detectors, the responsivity depends on the QE of the detector, which is the ratio of the basic "countable" events (such as photoelectrons or electron–hole pairs) produced by the incident photon to the number of photons incident, that is, the fraction of the incident photons that contributes to photocurrent. Thus, QE = 0.5 means that if 10 photons are incident on an "average" 5 photoelectrons (or electron–hole pairs) are generated. Thus, QE is a measure of a device's ability to convert the incident radiation to an electrical signal.

Spectral response: Each detector is sensitive to a spectral region within which only it can be used. Spectral response gives the variation of the responsivity with wavelength. It can be shown as a curve, with λ on the x axis and $R(\lambda)$ on the y axis, giving the absolute responsivity at each wavelength. However, it is more commonly depicted as a relative spectral response curve, in which the peak of the response curve is set equal to 100%, and all other points are relative to this value.

Noise equivalent power (NEP): A camera designer is concerned about the minimum radiation power that can be measured by the detector. The phenomenon of noise limits the measurement of everything in nature. Therefore, the minimum power a detector can measure depends on the noise that the detector generates and its responsivity. NEP is the amount of radiant power incident on a detector, which generates a signal current (voltage) equal to the noise current (voltage) of the detector, that is, the radiant power that produces a SNR of unity at the output of a given detector:

$$\text{NEP} = \frac{I_n}{R} \tag{5.2}$$

where I_n is the rms noise current (amperes) and R is the responsivity in amperes/watts. NEP is specified at a particular wavelength and temperature. As the noise is dependent on the bandwidth of the measurement, the NEP can only be given at a specific bandwidth and that bandwidth must be specified. To take care of this, manufacturers usually specify NEP as the minimum detectable power per square root of bandwidth. Therefore, the unit of NEP is watts/$\sqrt{\text{Hz}}$. It may be noted that the lower the value of NEP, the better the performance of the detector for detecting weak signals in the presence of noise.

It must be realized that even if we assume hypothetically that all the noise sources are eliminated the ultimate limitation to detector performance is "photon noise." This is due to the random arrival of photons from the very source of radiant energy under measurement and background radiation, if any. The photon noise–limited detector gives an SNR proportional to

$$(\text{QE} \times N)^{\frac{1}{2}}$$

where N is the number of photons arriving at the detector and QE the quantum efficiency.

Specific detectivity: Another way of specifying the detection capability of detectors is *detectivity*, D, which is the inverse of NEP. That is,

$$D = \frac{1}{\text{NEP}} \left(\text{W}^{-1} \text{ Hz}^{\frac{1}{2}} \right) \tag{5.3}$$

For most commonly used detectors, detectivity is inversely proportional to the square root of the detector area. Therefore, it is difficult to compare the performance of two detectors without knowing its active area. To provide a figure of merit that is dependent on the intrinsic properties of the detector, and not on how large the detector is, an area-independent figure of merit, specific detectivity D^* (pronounced D-star), is defined such that

$$D^* = \frac{\sqrt{A_d}}{\text{NEP}} \text{ cm Hz}^{\frac{1}{2}} \text{ w}^{-1} \tag{5.4}$$

D^* essentially gives the SNR at a particular electrical frequency and in a 1-Hz bandwidth when 1 W of radiant power is incident on a 1-cm² active area detector. The D^* is specified for the wavelength λ, the frequency f at which the measurement is made, and the bandwidth Δf (the reference bandwidth is frequently taken as 1 Hz). That is, D^* is expressed as $D^*(\lambda, f, \Delta f)$. When comparing different detectors, one with a higher value of D^* is better suited to measure weak signals in the presence of noise.

As the detectivity is inversely proportional to the square root of the detector area, the detector performance can be improved by reducing the area of the detector. However, the detector size is determined by other sensor performance requirements. A method to have the same "optical size" but keep the physical size low is to place a detector element in optical contact with a lens having a high index of refraction, referred to as an optically immersed detector (Jones 1962). This is achieved by placing the detector at the center of curvature of a hemispherical lens, which can produce an aplanatic image. In such a configuration, the apparent linear optical size in comparison with the physical size is increased by a factor of the refractive index, n. If one uses a hyperhemispherical lens, the advantage is n^2. (A hyperhemisphere can be generated by cutting a sphere at a distance R/n from the center, where R is the radius of the sphere.) Optically immersed detectors are normally employed for thermal infrared (IR) detection. Germanium ($n = 4$) is the generally used material for the lens.

Dynamic range: Dynamic range measures the range of signals that can be recorded by a detector. It is the ratio of the brightest signal that a detector can record to the lowest (darkest) signal. It is desirable that the detector has linear response within the dynamic range, that is, the output increases linearly with incident intensity.

Time constant: The time constant is a measure of how fast a detector can respond to an instantaneous change of input radiation. If a detector is exposed to a step input of radiation, and if the output of the detector changes exponentially with time, then the time required for the output voltage (or current) to reach 0.63 of its asymptotic value is called the time constant.

Another term used to describe the speed of detector response is rise time. Rise time is defined as the time difference between the point at which the detector has reached 10% of its peak response and the point at which it has reached 90% of its peak response, when it is irradiated by a step input of radiation.

After having familiarized ourselves with the various terms that characterize the performance of an electro-optical detector, we shall now review some of the basic principles involved in the working of the detectors.

5.6.2 Thermal Detector

Thermal detectors make use of the heating effect of electromagnetic radiation. The consequent rise in temperature causes a change in some electrical property of the detector, such as resistance, which is measured. The change in the electrical property is measured by an external circuit, which gives a measure of the incident radiant flux. The response of the thermal detector is only dependent on the radiant power that it absorbs and hence is independent of wavelength. However, in practice the wavelength characteristics of thermal detectors are limited by the absorption characteristic of the sensor material, which varies with wavelength, and the transmission characteristics of the window material that covers the detector. Some of the common thermal detectors include bolometers, thermocouples, and pyroelectric detectors.

In *bolometers*, the change in temperature caused by the radiation alters its electrical resistance, which is suitably measured. In its earliest form, a thin blackened platinum strip was used. Advanced bolometers use semiconductors. Compared to metals, semiconductors exhibit a much more pronounced resistance variation with temperature change. Semiconductor bolometer elements are called thermistors.

A *thermocouple* is formed by joining two dissimilar materials having different thermoelectric power. Two of the most commonly used materials are bismuth and antimony. One of the junctions is blackened to absorb the radiation, and the other junction is kept at a reference temperature. The difference in temperature between the junctions produces thermoelectric emf, which is a measure of the incident radiant power. To increase the sensitivity, a number of thermocouples are connected in series, called a *thermopile*.

When the temperature is below the Curie temperature, certain ferroelectric crystals exhibit spontaneous polarization with change in temperature. This is called the pyroelectric effect. A *pyroelectric detector* is basically a ferroelectric capacitor that is thermally isolated and exposed to incident radiation. When the incident radiation changes the temperature of the detector element, it exhibits spontaneous polarization, that is, opposite faces of crystallographic orientation exhibit opposite electric charges in response to temperature change. This can cause a displacement current to flow in an external circuit, which can be detected. Pyroelectric detectors respond only to changes in temperature. They have the maximum D^* and fastest response time compared to other thermal detectors.

Thermal detectors require a finite time to raise the temperature of the sensing element to the final value and hence a long time constant compared to photon detectors, wherein an electronic transition is virtually instantaneous on photon absorption. The time constant of thermal detectors is usually in milliseconds or more, and hence they are not suitable for high-resolution imaging.

5.6.3 Photon Detectors

In these types of detectors, also referred to as quantum detectors, photons (quanta of light energy) interact with the detector material to produce electrons/electron–hole pairs, which manifest as current or voltage. Depending on the detector material and the detection process, the photon should have a minimum energy, E_{min}, to release the charge carriers (electron/electron–hole pairs) from the atomic bonding forces:

$$E_{min} = h\nu = \frac{hc}{\lambda_c} \tag{5.5}$$

where h is Planck's constant, c the velocity of light, and λ_c the cutoff wavelength. For any radiation with wavelength above λ_c, the response of the photon detector drops to zero. These types of detectors respond to the rate of photon absorption. The spectral response of a quantum detector is shown in comparison with a thermal detector (Figure 5.14). There could be a little confusion in understanding the curve. Why should the responsivity increase

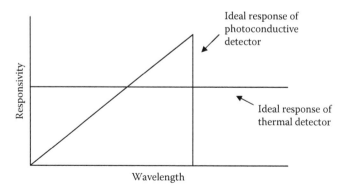

FIGURE 5.14
Idealized spectral response curve of a thermal and quantum detector for a constant energy input.

as the photon energy decreases? As the wavelength increases, the photon energy decreases; hence, the number of photons per constant input power increases. In a photon detector, as the response is dependent on the number of photons per second, responsivity increases as the wavelength is increased. On the other hand, as thermal detectors respond to the total energy absorbed, for a constant energy input response is independent of wavelength.

5.6.3.1 Photoemissive Detectors

These detectors are based on the photoelectric effect, where photons having energy above a certain threshold value generate free electrons from the surface of the detector material. The energy required by the electron to escape from the surface of the material depends on the material properties and is called the work function. The photoelectrons so emitted from the surface are accelerated to an anode and kept in vacuum at a positive potential with respect to the photoemissive surface, producing current in an external circuit (Figure 5.15). The detectors based on this principle, which are used in imaging cameras, are called photomultiplier tubes (PMTs) wherein the photoelectrons are multiplied by secondary electron emission through a number of plates (dynodes) kept at progressively increasing voltage and then collected by the anode. Because of this internal gain, the output signal is fairly large and can be handled for further processing without resorting to special low noise amplifiers. The spectral response of the photomultipliers depends on the photocathode material. Certain compound materials give a lower work function compared to a single metal. The silver-oxygen-cesium surface gives a work function of 0.98 ev, giving a λ_c of 1.25 μm. PMTs are available with spectral response extending from ultraviolet to near IR of the electromagnetic spectrum. The ATS-1 Spin-Scan CloudCover Camera, launched in 1966; ATS-3 Multicolor Spin-Scan Cloud Cover Camera, launched in 1967; and

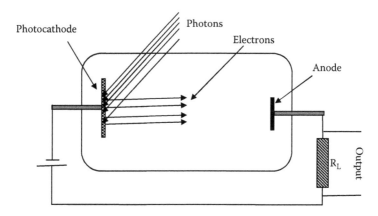

FIGURE 5.15
Schematics showing the functioning of a simple photoemissive detector. R, load resistance.

Landsat MSS are a few spaceborne Earth imaging cameras that use PMTs as detectors.

5.6.3.2 Photoconductive Detector

Photoconductive detectors are light-sensitive semiconductors whose resistance decreases with the amount of incident light. A photon of sufficient energy can excite an electron from valance band to conduction band. If E_g is the bandgap, the maximum wavelength for which this happens is given by

$$\lambda_c = \frac{1.24}{E_g} \tag{5.6}$$

where λ_c is in micrometers and E_g is in electron volts. The E_g for germanium is 0.66 eV (λ_c = 1.9 µm), for silicon 1.12 eV (λ_c = 1.1 µm), and for gallium arsenide 1.42 eV (λ_c = 0.87 µm). When the electrons are excited to the conduction band, the conductivity of the detector increases; this can be measured. Thus, photoconductive detectors can be considered to be light-sensitive variable resistors, whose conductivity varies with the number of photons absorbed. Although a photoconductive detector is made of homogeneous material, it can be either intrinsic (pure) or extrinsic (doped). The addition of an impurity into an intrinsic semiconductor (doping) results in energy levels that lie between conduction and valance bands of the intrinsic material, allowing detection of radiation at longer wavelengths. Thus, the bandgaps and hence the cutoff wavelength can be altered by suitable doping. Some of the commonly used photoconductive detectors for remote sensing include indium antimonide, indium gallium arsenide, and mercury cadmium telluride (MCT). MCT is of particular interest, as by adjusting the alloy composition it is possible to cover a wide range of wavelengths extending from the near IR to thermal IR. The INSAT VHRR uses MCT for the detection of thermal (10.5–12.5 µm) and water vapor (5.7–7.1 µm) bands (Joseph et al. 1994). To realize a practical detection system, the detector is connected to a resistor and a supply voltage is applied (Figure 5.16a). Depending on the variation of conductivity due to the changing incident radiation, the voltage across the resistor varies, which is measured by suitable amplifiers.

5.6.3.3 Photovoltaic Detector

When p and n extrinsic semiconductors are brought together, we have a p–n junction. An electric potential is formed at the junction because of the diffusion of electrons from the n-region to p-region and holes from the p-region to the n-region. When the photon flux of energy exceeding the bandgap energy E_g irradiates the p–n junction, electron–hole pairs are formed, which modify the junction potential. This effect is called the photovoltaic effect. Such a device can be operated either in photovoltaic mode or in photoconductive mode. In the photovoltaic mode, electron–hole pairs migrate to opposite

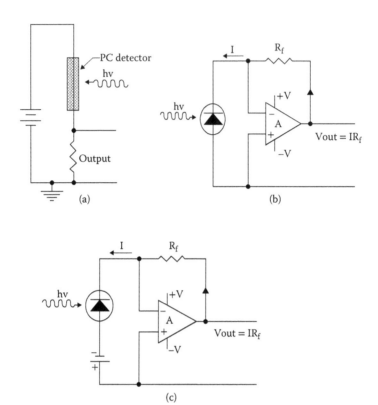

FIGURE 5.16
Typical configuration of photon detectors to measure signal generated. (a) Photoconductive detector. (b) Photodiode operating in photovoltaic mode. (c) Photodiode operating in photoconductive mode.

sides of the junction, producing a voltage. The voltage generated across the junction (open circuit) is logarithmically dependent on the radiation intensity. If the junction is short-circuited by an external conductor, current will flow in the circuit when the junction is illuminated and the current generated increases linearly (within limits) with intensity of radiation. In practice, the generated photocurrent is converted to voltage using a transimpedance amplifier (TIA). This mode of operation with zero bias is called photovoltaic mode (Figure 5.16b). The photovoltaic mode of operation is preferred for low-frequency applications (up to 350 kHz) as well as ultralow light level applications. Another advantage of photovoltaic operation is that the photocurrent in this mode has less variation in responsivity with temperature.

It is also possible to use a p–n junction to detect radiation by applying a bias voltage in the reverse direction, that is, when the p-side of the junction is made negative with respect to the n-side. This mode produces higher noise current compared to the photovoltaic mode of operation. But because the reverse bias increases the width of the depletion region, the junction capacitance is reduced, and therefore the reverse biased configuration can be operated at higher speed. This mode of operation of the photodiode is referred to as the photoconductive mode of operation (Figure 5.16c). Thus, for precise linear operation for low signal detection photovoltaic mode is preferred, whereas photoconductive mode is chosen where higher switching speeds are required. Photodiodes are usually optimized at the fabrication stage to operate in either photovoltaic mode or photoconductive mode.

Another semiconductor structure used in photodiodes to increase the frequency response is the PIN photodiode. The PIN photodiode consists of an intrinsic (lightly doped) region that is sandwiched between a p-type layer and an n-type layer. When this device is reverse biased, it exhibits an almost infinite internal impedance (like an open circuit), with an output current that is proportional to the input radiation power.

The spectral response of a photodiode can be tuned to different regions by choosing suitable materials to make the p–n junction. Silicon photodiodes are widely used in imaging systems for the spectral range from blue to near IR (~1 µm). Similarly, indium gallium arsenide (InGaAs) responds in about the 1 to 2 µm region. Indium antimonide (InSb) detectors can be fabricated to respond within the 1 to 5 µm spectral region. The most widely used detector in the thermal IR region (8–12 µm) is MCT. As mentioned earlier, its bandgap can be adjusted by varying the relative proportion of Hg and Cd to produce a peak sensitivity over a wide range in the IR region (Westervelt 2000). Figure 5.17 gives the spectral response of a few representative photon detectors.

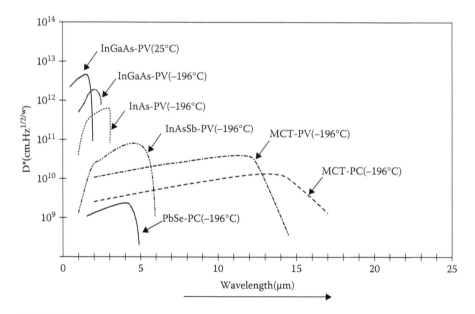

FIGURE 5.17
Typical spectral responses of a few photon detectors. The operating temperatures are shown in parentheses. (These are the typical responses of some of the detectors produced by Hamamatsu Photonics. From Hamamatsu technical information SD-12:- http://www.hamamatsu.com/resources/pdf/ssd/infrared_techinfo_e.pdf.)

5.6.4   Quantum Well Infrared Photodetectors

MCT detectors have been used for decades for detecting long-wave IR radiation in the 8–14 µm range. MCT at 77 K has a reported quantum efficiency exceeding 70% and detectivity beyond 10^{12} cmHz$^{1/2}$W^{-1}. Though moderate-size focal plane arrays have been realized using MCT, there are technical difficulties in realizing long arrays and yields are low. Quantum well infrared photodetectors (QWIPs) have emerged as an alternative for imaging in thermal IR. A quantum well is realized when a thin layer (on the order of a few nanometers) of a semiconductor like GaAs having a bandgap of approximately 1.4 eV is sandwiched between thick semiconductor layers of wider bandgap such as AlGaAs of bandgap 1.4–3.0 eV (value depends on the Al concentration) (Figure 5.18). The quantum effect creates a potential well that confines electrons or holes, which were originally free to move in three dimensions, forcing them to occupy a planar region. In a quantum well, the electrons and holes are still free to move in the direction parallel to the layers; however, they are confined to "discrete energy levels" in the direction perpendicular to the layers. The transitions between these discrete energy levels are used for IR wavelength detection. Quantum wells are doped such that electrons will occupy the ground state level until they are excited by incoming photons. When a bias is applied, the excited electrons will be swept away by the electric field, generating a photocurrent signal. Several quantum wells (sandwiched between barriers) are usually grown stacked on top of each other to increase photon absorption. By changing the thickness of the quantum well and the barrier height by choosing appropriate barrier material, the detection wavelength window can be tuned. Due to inherent dark current mechanisms (tunneling, thermal excitation, etc.), these detectors are also required to be operated at cryogenic temperatures (<120 K). The QWIP can be made to a high degree of precision by modern epitaxial crystal growth techniques. More details on QWIPs can be found in the work by Henini and Razeghi (2002). There is rapid growth in developing high-performance QWIP, with both area and linear arrays. Landsat 8 thermal imaging is carried out by quantum well detectors.

5.6.5   Operating Temperature

An ideal radiation detector should not generate any current in the absence of an incident radiation. However, this does not happen in a practical detector. In a semiconductor, at absolute zero temperature ($T = 0$ K), all electrons are bound to their parent atoms and therefore there is no flow of current. Thus, at the absolute zero temperature semiconductors behave like insulators. Above the absolute zero temperature, the average thermal energy of charge carriers increases as kT, where k is Boltzmann's constant and T is the temperature in kelvin. As the temperature increases more and more, electrons acquire sufficient energy to cross the bandgap. Such thermally exited electrons are undistinguishable from the photon-induced electrons. The current

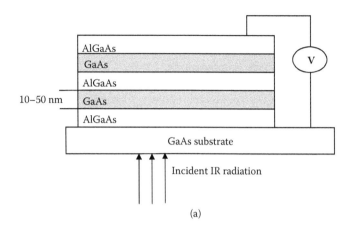

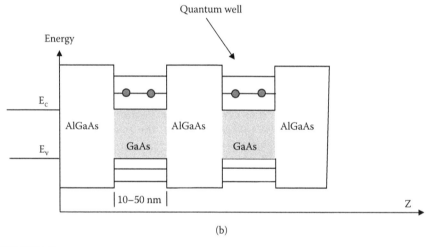

FIGURE 5.18
(a) Schematics of the heterostructure layout of a quantum well infrared photodetector. (b) Energy structure. AlGaAs acts as a barrier to create a potential well. Here, E_c denotes the conduction band edge and E_v denotes the valance band edge. (Babu Naresh, 2013, personal communication.)

so produced is called dark current, and its fluctuation is a major source of noise current. The situation is quite serious for detection at longer wavelengths, as detection of longer wavelengths requires a smaller energy gap. For example, silicon has an energy gap, E_g, of 1.12 eV with long wavelength cutoff, λ_c, at 1.1 μ. At 300 K, the thermal energy is 0.026 eV, which is only 2.3% of the energy gap. Consider a detector like HgCdTe with λ_c at 10 μ, for which E_g has to be 0.12 eV and the thermal energy at 300 K is about 22% of E_g. If we operate the same detector at 77 K, the kT value is 0.0066 eV, which is only 5.5% of E_g. Therefore, to reduce the thermal generation of carriers and corresponding noise, IR detectors have to be cooled for better performance. The temperature to which the detector is to be cooled depends on the wavelength region, and the method of cooling differs according to the operating temperature and overall system constraints.

For detectors operating at temperatures above 170 K, thermoelectric coolers are used. The operation of thermoelectric coolers is based on the Peltier effect, that is, the cooling of one junction and the heating of the other when electric current is passed through a circuit consisting of two dissimilar conductors. The effect is stronger in circuits containing dissimilar semiconductors. Many detectors for the 3- to 5-μm wave band are thermoelectrically cooled.

For the thermal IR (8–14 μ) region, the detectors are cooled to less than 100 K. For satellite-borne sensors, this is generally achieved by passive radiative coolers. The configuration of a three-stage cooler developed for cooling the MCT detector of INSAT VHRR I is shown in Figure 5.19 (Gupta et al. 1992). We shall briefly describe the design considerations of the radiative cooler. The three stages of the cooler are sun shields/housing, radiator, and patch. The patch is the radiating surface, which is radiatively coupled to the outer space. The detector is placed in thermal contact with the patch. The patch attains an equilibrium temperature depending on the radiative property of the patch surface; radiation inputs from the surroundings; power dissipation of the detector; and other conductive losses from wires connected, support studs, and so on. Therefore, while designing the cooler every possible effort must be taken to reduce the heat load reaching the patch. To minimize the radiative and conductive coupling, all the stages should be thermally isolated from each other. In addition, the radiative cooler should be mounted on the anti-Sun side of the instrument to have minimal external heat input. The radiator is the intermediate stage and

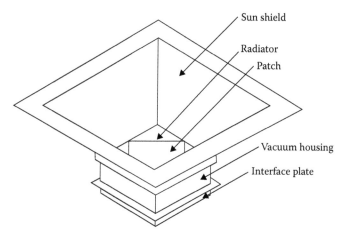

FIGURE 5.19
Schematics showing a passive radiation cooler employed for cooling detectors in satellites. (Courtesy of ISAC/ISRO; Gupta et al., *J. Spacecraft Technol.*, II, 1, 23–35, 1992.)

is coplanar with the patch. To reduce the heat load on the radiator, the surface facing the space is coated with a low solar absorptance and high emittance white paint. To avoid any direct Sun load to fall on the patch, a Sun shield assembly, a conical structure consisting of four trapezoidal panels, is mounted on the vacuum housing. The height and the cone angle of the Sun shield assembly are designed in such a way that no direct Sun load falls on the patch at any time of the year. The Sun shield surfaces facing the patch should have highly specular (mirror-like) reflecting surfaces with low solar absorptance and low emittance so that incident solar radiation is reflected back to deep space without scattering. This also shields the emitted IR load reaching the patch, thereby minimizing the thermal loads on the patch. The mechanical support and electrical wiring has to be judiciously designed to provide maximum thermal isolation of the patch. To facilitate testing on the ground, the patch/radiator assembly is attached to a vacuum housing. So as to maintain a stable operating temperature for the detector, the patch is designed to cool below the desired operating temperature and a thermal control (TC) system is required to control the temperature. This is achieved by attaching a heater coil to the patch and monitoring the temperature using a platinum resistance thermometer, whose output is used to control the heater to maintain the detector at the specified value and within the required range. In addition, decontamination heaters are incorporated to get rid of any deposited volatile contamination.

Passive coolers are generally used in Earth observation cameras, because of its reliability (no moving parts), virtually no power requirements (except for TC of detector), low mass, and production of no vibration that can lead to microphonics. The lifetime of passive coolers is limited only by surface contamination and degradation. Because of these advantages, passive coolers used to be the preferred choice for cooling detectors in the 80–100 K temperature range. However, the performance of passive coolers decreases rapidly as the cold temperature to be attained decreases below 70 K and the heat lift requirements are high. To address these limitations, active coolers (cryocoolers) are being successfully used in many space missions. Active coolers use closed thermodynamic cycles to cool at the cost of electrical input power. Early use of cryocoolers in space was for science missions to achieve temperatures below 10 K. Currently, a number of Earth observation missions are also using Stirling cycle coolers to keep the detector temperatures in the 60–80 K range. The Michelson Interferometer for Passive Atmospheric Sounding (MIPAS) and Advanced Along Track Scanning Radiometer instruments on board ENVISAT launched in 1999, along with MOPITT on the EOS platform, are some of the instruments that have successfully used Stirling coolers to cool the detectors (Jewell 1996). Many of the next-generation geostationary imagers such as GOES-R, METEOSAT third generation, and JAMI (Japanese Advanced Meteorological Imager) are also planning to carry active coolers for cooling the focal plane detector array (Ross and Boyle 2006).

5.6.6 Signal Processing

The output from a detector gives a continuous voltage/current signal proportional to the number of photons received by it. This signal also includes the noise contribution from the detector. In the case of IR detectors, there is a signal component associated with temperature and emissivity of the background targets in the FOV, such as optics, housing, and so on. In many cases, the contribution from the background could be very high compared to the signal. The major functions of signal processing electronics are to amplify the signal generated by the detector; provide a stable reference, referred to as DC restoration; provide a band limit; and digitize the output to convert the analog signal to digital for further formatting and transmission through the satellite communication channel. The electronics design and realization should ensure that their noise contribution does not dominate over the detector noise. Because the signal generated by the detector is very low, the preamplifier is kept very close to the detector. Preamplifier design depends on the type of detector.

Photovoltaic detectors generate current when photons are detected, which is converted to voltage. The zero-biased photovoltaic detector behaves like a high-impedance current source, and a low-impedance device has to be used to convert the current to voltage. A TIA is used to convert the current into voltage. A typical TIA circuit for photovoltaic operation is shown in Figure 5.20. The output voltage signal is the product of the short-circuited current from the photodiode (photon-generated current and dark/leakage current) and the feedback resistance value. TIA stability is sensitive to input capacitance due to detector p–n junction, connecting cable, and so on. A compensating capacitor, C_f, is added in the feedback to reduce gain peaking and protect against possible oscillation tendency. The C_f and the R_f decide the frequency response of the amplifier. The 3-dB cutoff frequency, f_c, of the signal is given by

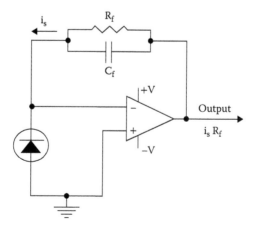

FIGURE 5.20
Schematics of the circuit of a transimpedance amplifier for photovoltaic operation.

$$f_c = \frac{1}{2\pi R_f C_f}$$ (5.7)

The thermal noise due to R_f is the dominating noise, and the SNR is proportional to $\sqrt{R_f}$. However, the maximum R_f value is limited by the bandwidth requirement. Also, the maximum value of feedback resistance is chosen such that the output does not get saturated for an expected maximum signal current. The actual input resistance of an operational amplifier is not infinite, and some bias current flows at the input terminal causing some offset. FET-input-type operational amplifiers have a much lower bias current compared to bipolar operational amplifiers and are the preferred choice. The other criteria for selecting an operational amplifier include bandwidth, drive requirement (length of the cable and capacitance), and noise performance.

Another type of detector used in optomechanical scanners is the photoconductive type. As discussed earlier, a photoconductive detector changes its resistance in response to the radiation it receives. So the signal extraction from a photoconductive detector depends on accurate measurement of the change in resistance of the detector due to the incident radiation. In principle, it can be measured by a simple biasing scheme, as shown in Figure 5.16a. These detectors usually have low impedance in the range of a few hundred ohms and have to be biased with a current value optimized by the vendor. As the bias noise/drift will be sensed as signal, a very stable and extremely low noise bias voltage has to be designed. Another approach to minimize the effect of drift of bias current is to use ratiometric measurement. The simplest form of ratiometric measurement is to use a bridge amplifier. The detector is placed in one of the arms of the bridge (Figure 5.21). R3 is the detector equivalent resistor used to balance the bridge. The bridge is balanced when the voltages at nodes A and B are equal. The values of the R1 and R2 resistors are decided based on the required bias current and are generally kept much higher in magnitude than the detector resistance. A low-noise op-amp can be used to sense the differential voltage. The bridge is designed to be balanced at a particular detector temperature referred to as the set temperature. The signal gain of this configuration is $R_f/R3$. The signal from these detectors is very low, and hence the preamplifiers for photoconductive detectors are generally designed to have a gain in the range of 500–1000. Because of the high gain of the preamplifier, any mismatch between the detector temperature and the set temperature leads to an imbalance in the bridge, which is reflected as the offset voltage at the output. This will be taken care of in subsequent circuits.

The preamplifier output generally contains signal as well as some offset component. The offset component could be due to various reasons such as electronics, detector dark current, drift, and so on. The offset component has to be removed and the signal has to be measured with respect to a known fixed level, usually when no radiation falls onto the detector, that is, dark

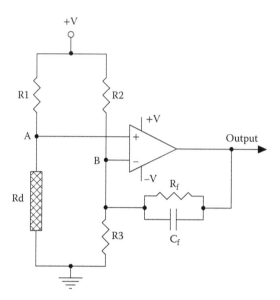

FIGURE 5.21
Simplified schematic of a bridge amplifier to measure photoconductive detector output.

level. The process is usually referred to as DC restoration. The most ideal dark level is when the detector is made to view deep space. This is possible when the scanning is done horizon to horizon as in the case of optomechanical scanners operating from a geostationary orbit such as METEOSAT, GOES, VHRR, and so on. Low Earth orbiting satellites carrying a rotating mirror, as in AVHRR, also view the space once in every rotation of the mirror. In the case of optomechanical scanners using an oscillating mirror (such as Landsat MSS and TM), at the completion of each imaging scan an obscuration shutter rotates to interrupt the normal optical path for each detector and in one position of the shutter a zero-radiance surface enters the detector's FOV (Fusco and Blonda 1986).

The DC restoration can be done either in the digital or in the analog domain. When dark space comes in the detector's FOV, electronics carries out a "clamping" action by estimating the background voltage and storing the value in a capacitor (analog mode) or converting it into a digital count value. When the target signal comes, stored background is removed to extract the true signal. This action is done ideally for every scan line. The action is similar to correlated double sampling (see Section 6.3.2). The DC-restored signal is amplified, and the gain is set such that the expected maximum signal gives the full-scale value of the analog to digital converter (ADC). Usually, a provision is provided to change the gain by command to vary the saturation radiance settings.

The spatial frequency component sensed by the detector depends on the spatial filtering carried out due to the combined MTF of the optics and detector aperture. The detector converts the spatial variation on the ground into the temporal variation in the electrical signal. For example, a wheat field or ocean surface is viewed as a low-frequency signal, whereas parallel highways or urban areas give high-frequency signals. A good transient response is important when viewing an abrupt change in reflectance such as land–water interfaces. Therefore, the electrical bandwidth also influences the spatial characteristics of the imagery. However, the bandwidth of the signal has to be limited to avoid aliasing. As per Shannon's criteria, the minimum sampling frequency should be twice the frequency content. The continuous electrical signal from the detector is usually sampled at every IGFOV in the spatial domain, which corresponds to an interval equal to the dwell time. If τ is the dwell time, then to satisfy Shannon's criteria the electrical bandwidth f has to be limited such that

$$f = \frac{1}{2\tau} \tag{5.8}$$

Therefore, a band-limiting filter is incorporated to avoid aliasing errors. However, such a filter also brings down the MTF, especially at the Nyquist frequency. To enhance the MTF, one can shift the cutoff to higher frequencies, but this apart from producing aliasing also increases the noise contribution. The design of the band-limiting filter is also important. A very sharp falloff is useful to reduce noise and also the aliasing effect; however, a sharp falloff may cause undesirable overshoot when the scene contains an abrupt transition of radiance (Norwood and Lansing 1983). To improve the MTF, systems are generally oversampled, that is, at intervals less than the dwell time, though this increases the data rate. The band-limited signal is sampled using an appropriate sample and hold amplifier and digitized using ADC with the number of bits consistent with the radiometric resolution requirements.

5.7 System Design Considerations

A system designer is given a set of performance requirements to be realized based on the end applications for which the data from the instrument are used. The system designer should be able to achieve this goal in the most efficient way, that is, by optimizing weight, volume, power consumption, and so on, finally leading to a design meeting the budget and schedule constraints. We shall now work out the interrelationship between various parameters of an optomechanical scanner to choose an optimal design. The performance goals projected to the designer generally include the following:

1. Spatial resolution r
2. Swath-FOV Ω

3. Number of spectral channels and their central wavelength
4. Spectral bandwidth $\Delta\lambda$
5. Minimum detectable radiance or temperature difference
6. Satellite altitude h

For a spatial resolution r (i.e., the footprint of the detector on the ground) from a satellite altitude h, the IFOV β is given by

$$\beta = \frac{r}{h} \tag{5.9}$$

The scan mirror scanning period should be such that the time taken, t, by the mirror to scan one line and come back to the start should be equal to the time taken by the subsatellite point to move a distance r. If v is the subsatellite velocity,

$$t = \frac{r}{v} = \frac{\beta h}{v} = \frac{\beta}{(v/h)} \tag{5.10}$$

If n detectors are used along track, the scan mirror scans n lines at a time and hence

$$t = \frac{n\beta}{(v/h)} \tag{5.11}$$

The time t consists of the time for forward motion, t_f, of the mirror to cover a swath of Ω rad and the time, t_r, to return back to the start position. Of these, only t_f is useful in imaging (when imaging is done only in forward scan as in Landsat MSS); t_f/t is called the scan efficiency S_e. During t_f, the sensor views (Ω/β) pixels. Therefore, the dwell time τ is given by

$$\tau = \frac{t S_e}{\Omega/\beta} = \frac{n\beta^2 S_e}{(v/h)\Omega} \tag{5.12}$$

As the signal collected is proportional to the dwell time, the sensor designer tries to maximize the dwell time within practical engineering limitations. As β and Ω are generally the required specifications and v/h is decided by the satellite orbit, the only variable available to the designer is the scan efficiency, S_e, and the number of detectors along track. You will now appreciate why Landsat MSS had six detectors along track for each band. In the case of TM, due to higher spatial and spectral resolution, this was increased to 16 and, in addition, the scan efficiency was improved by using both forward and reverse scans for imaging.

Another important system parameter is data rate. If there are m spectral bands and each band output signal is digitized to b bits, the total number of bits to be transmitted during the integration time is n ($m \times b$). Therefore, the data rate is

$$\frac{n(m \times b)}{\tau} = \frac{mb(v/h)\Omega}{\beta^2 S_e} \text{ bits / s} \tag{5.13}$$

This assumes that across track is sampled at every IFOV. However, one usually oversamples to get a better MTF. For example, Landsat MSS is oversampled 1.41 times. In addition, there will be other overhead due to housekeeping information and so on. The aforementioned equation shows that the data rate is inversely proportional to the square of the IFOV. That is, when the resolution is doubled (i.e., IFOV halved) the data rate increases four times. A higher data rate is one of the major concerns in realizing high-resolution imaging systems.

Amplifier bandwidth Δf is given by

$$\Delta f = \frac{1}{2\tau} = \frac{(\frac{v}{h})\Omega}{2n\beta^2 S_e} \tag{5.14}$$

However, if oversampled in the across-track direction, one can broaden the bandwidth and improve the MTF.

Ultimately, our interest is to measure as small a radiance as possible. This depends on the energy collected by the system and the noise from various sources. The SNR is an important parameter, contributing to the quality of the image. The SNR depends on the energy reaching the detector and the noise characteristics of the detector used. We have seen in Chapter 3 that the power received at the detector is (Equation 2.14)

$$\Phi_d = \frac{\pi}{4} O_e \Delta\lambda \, L_\lambda \beta^2 D^2 \text{ watts}$$

The signal current I_s generated depends on the responsivity of the detector.

The next task is to evaluate the noise contribution. The contribution to noise comes from various sources. These include the following:

Dark current: The current through the photodetector in the absence of light. Because of the statistical nature of the carrier generation to produce dark current, its fluctuation results as noise.

Thermal noise, also called Johnson noise or Nyquist noise, is the noise generated by the thermal motion of the electrons inside an electrical conductor. This happens regardless of any applied voltage. For an electrical conductor with a bandwidth Δf, the mean square thermal noise current (i_{th}^2) is given as follows:

$$i_{th}^2 = \frac{4kT\Delta f}{R} \tag{5.15}$$

where k is Boltzmann's constant (joules per kelvin), T is the resistor's temperature (kelvin), and R is the effective load resistance. A higher value of R reduces the thermal noise but increases the time constant, thereby reducing the bandwidth, and hence requires a judicious choice of the load resistor.

Generation–recombination (G-R) noise is caused by fluctuations in the electron–hole generation rate and recombination rate. G-R noise is unique to photoconductive detectors.

1/f noise: The mechanisms that produce $1/f$ noise are not well understood. Generally, it is not of much concern if the lower cutoff of the electronics bandwidth is above a few hundred Hz.

Shot noise (photon noise) is due to the random arrival rate of photons from the source of radiant energy under measurement and background radiation. Even if all other noise sources are eliminated, photon noise would set the ultimate limit for detector performance. Thus, an ideal detector system should have all other noises much smaller than the photon noise; then, one has a photon noise–limited detector system.

Preamplifier noise: Depending on the preamplifier chosen, the noise current of the preamplifier also contributes to the overall noise current.

In a photodiode, for low signal levels the Johnson noise and the preamplifier noise are the dominant contributors to the noise current (NASA 1973).

As all these noise generation mechanisms are independent, the total noise current I_n can be found as the square root of the sum of the squares of these independent noise terms:

$$I_n = \sqrt{(i_1^2 + i_2^2 + i_3^2 + ----)\Delta f} \tag{5.16}$$

$$I_n = i_n \Delta f^{\frac{1}{2}} \tag{5.17}$$

where i_n is the root sum square of the current sources contributing to the noise.

The SNR can be found from the ratio of the signal photocurrent to the noise current. The signal current I_s is given by

$$I_s = \frac{\text{power } \Phi \text{ received at}}{\text{the detector (watts)}} \times \frac{R \text{ the mean detector responsivity}}{\text{over } \Delta\lambda \text{ (ampere/watts)}}$$

Substituting for power Φ received,

$$I_s = \left[\frac{\pi}{4} O_e \, \Delta\lambda \, L_\lambda \, \beta^2 \, D^2 \right] R \tag{5.18}$$

$$\text{SNR} = \frac{I_s}{I_n} = \frac{\dfrac{\pi}{4} O_e \, \Delta\lambda \, L_\lambda \, \beta^2 D^2 R}{i_n \Delta f^{1/2}} \tag{5.19}$$

Substituting for Δf from Equation 5.14 and rearranging,

$$[\text{SNR}]_{\text{PD}} = \left[\frac{\pi}{2\sqrt{2}} \frac{L_\lambda \Delta\lambda \sqrt{S_e} O_e}{i_n \left(v/h \right)^{1/2}} R \right] \frac{\beta^3 D^2 \; n^{\frac{1}{2}}}{\Omega^{\frac{1}{2}}} \tag{5.20}$$

The quantities within the bracket are constant for a given satellite height, detector characteristics, and scene radiance. The spectral characteristics are generally theme dependent, and hence $\Delta\lambda$ is specific for a mission irrespective of spatial resolution. One normally optimizes the optics diameter and scan efficiencies within engineering limitations to get the best performance. To understand the design constraints to achieve higher spatial resolution, Equation 5.20 can be represented for a fixed swath (i.e., Ω constant) as follows:

$$[\text{SNR}]_{\text{PD}} = K\beta^3 D^2 \; n^{\frac{1}{2}} \tag{5.21}$$

where K is a constant. Equation 5.21 gives the interrelation between various instrument parameters. Similar relationships when photomultiplier and photoconductive detectors are used as detectors have been worked out by the National Aeronauts and Space Administration (NASA 1973).

To evaluate how the optics diameter D and the number of detectors are affected as the resolution is changed, keeping all parameters the same, it is convenient to express the right-hand side of the equation as $\beta \, D^{\frac{2}{3}} \, n^{\frac{1}{6}}$. Thus, if we improve the resolution by a factor of two (β halved) keeping the number of detectors the same, to have the same radiometric quality (i.e., same SNR) the diameter of the optics has to be increased 2.8 times. Figure 5.22 gives the aperture diameter to achieve increased resolution with TM as a reference. Using Equation 5.21, Joseph (1996) has shown that if one were to realize SPOT HRV camera, IFOV and FOV using an optomechanical scanner with TM as a reference and keeping the number of detectors the same as in TM, the diameter of the optics needed is 69 cm compared to the 47 cm of the actual SPOT camera operating in pushbroom mode. However, if we decide to keep the optics diameter the same as SPOT, to achieve the same SNR as TM we have to use 128 detectors against the 16 used in TM. This shows the complexity of using optomechanical scanners to achieve higher resolutions. However, optomechanical scanners will still dominate when one requires a wide swath of coverage. Another area in which an optomechanical scanner has an advantage is that it can image in any spectral band extending from violet to far IR, as discrete detectors operating in these ranges are well developed but long detector arrays are still in the developmental stage for longer wavelengths.

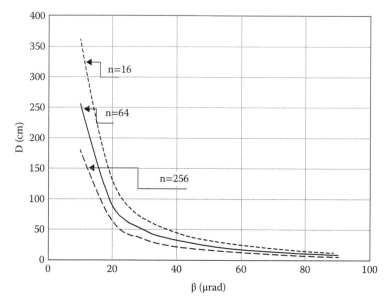

FIGURE 5.22

Aperture diameter D versus instantaneous field of view (β) for optomechanical scanners employing photodiode detectors for different values of n (relative to Thematic Mapper; $\beta D^{\frac{2}{3}} n^{\frac{1}{6}} = $ constant).

As a typical example of an optomechanical scanner, we shall describe the Landsat Enhanced Thematic Mapper, which is the highest spatial resolution Earth imaging camera operating in the whiskbroom mode.

5.8 Enhanced Thematic Mapper Plus

Optomechanical scanners have been used for meteorological observation since the 1960s. The multispectral scanner system on board NASA's Earth Resources Technology Satellite Landsat 1, popularly known as MSS, was the first operational satellite-borne optomechanical scanner specifically designed for civilian Earth resources applications. The TM is an advanced second-generation optomechanical multispectral scanner first carried on board Landsat 4. TM provides seven spectral bands covering visible, near IR, shortwave IR, and thermal IR spectral regions with a 30-m resolution in the visible, near, and shortwave IR bands and 120-m resolution in the thermal IR band. Apart from improved spatial and spectral resolution, the TM provides a factor of two improvement over MSS in radiometric sensitivity. The follow-on to Landsat 5, the Landsat 7 satellite carried an improved version of

TM called Enhanced Thematic Mapper Plus (ETM+) with an additional band (panchromatic) providing a 15-m resolution covering the spectral region from 0.5 to 0.9 μm and an improved resolution (60 m) in the thermal band. However, the basic instrument design of TM and ETM is similar.

Considerable innovation in the electro-optical design was required to achieve such improvements (Blanchard and Weinstein 1980). TM/ETM+ uses an F/6, RC telescope with an aperture of 40.6 cm. Silicon photodiode arrays are used to detect visible and near IR bands. Cooled indium antimonide (InSb) photodiodes are used to detect the middle IR channels, whereas the thermal band uses cooled MCT detectors. The detectors are located in two focal planes. A monolithic silicon array for five bands is located at the prime focal plane: band 1 through band 4 and the pan band. The arrays for bands 1 through 4 contain 16 detectors divided into odd–even rows. The array for the pan band contains 32 detectors, also divided into odd–even rows. A relay optics transfers the radiation from the prime focal plane to the cold focal plane. At the cold focal plane, the detector arrays for bands 5, 6, and 7 are located, which are cooled by a radiative cooler operating at a nominal temperature of about 90 K that can be adjusted to one of three set points (90, 95, or 105 K) by a heater at the back of the substrate. The thermal IR band contains eight MCT detector elements. Band 5 and band 7 contain 16 indium antimonide photovoltaic arrays. The nominal spatial resolution of bands 5 and 7 is the same as that of bands 1 through 4. The input stages and the feedback resistors of the preamplifier for band 5 and 7 photovoltaic detectors are mounted with the detectors on the cold stage of the radiative cooler. The band 6 preamplifier and the remainder of the band 5 and band 7 preamplifiers are mounted adjacent to the radiative cooler. The spectral filters that primarily decide the spectral bandwidth are located immediately in front of each detector array. The EFL with the detector size gives an IFOV of 42.5 μrad for the very near IR and shortwave IR bands, giving a ground resolution of 30 m at the nadir. The pan gives 15-m resolution, whereas the thermal band provides 60-m resolution. The layout of the detectors in the focal plane is shown in Figure 5.23. As seen from the figure, at any instant the actual ground location seen by each band's detectors is not the same due to horizontal spacing of detector rows within and between bands. That is, there is no inherent band to band registration at the instrument level and this is taken care of in the data processing to produce a band-registered data product.

One of the major challenges in realizing ETM+ is the design of the scan assembly. To improve the scan efficiency, unlike the MSS, both forward and reverse scans of the mirror are used for collecting data. Thus, the TM generates the 185-km swath imagery using a bidirectional scan mirror, which on descending passes alternately scans west to east (forward scan) and then east to west (reverse scan). The TM scan mirror assembly consists of a 406 × 503 mm flat beryllium mirror supported by flexural pivots on either side onto a fixed frame. An electromagnetic torquer produces a torque in response to an applied electrical control signal, which oscillates the scan mirror ±3.85°

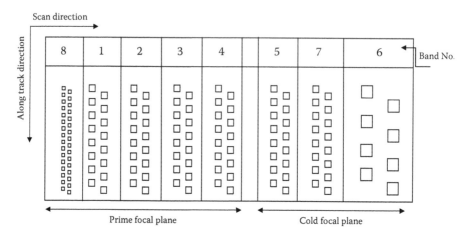

FIGURE 5.23
Focal plane detector layout of Enhanced Thematic Mapper Plus. Bands 1–5 and band 7 have 16 detectors each. Band 6 has 8 detectors, and band 8 has 32 detectors. The spectral bands values in micrometers are as follows: (1) 0.45–0.52, (2) 0.53–0.61, (3) 0.63–0.69, (4) 0.78–0.90, (5) 1.55–1.75, (7) 2.09–2.35, (6) 10.4–12.5, (8) 0.52–0.90.

with respect to the housing, thereby causing the optical line of sight to sweep ±7.70° back and forth to cover a 185-km swath across the subsatellite track. The motion of the mirror in each direction is stopped by two leaf spring bumpers preloaded against a rubber stop to reduce the force on the flexural pivots during turnaround. To achieve a linear angular motion during the data collection period, torque is applied only during the turnaround period (Blanchard and Weinstein 1980). To control the scan mirror motion and to aid data reduction, an accurate knowledge of the angular position of the scan mirror assembly is required. A scan angle monitor is attached to the back of the scan mirror, which provides the angular position of the scan mirror and also generates timing pulses to the scan mirror control electronics at the beginning, middle, and end of each Earth viewing active scan (USGS 2006).

As we discussed earlier for the MSS ground pattern (Figure 5.5), due to orbital motion the detector's line of sight is not aligned at right angles to the subsatellite track. In the case of TM, because of the bidirectional scanning the cross-track scan lines produce a zigzag pattern with overlap at one end of the scan and underlap at the other end (Figure 5.25a). The maximum gap at the edge of the swath is the distance traveled in two active scans plus one mirror turnaround period (USGS 2003). To take care of the overlap and underlap in ground coverage between successive scans, a scan line corrector (SLC) is introduced in the optical path. The SLC consists of two parallel mirrors set at an angle on a shaft and rotates about an axis normal to the axis of the scan mirror in a sawtooth fashion (NASA 2003). The SLC is positioned behind the primary optics (Figure 5.24) and compensates for the along-track motion of the spacecraft that occurs during an active cross-track scan of the mirror. Thus, the SLC enables contiguous data collection without overlap and underlap (Figure 5.25b).

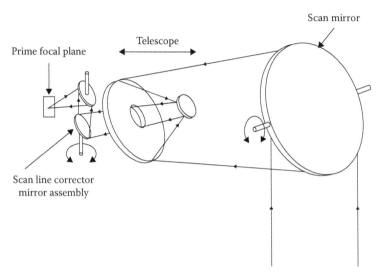

FIGURE 5.24
Schematics showing scan line corrector assembly in the Thematic Mapper optical path. (Adapted from NASA Landsat Handbook 3.2, 2011, http://landsathandbook.gsfc.nasa.gov/payload/prog_sect3_2.html, accessed on May 14, 2014.)

To monitor the radiometric stability of the instrument, ETM+ carries three calibration devices: an internal calibrator consists of two tungsten lamps, a black body, and a shutter that oscillates in front of the primary focal plane. There are two external calibration schemes, which direct sunlight into the ETM+. In one case—partial aperture solar calibrator—an additional optics covering a reduced aperture directs sunlight to the instrument. Viewing this when the Earth underneath is dark is used to monitor the stability of the reflective channels. The other external calibration scheme has a diffuser panel deployed in front of the instrument, which reflects the Sun's radiation into the telescope at an appropriate position in the orbit. By knowing the reflectance of the panel and the illumination geometry, radiometric calibration of the reflective channels can be carried out (Markham et al. 1997).

The ETM+ can be operated in two gains. The gain setting is chosen by ground command. When surface brightness is high the low gain mode is used, and the high gain mode is chosen when surface brightness is lower. Hence, the instrument can collect data with full radiometric resolution without saturation. The data is digitized to 9 bits per pixel, of which 8 bits are transmitted.

As mentioned earlier, ETM+ has the highest spatial resolution for a spaceborne Earth observation camera operating in whiskbroom mode. The limitation is primarily due to the inherent limitation of dwell time because of the scan technique. A camera operating in pushbroom mode is the way to realize a high-resolution Earth observation camera. We shall discuss this in Chapter 6.

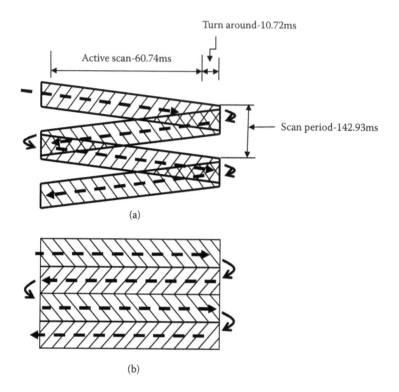

(a)

(b)

FIGURE 5.25

Enhanced Thematic Mapper Plus scan line projection on ground. (a) Without scan line corrector. (b) After correction using scan line corrector. (Adapted from NASA Landsat Handbook 3.3, 2011,http://landsathandbook.gsfc.nasa.gov/payload/prog_sect3_3.html, Accessed May 14, 2014.)

References

Abel, I. R.and B. R. Reynolds. 1974. Skylab Multispectral Scanner (S-192)—Optical design and operational imagery. *Optical Engineering* 13(4): 292–298.

Blanchard, L. E. and O. Weinstein. 1980. Design challenges of the Thematic Mapper. *IEEE Transactions on Geoscience and Remote Sensing* GE 18:146–160.

Feng, X., J. R. Schott, and T. W. Gallagher. 1994. Modeling the performance of a high-speed scan mirror for an airborne line scanner. *Optical Engineering* 33(4):1214–1222.

Fusco, L. and P. N. Blonda. 1986. ESA-Earthnet experience in high resolution sensor performance. *SPIE* 660: 35–44.

Gupta, P. P, S. C. Rastogi, M. Prasad, H. S. Dua, and A. Basavaraj. 1992. Development of passive cooler for Insat-2 VHRR. *Journal of Spacecraft Technology* II(1): 23–35.

Harnisch, B., A. Pradier, M. Deyerler, B. P. Kunkel, and U. Papenburg. 1994. Development of an ultra-lightweight scanning mirror for the optical imager of the second generation METEOSAT (MSG). *Proceedings of SPIE* 2210: 395–406.

Harnisch, B, B. Kunkel, M. Deyerler , S. Bauereisen, and U. Papenburg. 1998. Ultra-lightweight C/SiC mirrors and structures. ESA Bulletin 95.

Henini, M. and M. Razeghi. 2002. *Handbook of Infrared Detection Technologies*. Elsevier, Oxford, United Kingdom.

Hollier, P. A. 1991. Imager of METEOSAT second generation. *Proceedings of SPIE* 1490: 74–81.

Jewell, C. I. 1996. Cryogenic cooling systems in space, 1996. *Proceedings of the 30th ESLAB Symposium*, Noordwijk, the Netherlands, ESA SP-388.

Jones, R. C. 1962. Immersed radiation detectors. *Applied Optics* 1(5): 607–613.

Joseph, G., Iyengar, V. S., Rattan, R. et al. 1994. Very-high resolution radiometers for INSAT-2. *Current Science* 66(1): 42–56.

Joseph, G. 1966. Imaging sensors for remote sensing, *Remote Sensing Reviews* 13(3–4):257–342.

Joseph, G. 2005. *Fundamentals of Remote Sensing*, 2nd edition, Universities Press (India) Pvt. Ltd , Hyderabad, India.

Krishna, M. P. and Kannan. 1996. Brushless DC limited angle torque motor, *Proceedings of the 1996 International Conference on Power Electronics, Drives and Energy Systems for Industrial Growth* 1: 511–516.

Lansing, J. C. and R. W. Cline. 1975. The four and five band multi-spectral scanners for Landsat. *Optical Engineering* 14(4): 312–322.

Markham B. L.,J. L. Barker, E. Kaita, and I. Gorin. 1997. Landsat-7 Enhanced Thematic Mapper Plus: Radiometric calibration and prelaunch performance. *Proceedings of SPIE* 3221: 170–178.

NASA. 1973. *Advanced Scanners and Imaging Systems for Earth Observation*, Working Group Report, NASA/GSFC SP 335.

NASA. 2003. *Landsat 7 Science Data Users Handbook*. http://landsathandbook.gsfc .nasa.gov/pdfs/Landsat7_Handbook.pdf (accessed on May 14, 2014).

NASA. 2009. http://goes.gsfc.nasa.gov/text/GOES-N_Databook_RevC/Section03 .pdf (accessed on May 14, 2014).

NASA Landsat Handbook 3.2. 2011. http://landsathandbook.gsfc.nasa.gov/payload /prog_sect3_2.html (accessed on May 14, 2014).

NASA Landsat Handbook 3.3. 2011. http://landsathandbook.gsfc.nasa.gov/payload /prog_sect3_3.html (accessed May 14, 2014).

Norwood, V. T. and J. C. Lansing. 1983. Electro-optical imaging sensors. In Chapter 8 of *The Manual of Remote Sensing*, 2nd edition. American Society of Photogrammetry.

Photonics Handbook. http://www.photonics.com/Article.aspx?AID=25113 (accessed on May 14, 2014).

Ross, R. G. Jr. and R. F. Boyle. 2006. An overview of NASA space cryocooler programs—2006. http://trs-new.jpl.nasa.gov/dspace/bitstream/2014/40122/1/06-1561 .pdf (accessed on May 14, 2014).

Starkus, C. J. 1984. Large scan mirror assembly of the new thematic mapper developed for LANDSAT 4 earth resources satellite. *Proceedings of SPIE* 0430: 85–92.

Suomi, V. E. and R. J. Krauss. 1978. The spin scan camera system: Geostationary meteorological satellite workhorse for a decade. *Optical Engineering* 17: 6–13.

USGS. 2003. Scan Line Corrector Theoretical Basis, version 1.1. http://landsat.usgs
   .gov/documents/SLCOff_Processing_ATBD.pdf (accessed on May 14, 2014).
USGS. 2006. Bumper Mode Theoretical Basis Version. http://landsat.usgs.gov/
   documents/BumperModeATBD.pdf (accessed on May 14, 2014).
Westervelt, R. 2000. Imaging Infrared Detectors II. http://www.fas.org/irp/agency/
   dod/jason/iird.pdf (accessed on June 17, 2014).

6

Pushbroom Imagers

6.1 Introduction

We have seen in the previous chapter the limitations of an optomechanical imaging system in generating high spatial resolution imagery from space platforms. The limitation primarily arises from the mode of imaging on an instantaneous geometric field of view (IGFOV) by IGFOV basis, thereby limiting the time to collect the radiation. The situation can be substantially improved if data are collected for an entire strip by imaging the complete swath at a time, since all the pixels of the strip get the integration time corresponding to the time taken by the satellite to move through one IGFOV. Such a mode of operation is possible due to the development of long array detectors.

6.2 Principle of Operation

A strip of the terrain is focused by an optical system, say a lens, onto the linear detector. At any given instant of time, only those points on the ground that lie in the plane defined by the optical center of the imaging system and the line containing the sensor array (the optical center is the imaginary point at which all the rays intersect are imaged). This plane may be referred to as the instantaneous view plane. The radiation from the strip on the ground is received simultaneously by every detector element of the sensor and each detector element produces electrons proportional to the radiant flux received by that detector element and the duration for which the detector is exposed, also referred to as the integration time. Depending on the number of elements in the detector, one line of information generates that many picture elements. When such a system is mounted on a moving platform (usually such that the array length is at a right angle to the velocity vector of the platform), as the platform advances the view plane sweeps out a strip of the terrain that is continuously projected onto the charge-coupled device (CCD) array. Normally, the detector is exposed for a duration equal to the time taken by the subsatellite point to move through one IGFOV, referred to as the dwell time. Thus, due to the platform motion every successive exposure produces contiguous image strips. Hence, a two-dimensional image is produced—the linear array detector producing one image line across track (XT) and successive image strips along track (AT) by the motion of the platform (Figure 6.1). When compared to optomechanical scanning systems, the scan mirror is avoided and instead the detector array is used to produce one scan line of information. This mode of scanning is referred to as *pushbroom scanning*. In this case, the total time taken by the subsatellite point to move through one ground resolution element is available for integration of the signal for all the pixels along the linear array. The dwell time for a pushbroom scanner is given by

$$\tau_p = \frac{\beta}{v/h} \qquad (6.1)$$

where β is IFOV, v is the velocity of the subsatellite point, and h is the satellite height.

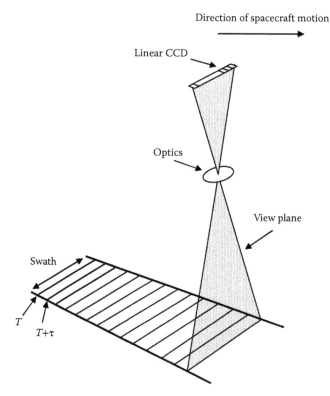

FIGURE 6.1
Illustration of the pushbroom scanning technique. τ is the integration time.

Here, the scanning is done electronically, and the scan efficiency is 1. If τ_0 is the dwell time for an optomechanical scanner for $n = 1$, using Equation 5.12, we have

$$\frac{\tau_p}{\tau_o} = \frac{\beta}{v/h} \frac{\left(v/h\right)\Omega}{\beta^2 S_e} = \frac{\Omega}{\beta S_e} \tag{6.2}$$

$\frac{\Omega}{\beta}$ represents the number of detector elements on the linear array (XT pixels). Thus, the dwell time is considerably longer in a pushbroom camera compared to an optomechanical scanner, thereby improving the signal to noise ratio (SNR). Hence, for the same aperture and S/N as that of an optomechanical scanner, higher spatial/spectral resolution imagery can be produced with the pushbroom scanning technique. The absence of mechanical motion increases the reliability and makes the system more compact. It is worthwhile to mention that currently all high-resolution imaging systems are based on this principle.

6.3 Linear Array for Pushbroom Scanning

Apart from being used in spaceborne cameras, pushbroom scanning technology has many commercial applications as in page readers. The pushbroom scanning technique could only be realized because of the availability of photosensitive linear arrays. Now various types of solid-state linear arrays are available operating in the visible to infrared (IR) region. We shall now discuss the operation of the linear arrays usually used for spaceborne pushbroom imaging.

6.3.1 Charge-Coupled Devices

The concept of solid-state image sensors were experimented from early sixties (Weimer et al. 1967). A practical imaging solid-state camera with acceptable quality was possible only with the invention of the CCD by Willard S. Boyle and George E. Smith in 1969 for which they were awarded Nobel Prize in 2009. The important feature of the invention is the process of efficiently transferring electrical charge along the surface of a semiconductor from one storage capacitor to the next. CCDs are used in a number of areas other than imaging such as analog delay lines, to perform coherent analog signal processing, and memory functions.

The detection of photons by a CCD is same as that of a photodiode; that is, it converts the light (incoming photons) into electrons, which are stored as electrical charge in a capacitor formed by the depletion region. CCDs operating in the visible and near-infrared (VNIR) region use suitably doped silicon as the photosensitive element. A number of such photosensitive elements can be placed in close proximity and organized either as a two-dimensional array (area array) or as a linear array. The photodiodes are isolated as individual pixels by nonconducting channel stops and biased gate electrodes. When a scene is focused onto such a device, each capacitor accumulates an electric charge proportional to the number of photons received at that location. This is the first step in capturing the image. The limit to the number of electrons that a pixel can hold is called well capacity and is expressed in units of electrons. The well capacity is set by the pixel geometry and the controlling electronic structures. Well capacity is a significant contributor to the dynamic range of the radiation that the CCD sensor can measure.

We shall now describe the processes involved in extracting the signal using a linear array. Located adjacent to the image sensing elements (photosites) is a structure called a *transfer gate*. By applying suitable voltages to the transfer gate at the end of the integration period, the charge packets accumulated in the image sensing elements are transferred out in parallel to a shift register called the *transport shift register* or readout register. At the transport shift register, the charge corresponding to each pixel is stored in a separate capacitor. By manipulating various gate voltages, the charges in each of the capacitors are transferred to the adjacent capacitor sequentially and finally to the output amplifier (Figure 6.2). Thus, the charges from each pixel, one at a time, reach the charge sensing output amplifier located at one end of the readout register. The amplifier converts the charge to voltage, thereby producing a sequence of voltage pulses with amplitudes proportional to the

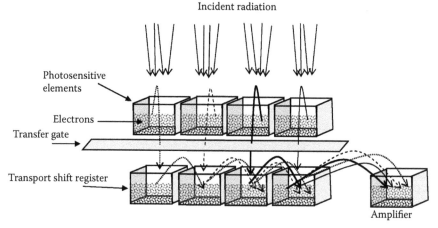

FIGURE 6.2
Illustration of working of a linear charge-coupled device (CCD). The charge accumulated at the photosite is transferred to transport shift register. The charges in the transport shift register are sequentially transferred from one capacitor to the other, finally reaching the amplifier one after the other from each pixel.

integrated photons incident on the photosites. We shall deal with the output signal characteristics in Section 6.4.

In the transport shift register as the charges move serially from one capacitor to the next capacitor, the total charge collected should be faithfully transferred in each transfer. However, in practice, some charge is "left behind" during each transfer (Brodersen et al. 1975). The efficiency with which the charge transfer takes place is called *charge transfer efficiency* (CTE) and ideally it has to be close to 100%. The charge from the capacitor located farther away from the output amplifier is maximally affected since it has to undergo many transfers. Therefore, the effect of CTE on loss of charge is not the same for all the pixels. The effect of reduced CTE is to decrease the overall modulation transfer function (MTF) of the line image in a nonlinear fashion along the length of the array. One method to improve the CTE is by reducing the clock frequency, thereby allowing enough time for the carriers to transfer to the next capacitor. Of course, if the number of transfers is reduced, the effect due to CTE will also get reduced. In practice both these can be achieved by having two separate transport registers on either sides of the pixels, wherein one transport register is connected to the odd pixels and the other to even pixels (Figure 6.3). As the spatial resolution increases the integration time is reduced and readout clock frequency increases. To take care of the increased speed requirement, the transport registers on both sides can be further divided, as in the case of the 12K CCDs used in the Indian Remote Sensing (IRS) LISS-4 camera, which has four transport shift registers each for odd and even pixels.

The MTF of the CCD decreases as the sensing wavelength increases (Figure 6.4). Therefore, imagery taken at longer wavelengths will have comparatively poor quality.

Another issue of concern in using CCD for imaging is blooming. Each pixel storage capacitance can hold only up to a certain maximum number of electrons. When the CCD is exposed to a bright source and the photoelectrons generated exceed the maximum well capacity, the excess charges then spread into neighboring pixels, causing them to convey erroneous values.

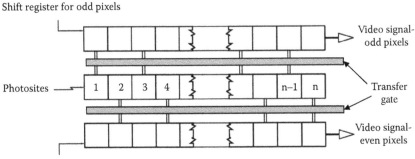

FIGURE 6.3
Schematics showing the organization of a linear CCD with two transport shift registers.

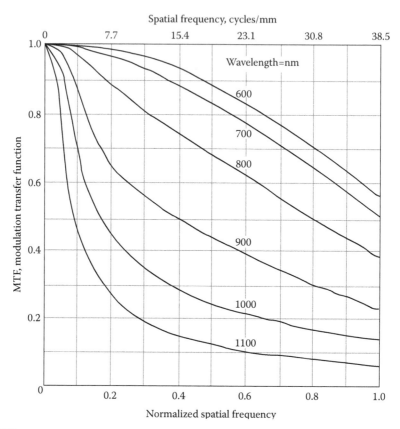

FIGURE 6.4
Modulation transfer functions for narrow band illumination source for a 13 μm pixel device (Fairchild datasheet for CCD143A).

This phenomenon of spillover of electrons from a saturated pixel to the adjacent pixels is called blooming and appears as a white streak or blob in the image. Thus, blooming affects both quantitative and qualitative imaging characteristics. Some CCDs are designed with antiblooming (AB) structures. AB structures bleed off any excess charge into a specifically designed sink before they can overflow to the adjacent pixel and thereby stop blooming. AB provision is useful when one has to observe targets with large dynamic range. However, AB structures can reduce the effective quantum efficiency and introduce nonlinearity into the sensor and hence AB protection is not the preferred option for most scientific CCDs (Janesick 2001).

6.3.2 CMOS Photon Detector Array

Complementary metal oxide semiconductor (CMOS) photodiode imaging arrays were the first to hit the market in the 1960. With CCDs entering into the market in 1970s, CCDs dominated vision applications because of their superior dynamic range, low fixed pattern noise (FPN), and better sensitivity to light. However, with the advent of CMOS technology, CMOS imaging arrays are coming back as a viable alternative to CCD, at least in certain imaging applications. In both cases, each pixel accumulates signal charge proportional to the incident illumination. The basic difference is how the charge is transported to the outside world for further processing. CCD transfers charge packet from each pixel sequentially to a common output amplifier, which converts the charge to a voltage, buffered, and is available off-chip as an analog signal. In the case of CMOS, the charge-to-voltage conversion takes place in each pixel. Each pixel is multiplexed and the output from each pixel is connected successively to a common output amplifier (Figure 6.5). (Since each pixel has an active circuit, this is also referred as active pixel sensor or APS.) Using the CMOS process it is possible to integrate many additional features on the chip along with photon detection. These include timing and control, clock and bias generation, autoexposure, automatic gain control, gamma correction, analog to digital conversion, and so on, thus providing a readout integrated circuit (ROIC) along with the photon detection function. Hence, we get a digitized output providing a robust, noise-immune interface for further processing off the chip. Thus, it is a complete focal plane chip with only command signals and power as inputs to the chip and digital image data available as output.

Since a portion of the pixel surface area is used to accommodate amplifier and other readout circuits, the fill factor of an APS is substantially lower than that of a CCD (fill factor is a measure of the fraction of the pixel area that is light sensitive). One method to improve fill factor is by using a microlens to focus the incoming illumination onto the light-sensitive part of each pixel. However, the use of a microlens increases system complexity and cost (Gamal and Eltoukhy 2005). Because of combined functionality, miniaturization, and simplification of the sensor electronics, it is possible for a CMOS APS to offer lower system power, increased functional flexibility, and higher system reliability all of which lead to reduced system assembly and test cost. The APS also has an advantage of reduced fabrication cost since a

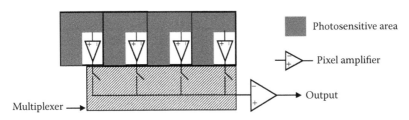

FIGURE 6.5
Schematics showing the organization of an active pixel linear array. Charge generation takes place at the photosensitive area of photodiode. Charge-to-voltage conversion is carried out for each pixel by amplifiers. A multiplexer successively connects amplifiers to a common bus to transfer the data to the outside.

proven high-yield CMOS fabrication technology can be adopted. Therefore, CMOS-based detector arrays are extensively used in commercial applications. Nevertheless, CCDs are the preferred option for scientific imaging due to their higher SNR, low photoresponse nonuniformity (PRNU), low FPN, and low dark current (Kevin Ng).

6.3.3 Hybrid Arrays

Monolithic detector arrays—that is, photon sensor and readout on the same substrate—are generally silicon based. Silicon-based devices such as CCD and CMOS arrays do not respond to wavelengths beyond about 1 μm (Figure 6.6). Therefore, one technique to construct an IR detector array is to separate the two functions (photon sensing and readout circuitry) and create a hybrid detector array. In this scheme, the radiation detectors are made of suitable IR material and can be wire bonded on a silicon chip or the entire detector array can be bump bonded to a silicon readout chip.

Many detectors such as mercury cadmium telluride, indium antimonide, and platinum silicide have been used for shortwave infrared (SWIR) detection. However, they require cooling far below room temperature, some even

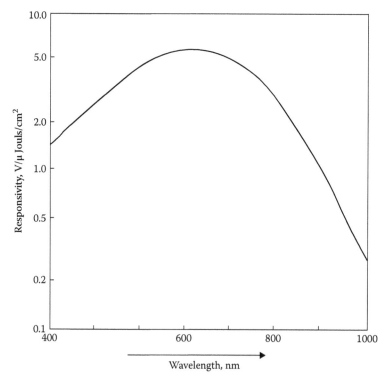

FIGURE 6.6
Typical spectral response of a silicon CCD (Fairchild CCD143A, Fairchild datasheet).

to cryogenic temperatures, so that the dark current is within acceptable limits. An indium gallium arsenide (InGaAs) linear array is a possible choice for operating in a near room temperature environment. $In_xGa_{1-x}As$ is a compound semiconductor and by varying "x," the optical, electrical, and mechanical properties of InGaAs can be varied. Thus, it is possible to design InGaAs detectors with a response between 0.9 and 1.7 μm. As the temperature is reduced, in general, the upper cutoff shifts to a lower wavelength. A typical value is about an 8 nm shift to shorter wavelength for every 10°C cooling (Guntupalli and Allen 2006).

6.4 CCD Signal Generation and Processing

Figure 6.7 gives a simplified block diagram of single-channel CCD camera electronics. For the operation of a CCD, different bias voltages and a number of synchronized clocks are required. Clocks for the CCD and the pulses required for various off-the-chip CCD signal handling like the analog-to-digital converter (ADC) are controlled by customized timing and control logic. Synchronized clocks are generated using a master clock and a sync pulse so as to maintain the required phase relationship. The external clocks applied to the CCD generally include the transfer clock (Φ_X), applied to the transfer gate to move the accumulated charge from the photosensor elements to the transport shift register; the transport clock (Φ_T), applied to the gates of the transport shift register to move the charges to the charge detector amplifier; and the reset clock (Φ_R), applied to reset the sense amplifier after reading every pixel. The clock characteristics such as rise and fall times, voltage level, and the relative phase relationship should be strictly followed as per the manufacturer's data sheet. The device used for interfacing the clock to the CCD should meet the requirements of speed, voltage swing, and capacitive load drive.

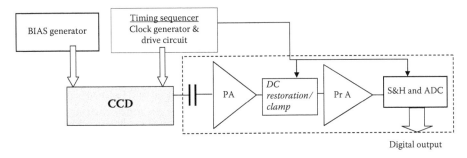

FIGURE 6.7
Schematics showing the architecture of a single-band CCD camera. The blocks in the dotted box perform off-the-chip signal processor functions. PA, preamplifier; PrA, programmable amplifier.

Low-resolution cameras like IRS LISS-1 have low speeds (5.2 MHz) and timing logic is designed using the TTL/LSTTL logic family. On the other hand, high-resolution cameras like the IRS Cartosat series with meter resolution require emitter-coupled logic (ECL) devices to operate at 105 MHz. At higher speeds, the power increases due to parasitic capacitance. A simple model for alternating current (AC) power W is given by (Texas Instruments 1997)

$$W = CV^2 f \qquad (6.3)$$

where
C = total capacitive load
V = peak output voltage swing
f = clocking frequency

As can be seen, the power consumed is highly dependent on the voltage swing. Miniaturized designs are now made with low voltage differential signaling and field programmable gate array (FPGA) devices. As FPGA core logic operates at 2 V or lower, power consumption is minimized at higher speeds. For operation of CCDs, voltages higher than that available from logic families are required. Therefore, the voltage levels of clocks are further level translated by specialized clock driver devices to interface with the CCD.

The bias voltages should have a very low noise since any noise on the bias voltages can reflect on the output video signal. This is particularly challenging for spaceborne systems where direct current (DC) to DC converters are used for generating various bias voltages. Therefore, extensive filtering of the power supply lines and judicious grounding schemes are required to minimize noise coupling.

6.4.1 CCD Output Signal

In this section, we shall discuss the nature of the output signal from the CCD. Figure 6.8 shows the conceptual schematics of the output stage of a CCD. We have seen that the electrons produced in each pixel are transported by the transport shift register to an amplifier located at the end of the shift register, where the electrons (which represent a charge Q) are converted to voltage. The charge measurement is accomplished by dumping the charge from each pixel onto a "sense capacitor" C_s located at the end of the transport shift register. A voltage V is developed across the sense capacitor according to the relation $V = Q/C_s$. The voltage so generated is taken off the chip through a source follower. The output of the amplifier gives an amplitude-modulated voltage in direct proportion to the number of charges stored in each pixel.

The process begins by closing the field-effect transistor switch "S" with a reset clock (Φ_R) to precharge the sense capacitor C_s to the reference voltage V_r. Initially there is a "reset feedthrough" due to the parasitic capacitive coupling of the switch to the output of the amplifier. This is followed by the reference level. The signal charge from the readout shift register is then

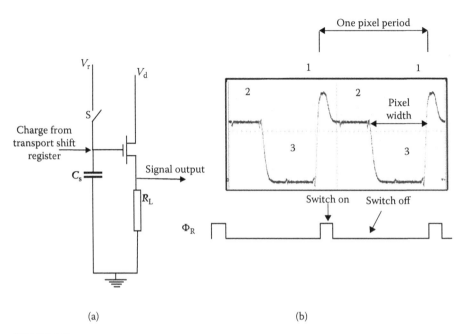

(a)   (b)

FIGURE 6.8
(a) Schematics of CCD output charge detection amplifier. V_r, reference voltage; V_d, drain voltage; C_s, sense capacitor; R_L, load resistor; S, switch. (b) CCD output voltage waveform 1—reset feedthrough, 2—reference level + V_r, and 3—pixel voltage level. The difference between 2 and 3 gives the voltage due to the charge accumulated at the pixel photosensitive area. The width of 3 gives the video duration.

transferred onto the node of C_s, whose potential changes linearly depending on the amount of the signal charge delivered. Since the electron charge is negative, the photon-generated signal (hereafter referred to as the video signal) goes negative from the reference level. The signal so generated is made available through a MOSFET source follower for off-the-chip processing. In summary, the output of the CCD is a sequence of stepped DC voltages for each pixel, consisting of reset feedthrough, reference level, and pixel video level. Another property of relevance is the pixel period, which decides the readout rate and hence the data rate.

When there are multiple readout shift registers in a CCD, they may be combined on the chip so that a single output is available to the outside world or each shift register may have its own output amplifiers producing multiple video outputs.

6.4.2 Off-the-Chip Signal Processing

The goal of the signal processing electronics is to convert the video signal at the output of the CCD to digital data of required resolution (number of bits). To faithfully reproduce the quality of the signal leaving the CCD, it is necessary to consider the design and choice of components, including the physical layout and grounding scheme of the off-the-chip signal processing electronics until the signal leaves the ADC. In doing so, the architecture of the processing electronics should maintain the linearity and no further noise should be added. For the purpose of discussion let us consider that the total video data of a linear array CCD is available from one output amplifier.

The DC level of the CCD output signal is usually several volts, while the actual video signal is in millivolts. To avoid the reference level saturating the successive stages, the output from the CCD is coupled through a capacitance to a preamplifier. Since the preamplifier has to amplify the low level signal from the CCD, the preamplifier should be chosen such that the noise from the preamplifier should be much less than the noise in the signal and the first stage should have enough gain so that the noise introduced by subsequent stages is also negligible. However, the gain should not be so high that the voltage swing at the output saturates the amplifier and also the frequency response should not be compromised in a gain-bandwidth trade-off.

As we discussed earlier, the output signal from a CCD is in the form of amplitude-modulated pulses. Therefore, apart from having low noise, the amplifier design should preserve the CCD output signal pulse characteristics so that the bandwidth, slew rate, settling time, and so on, are commensurate with ADC resolution. The duration of the video signal (Figure 6.8b) depends on the readout clock frequency, which in turn depends on the integration time and the number of pixels the transport shift register has. For example, LISS-1 with an integration time of 11.2 ms and 1024 active pixels per transport register has a video duration of 5.4 µs, which is approximately half the pixel period. The LISS-4 with 0.877 ms integration time and 1500 active pixels per transport shift register has a video duration of 0.28 µs. The settling time of the amplifier should be such that the amplifier output should attain the full video signal within one least significant bit of ADC before the conversion starts. To reduce the capacitance load seen by the CCD output stage, the cable between the CCD and the preamplifier should be kept to a minimum. Therefore, the preamplifier and all circuits that connect to the detector are located on a single board, with a robust ground plane, and are housed in a detector electronics (DE) package, which is placed close to the CCD.

As mentioned above, the output of the CCD is a sequence of stepped DC voltages for each pixel. The difference between the reference level and the video level gives the actual signal produced due to the photons falling on the pixel, which is the information of interest (Figure 6.8b). Since the CCD output is coupled to the preamplifier through a capacitance, the preamplifier output does not have a defined DC level and hence a reference level has to be established—a method referred to as DC restoration. The procedure involves taking two samples during the pixel period of the CCD signal—one at the reference level and the other at the video level—and subtracting them, that is, the difference between the reference and video levels of the CCD signal. This process is called *correlated double sampling* (CDS). This process also eliminates any noise source that is correlated to the two samples. A slowly varying noise source that is not correlated can also be reduced in magnitude. The CDS essentially functions as a "differential-in-time amplifier," which takes two samples separated in time at the input corresponding to reference level and video level and generates the difference between them at the output of the amplifier. Thus, CDS establishes a DC reference and also reduces some of the noise components in the CCD signal. CDS is one of the important steps in processing the CCD signal and there are a number of methods to achieve the CDS function (Wey and Guggenbuhl 1990). We shall discuss some of the CDS schemes.

In the clamp-sample approach, the preamplifier output is connected to a clamp capacitance C_c, which is connected to a "clamp switch" as shown in Figure 6.9a. The clamp switch is closed during the pixel reference period thereby grounding one side of the capacitor. At that time, the capacitor will be charged to the reference level. After the reference level duration, the clamp switch opens and the video signal passes through the capacitance C_c to a buffer amplifier. By passing through C_c, the video signal level gets subtracted by the amount of charge previously stored during the time the capacitance was grounded. Thus, the circuit functions as

1. Sample and hold (S&H) for reference level

2. Translation of AC-coupled reference level to 0 V, lower end of ADC (input corresponding to all zeroes)

3. Generation of difference of video and reference levels

For proper functioning of the circuit, the selection of the switch is very important. The switch should have very fast switching speeds and low ON resistance and parasitic capacitance and high OFF resistance. This technique of realizing CDS with discrete components like operational amplifiers and switches is best suited when the video period is >500 ns.

Another possible topology for CDS is using two S&H circuits. Here, the signal coming from the CCD is applied to the two S&H circuits, with their outputs connected to a difference amplifier (Figure 6.9b). At time T_1, the sample and hold S/H_1 goes into the hold mode, sampling the reference level including the noise. This voltage is applied to the noninverting input of the difference amplifier. At time T_2, the S&H will sample the video level and the output is applied to the inverting input of the difference amplifier. The output voltage of the difference amplifier gives the voltage level corresponding to the pixel radiance after removing the correlated noise.

Another alternative configuration called digital double sampling (DDS) is considered especially for high-resolution systems (Mehta et al. 2006). The block schematic of the DDS video processor is shown in Figure 6.9c. The DDS configuration uses digitization of the signal during both reference and video periods for true video extraction. The digitizer usually used is either

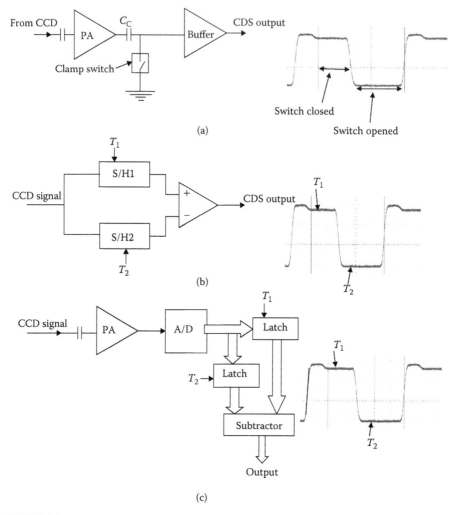

FIGURE 6.9
Schematics showing different schemes to implement CDS: (a) clamp-sample approach, (b) using two sample and hold circuits and a difference amplifier, and (c) using digital sampling.

a flash type ADC or it contains a high-speed track and hold preceding the digitizer. A digitizer with 12 bits is used to minimize the digitization error due to double sampling and also to cater to a large signal swing (up to 150%) expected after AC coupling based on scene modulation.

When the CDS output is analog (Figure 6.9a and b), the signal has to be converted into digital data for further down the line processing and transmission to the receiving station. This is accomplished by a S&H and ADC. A programmable amplifier follows the CDS so that depending on the scene radiance, CCD signal amplitude matches the full-scale voltage of the ADC. The S&H and ADC requirements are worked out based on the radiometric resolution, time available for signal processing, and the data transmission capability. The major parameters of S&H are acquisition time, hold time, sampling accuracy, droop rate, and interface compatibility for control signals. Similarly, the major parameters of interest for ADC are resolution, conversion time, linearity (differential nonlinearity and integral nonlinearity), missing code, noise, power, and so on. The total signal processing function can be carried out by an application-specific integrated circuit (ASIC). The advantages of an ASIC approach are reduced camera size and power, and increased reliability. Special ASICs (analog front end [AFE] devices) are now available from many semiconductor manufacturers to carry out off-chip signal processing that also have some flexibility to meet the user requirements (Patel et al. 2012).

6.5 Spaceborne Pushbroom Cameras

The first mission to use CCD in a spaceborne camera to image the Earth was the second KH-11 (Keyhole), a reconnaissance satellite launched on June 14, 1978. The satellite camera carried an 800 × 800 pixel CCD area array. The first civilian camera using the pushbroom technique to be flown in a satellite was the Modular Optoelectronic Multispectral Scanner (MOMS) developed by the German Aerospace Research Establishment, which was part of Shuttle Mission STS-7 launched in 1983. MOMS had two spectral bands, at 0.575–0.625 µm and 0.825–0.975 µm, to produce 20 m resolution images from the shuttle orbit. The first CCD-based pushbroom scanner on an unmanned Earth-observing spacecraft was on the French satellite SPOT-1 launched in 1986. This was followed by the launch of the Indian Space Research Organisation's (ISRO) IRS-1A in 1988. We shall now discuss the design considerations for a spaceborne Earth observation pushbroom camera system taking the IRS LISS cameras as examples.

6.6 IRS Cameras: LISS-1 and -2

To realize a camera system, depending on the end use, the designer is given a set of specifications, which the camera should meet. These include IFOV/IGFOV and swath in the spatial domain; number of spectral bands, central wavelength, and bandwidth in the spectral domain; saturation radiance, radiometric resolution, and number of digitization bits in the radiometric domain; and at the overall system level performance criteria such as MTF S/N, and band-to-band registration (BBR). Of course, the camera has to be realized within the given budget and delivered to the satellite project within a specified time frame. The design should be such that the camera should perform satisfactorily under various specified environmental conditions. The camera design should optimize the weight, volume, and power requirements.

For the first operational Earth observation system of ISRO, it was decided to have two imaging systems, one with a resolution similar to Landsat MSS and the other with a spatial resolution close to Landsat Thematic Mapper (TM). Such a combination gives continuity of data to those users who were using MSS and TM. Both the cameras were to have the same spectral characteristics. The imagery from the same platform with the same spectral bands will enable one to study the effect of spatial resolution for some specific theme. Therefore, it was decided to have two cameras, one with a spatial resolution around 75 m (LISS-1) and the other around 30 m (LISS-2), the exact value to be decided based on engineering considerations. The swath shall not be less than 120 km to achieve a temporal resolution of less than 25 days.

The next important aspect is to decide on the spectral domain. ISRO studied the spectral signature to understand the spectral characteristics of various objects. Aerial data were collected over a number of test sites covering various themes using ISRO MSS and the Bendix multispectral scanner. Based on the statistical analysis, a set of bands was recommended (Majumder et al. 1983; Tamilarasan et al. 1983). Another crucial factor is that the bands should be outside the strong absorption bands of the atmospheric gases. Table 6.1 gives the central wavelength (λ_c) and the spectral bandwidth $(\Delta\lambda)$ of the selected bands for IRS in comparison to Landsat MSS/TM and SPOT. It is worth noting that all the IRS cameras are designed such that the

TABLE 6.1

Central Wavelength (λ_c) and the Spectral Bandwidth $(\Delta\lambda)$ of IRS, Landsat MSS/TM, and SPOT

MSS		TM		SPOT HRV (XS)		IRS	
λ_c	$\Delta\lambda$	λ_c	$\Delta\lambda$	λ_c	$\Delta\lambda$	λ_c	$\Delta\lambda$
–	–	485	70	–	–	485 (B1)	70
550	100	560	80	545	90	555 (B2)	70
650	100	660	60	645	70	650 (B3)	60
750	100	–	–	–	–	–	–
950	300	830	140	840	100	815 (B4)	90
–	–	1650	200	–	–	–	–
–	–	11,450	2100	–	–	–	–
–	–	2215	270	–	–	–	–

Note: All values are in nanometers.

designated bands B1, B2, and so on, will always have the same central wavelength and bandwidth subject to the manufacturing tolerance of the filters. This makes it easy for intercomparison or merging of data from different cameras/missions.

We shall now consider how to make an optimal choice of various camera subsystems to meet the above requirements. At the time of designing the IRS-1A camera, linear array CCDs were used mostly for commercial applications on the ground. Instead of getting a CCD fabricated to our specifications it was decided to select a CCD available in the market that closely met our requirements and qualify the device for space use. Based on an extensive market survey the CCD chosen was the Fairchild 143A, with 2048 photosensitive active pixels, which is a second generation device having overall improved performance compared to the first generation devices. The improvements include a higher sensitivity, an enhanced blue response, and a lower dark signal. These devices can be operated up to 20 MHz. The photo element size is 13 μm by 13 μm with 13 μm pitch. The focal length "*f*" of the optical system depends on the detector pixel size "*a*," the height of the spacecraft "*h*," and the required IGFOV "*x*," which defines the spatial resolution. Referring to Figure 2.15 we find that focal length is given by

$$f = \frac{a \times h}{x}$$

As an example let us consider the focal length requirement for a system similar to LISS-1. The CCD chosen has a pixel size of 13 μm; for a satellite height of 900 km and spatial resolution of 75 m, the focal length works out to be 156 mm. (Note that these are indicative examples, not the actual values for LISS-1, which will be given later.) The next important parameter to be defined is the field of view (FOV), which depends on the required swath of the imaging strip and the satellite height. At the time of project initiation, CCD arrays with maturity of production were limited to 2048 elements. Therefore, in the present example the swath is 2048.

The FOV θ of the optics can be found from the simple trigonometric relationship as shown in Figure 6.10.

$$\theta = 2\tan^{-1}\left(\frac{77}{900}\right) = 9.8°$$

The actual values for IRS-1A LISS-1 are 162.2 mm equivalent focal length (EFL) and 9.4° FOV. We have discussed in Chapter 3 various optical systems and their relative merits. Though it is possible to design a reflective telescope of the three-mirror anastigmatic (TMA) type as discussed in Section 3.2.2, for a 162.2 mm focal length imaging system a refractive optical assembly makes the system more compact.

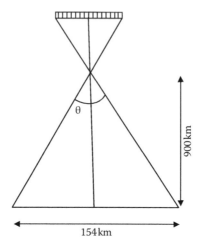

FIGURE 6.10
Schematics showing θ, the field of view of a camera using a linear array detector. Satellite altitude of 900 km and 154 km swath is considered.

6.6.1. Focal Plane Layout

Once we have selected the type of imaging optics, the next task is how to accommodate the CCDs in the focal plane. The scene energy reaching the focal plane of the collecting optics has a broad spectral coverage depending on the scene spectral characteristics and the transmission characteristics of the optical system. So we have to split the energy received at the focal plane into four bands and relay to the appropriate detectors. In conventional spectrometers this is achieved by using dispersive elements like prisms/gratings (refer to Chapter 8). Such systems have been used in some of the multispectral scanners (MMSs). However, detectors with band-limiting filters are used in most modern spaceborne systems. We shall now discuss various focal plane options for generating a four-band multispectral imaging camera.

6.6.1.1 *Single Collecting Optics Scheme*

Can we use a single lens to accommodate all four bands? If so, what are the options? The first simple option one could imagine is to put all 4 CCDs side by side in the focal plane. During the time of development of IRS only single line CCDs were available. The Fairchild 143A CCD packaging outline is given in Figure 6.11a. Though the active width of CCD is only 13 microns, when packaged the width of the packaged CCD is 15.49 mm. That is about 1200 pixels. Thus if the CCDs are placed side by side across track (AT), the first and the fourth CCD will be separated by about 3600 pixels (Figure 6.11b). Simple geometrical consideration shows that to accommodate such a system for LISS-1 (in the example given earlier) will require a field correction of ±8.5° in the AT direction. In addition, each CCD will have a different look angle. Moreover, the lens has to be corrected

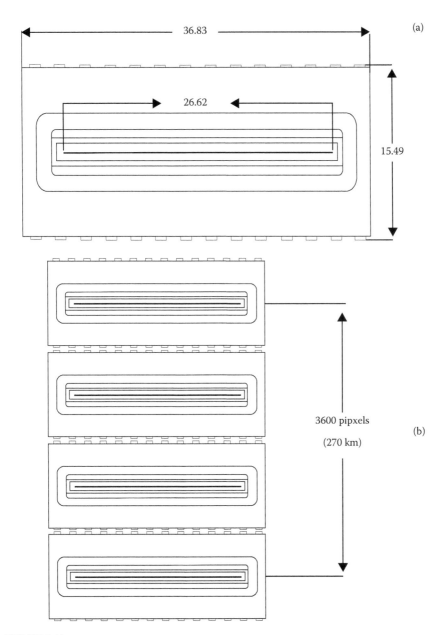

FIGURE 6.11
(a) CCD 143A packaging outline (*source:* Fairchild datasheet). The numbers are dimensions in millimeters. The active CCD length is only 26.62 mm. (b) Schematics showing positioning of four CCDs side by side in the focal plane of a single lens.

to cover a wavelength region extending from visible to near IR. Such broad wavelength coverage will bring down the quality of the lens performance, especially the MTF. The most critical issue is the time difference between the bands for imaging a strip on the ground. For example the separation between the extreme bands, which is about 270 km on the ground, corresponds to a delay of about 42 s between Band 1 and Band 4 data acquisition. To have the required BBR even if one is able to correct for such a large spatial displacement, a large delay in acquisition can change the atmospheric and even some surface conditions (e.g., canopy change due to wind), which is not acceptable for data interpretation.

Another configuration by which a single lens assembly can be used for multispectral imaging is to optically "split" the focal plane using a prism with a number of dichroic beam splitters as discussed in Chapter 5. A possible scheme for the spectral separation is shown in Figure 6.12. This concept

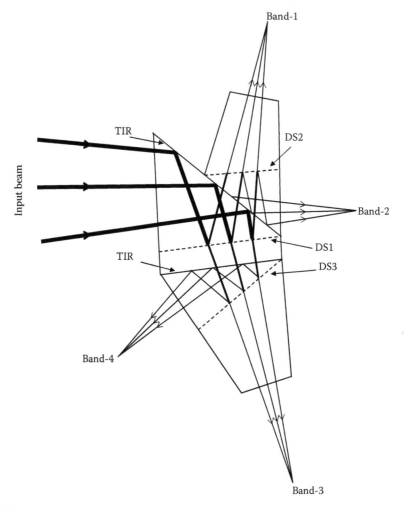

FIGURE 6.12
Schematics showing a beam splitter assembly to generate four focal plane locations. DS, dichroic beam splitter; TIR, total internal reflecting surface. DS1 transmits wavelengths greater than 0.62 μm and reflects the radiation less than 0.62. DS2 and DS3 further split the radiation to produce four focal planes.

has been used in the SPOT camera. However, for IRS this concept was discarded since for such a scheme the lens has to be corrected for a broad wavelength region. Since the beam splitter is in the converging beam, the design of the lens becomes more complex and the image quality poorer than having a lens for each band. It may be mentioned that with the advancements in technology, currently it is possible to have well-aligned multiple arrays on the same substrate with spectral filters overlaid. Such detector arrays can be used at the focal plane without any further arrangements though the optics have to be corrected for the full spectral range.

6.6.1.2 Multiple Lens Option

Probably the most straightforward method is to use separate collecting optics for each band. This is practically viable only if the aperture is not too large as in the present case. In this scheme the lens design is much simpler since there is no beam separating blocks in the converging beam. Also each lens needs to be corrected only for a narrow wavelength region. Therefore, one can get the best optical performance for all the bands. In a single lens option, any damage/performance degradation to the lens during launch jeopardizes all bands, whereas when we use multiple optics even if one lens is damaged other bands are not affected. The realization of each band, including the focal plane alignment can be done in parallel and hence can have better schedule management. However, in this case the lens design and fabrication are very stringent to ensure that the focal lengths of lenses of all bands of a camera are well within a certain limit, and the nature of distortions for all lenses has to be identical, so as to ensure BBR. In addition, the optical axis orientation shall be stable within a specified tolerance over the operating temperature range and other environmental conditions such as shock and vibration. We discussed these issues in Chapter 3. Based on a detailed system analysis on various aspects such as image quality, alignment requirements, and work scheduling it was decided to go for the multiple lens option for both LISS-1 and LISS-2. In the case of LISS-1 a single 2048 element CCD could give the required swath. Thus, LISS-1 had four matched lenses each having the desired spectral filter with CCDs placed at the focal plane (Figure 6.13).

In the case of LISS-2, since the spatial resolution is better by a factor of two, to cover the same swath we require a CCD with twice as many pixels compared to that used in LISS-1, that is 4096 elements. However, at the time of design such CCDs were not available and hence it was decided to configure LISS-2 using 2048 element CCDs. We could have used optical butting (see Section 6.9) to generate 4096 elements using two 143A devices. However to reduce the complexity and to have better image quality, it was decided to use two cameras each with half the swath of LISS-1. However, the LISS-2 lenses were corrected for the full FOV as in LISS-1 for future use when CCDs of the required number of pixels to cover the full swath are available. For LISS-2, the CCDs were aligned in one camera (LISS-2A) in one-half of the field and in the other (LISS-2B) toward the other half as shown in Figure 6.13.

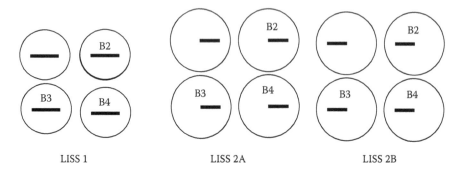

LISS 1   LISS 2A   LISS 2B

FIGURE 6.13
Focal plane arrangement of CCDs for LSS-1 and LSS-2. The circle represents the focal plane of the lens. The central line represents CCD position in the focal plane.

When projected on the ground, the swath of each camera has a predetermined overlap so that the imageries from LISS-2A and LISS-2B could be merged to produce a combined swath close to that of LISS-1.

The LISS-1 and LISS-2 lenses are of a similar design to the double Gauss type. The filter is placed in front of the first powered element, which if necessary can be changed to accommodate another band within the lens design range. Both LISS-1 and LISS-2 lenses have an f/no. of 4.5. Considering the required saturation radiance and the CCD response, an f/8 system would have been adequate for LISS-1. But the diffraction-limited performance of such a slow system would not have been acceptable. Thus, considering various aspects such as diffraction-limited performance, design/fabrication complexities, and size, an f/4.5 system was found to be optimum. Transmission of the neutral density (ND) filter kept at the entrance of the lens, which also reduces the thermal gradient in the subsequent elements, can be adjusted so that the maximum radiance does not saturate the CCD elements. To meet the stability requirements of a set of four lenses in a camera, so as to ensure band-to-band registration, is a major challenge to the design, fabrication, and test of the lenses. These aspects have been dealt with in Chapter 3.

6.6.2 Mechanical Design

Each camera has three packages:

1. Electro-optics module, which accommodates the collecting optics, detector assembly, and DE box containing the preamplifiers and CCD clock drivers.

2. The package that houses all the electronic circuits, such as video processor, clock generator, and calibration logic.

3. The power package, which takes the power from the satellite raw bus and generates various voltages required for the operation of the camera.

The mechanical design has to take into account various environmental conditions encountered by the payload during storage, launch phase, and in-orbit operation. Structural analysis is carried out to ensure that the fundamental frequency is above a value required by the satellite system. Each band is assembled as an independent structure so that band level testing is possible. Each lens is mounted on a barrel of invar with three flanges. To the front flange, the lens assembly is fixed and to the back-end the detector head is attached, thus forming a single band assembly. One of the challenges is to ensure the CCD location is stable with reference to the lens flange. Mounting the CCD directly onto a printed circuit board (PCB) and mounting it on the flange would have produced changes in the CCD plane due to warping of the PCB. Therefore, the CCD device is sandwiched between two metallic surfaces, which are mounted onto the detector head flange. Such an arrangement also helps for thermal management of the CCD. The central flange of the barrel located at the centre of gravity (CG) is used to mount the single band assembly to a bracket carrying all four bands (Figure 6.14). This bracket is mounted onto a base plate for mounting onto the satellite structure. In the initial tests, it was noticed that base plate mounting stresses distorted the band assembly plate, thereby disturbing the band-to-band alignment. To take care of this, the front and back-end flanges were constrained by suitable coupling plates. Nevertheless, the planarity of the mounting locations on the satellite needs to be maintained within a tight tolerance.

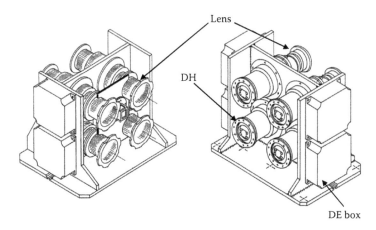

FIGURE 6.14
IRS-1A/B LISS-1 and LISS-2 camera mechanical configuration of electro-optics (EO) module. DH, detector head carrying CCD; DE box, detector electronics box housing circuits to drive CCD and preamplifier. (Reproduced from Joseph, G., *Fundamentals of Remote Sensing*, 2nd edition, Universities Press (India) Pvt Ltd, 192, 2005. With permission.)

6.6.3 Electronics

In addition to meeting the system requirements, the electronics are designed to have maximum reliability, which ensures that a single point failure does not affect the multispectral imaging capability. The same basic design is used for all three cameras (LISS-1, LISS-2A/B) and functionally modular so that testing becomes easier. The components are chosen from the preferred parts list of the IRS project and when it is necessary to choose a component not available in the preferred parts list they were qualified as per the set procedure of the project. As a general guideline where speed is not critical CMOS is used and when higher speed is required LSTTL/ECL is used depending on the speed.

To generate good quality imagery, proper design of the circuit and the PCB layout is essential. Isolation of power supply noise from the signal is a major challenge. The signal at the output of the CCD is first amplified by suitable low noise amplifiers and DC restored by clamping the DC level to the reference level obtained from the CCD device. The video signal generated by each CCD is digitized by suitable ADC. Each spectral band has independent ADC. The data are digitized to seven bits to have adequate radiometric resolution and consistent with data transmission capability. Necessary clock signals for the functioning of the CCD are generated by suitable logic. The device used for interfacing the clock to the CCD (clock driver) should have adequate frequency response and should be able to drive the required capacitance load. The preamplifier, clock driver, and power filters are housed in a separate box (DE box) and mounted on the electro-optics module so as to be as close to the CCD as possible. There is a provision to select four gains.

An in-flight calibration (IFC) system is included to assess the stability of the overall performance of the camera in the orbit. To achieve maximum reliability, it was decided not to use any moving mechanisms in the path of the camera since a failure of any moving mechanism will lead to mission failure. The IFC uses two light-emitting diodes (LEDs) for each band that illuminate the CCD from the two sides. By varying the LED current and using the on–off combination of the two LEDs, it is possible to generate 12 nonzero intensity levels. Though such a scheme does not include the optics, the IFC was found to be very useful to check the payload performance during various stages of payload testing and to access the in-orbit stability of the CCD and associated circuitry.

6.6.4 Alignment and Characterization

Once various components/subsystems, such as lenses, detectors, and mechanical structure have been received as per the projected requirements, the task of realizing a "space-worthy" camera begins. For this purpose, a number of alignment/characterization systems have to be developed, some of which may be specific to the camera under development. We shall describe the general approach to align and characterize that could be applied to any camera system. The activities primarily involve the camera focusing, that is, placing the detector in the focal plane of the lens assembly (optics), detector alignment for band-to-band registration, MTF evaluation, and radiometric performance assessment. These are generally carried out with the help of a scene simulator.

The scene simulator is essentially a collimator at whose focal plane different targets can be placed to simulate an object space at infinity. An off-axis parabolic mirror is generally used as a scene simulator. An off-axis parabola is a segment cut out of a large parent parabola. Theoretically, a concave parabolic surface will form a highly collimated beam of light from a point source placed at its focus. An off-axis mirror has an unobstructed aperture, unlike in the conventional Ritchey–Chretien (RC) design, where the secondary mirror obstructs rays coming onto or reflected from the primary mirror. Since a collimator is an all-reflective system, the performance is completely achromatic, and thus it can be used at any wavelength without realigning or refocusing. Many manufacturers do make collimators with standard off-axis parabolic mirrors as off-the-shelf products. One could also get custom-built mirrors to suit one's requirements. The mirror manufacturers can also supply suitable mirror mounts to support the mirror without stress.

Once you have the off-axis mirror, the next task is to locate its focus precisely so that a point source at the focus gives a collimated beam. There are a number of techniques to achieve this. However, a simple method is to use a lateral shear interferometer (Grindel). The testing device consists of a high-quality optical glass with extremely flat optical surfaces having a slight angle between the two surfaces. When a plane wave is incident at an angle of 45° to the shear plate, it is reflected from the front and rear surfaces. The two reflections are laterally separated due to the finite thickness of the plate and by the wedge (Figure 6.15). This separation is referred to as lateral shear. The shifted wavefronts interfere resulting in fringes in the overlapping area of the two wavefronts. The wedge of the plate is aligned normal to the shear direction. For a collimated beam, the fringes would run exactly parallel to the direction of shear (horizontal direction for shear is generally preferred). The interference fringes are inclined with respect to the direction of the shear when the collimation is imperfect. A scene simulator should have provision to place various targets in the focal plane of the collimator and move them precisely in the x, y, z directions.

6.6.4.1 Focusing of the Camera System

The first task is to place the detector at the focal plane of the lens. The focusing exercise is carried out using the scene simulator with a bar target at its focal plane simulating an object at infinity. The bar target consists of alternate bright and dark lines, illuminated with uniform white light, and having

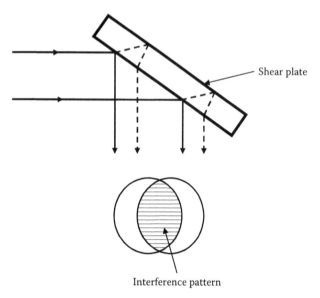

Shear plate

Interference pattern

FIGURE 6.15
Schematics showing interferogram of a lateral shear interferometer.

a spacing corresponding to the Nyquist frequency at the camera focal plane. (The Nyquist frequency is one-half of the reciprocal of the center-to-center pixel distance of the detector array.) The bar pattern is imaged onto the detector array by the optics of the camera under test. When properly aligned, the alternate bright and dark bars fall on alternate pixels of the detector array, that is, if the dark bars fall on the odd pixels of the array, the even pixels are illuminated by the bright bars. The minimum and the maximum count obtained at alternate pixels of the detector are used in the following expression to compute the square wave response (SWR) of the system:

$$SWR(\%) = \frac{\text{Maximum count} - \text{Minimum count}}{\text{Maximum count} + \text{Minimum count}} \times 100$$

The tube length connecting the lens and detector head is fabricated to be slightly less than the theoretically required distance between the lens and the detector head assembly. To place the detector at the focal plane of the lens, initially a spacer of required (estimated) thickness is placed either at the interface of the lens and the camera structure or at the detector–structure interface.

The focusing exercise is to find out the correct thickness of the spacer so that the CCDs are at the focus of the lens in the operating condition, that is, in vacuum. The target at the collimator focal plane is subjected to controlled incremental movements along the optical axis of the collimator. At each step, the SWR of the system is measured by recording the minimum and maximum counts as obtained at alternate pixels of the detector. If the detector is not at the right focus, the plane of maximum SWR lies away from the initial location indicated by the "0" in Figure 6.16. The position of the target at the collimator that gives maximum SWR corresponds to the position where the detector should be placed for best focus. The change at the lens focus Δf_L due to a change of Δf_C of the target in the focal plane of the collimator is given by

$$\Delta f_L = \left(\frac{f_L}{f_C} \right)^2 \times \Delta f_C \tag{6.4}$$

where f_L is the focal length of the lens assembly and f_C is the focal length of the collimator. This relationship can be used to locate the position of the detector in the focal plane of the lens and hence the actual thickness of the spacer required. The SWR is measured at the center and at the extreme field angles covering the entire FOV of the camera, and the correct shim thickness is determined after taking into account the total FOV. The measurement covering the total FOV of the system ensures that the camera performance is optimized for all the field angles in terms of image contrast. As discussed in Chapter 3, in the actual operating condition (in space) the location of the focal plane of the lens is different from the laboratory (air) environment due to change in the refractive index from air to vacuum. Therefore, in the laboratory, CCDs have to be offset from the focus in air to match the vacuum focus and the final spacer thickness is chosen accordingly. Once the camera is optimized, it is tested in vacuum at different temperatures to ensure that it

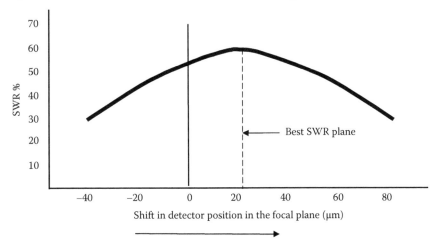

FIGURE 6.16
Modulation transfer function (MTF) variation with the detector position in the focal plane. "0" represents the initial location of the detector. The detector is normally placed at the plane where SWR is maximum.

delivers the optimum performance in space. The procedure can be generalized and can be used for focusing all camera systems.

When the camera is mounted on the satellite deck, the CCDs are to be at right angles to the velocity vector. A reflecting cube is fixed onto the electro-optics module and with respect to which the CCDs are aligned. The camera viewing axes in the roll (velocity vector), pitch (along the CCD), and yaw (optical axis) directions are thus defined with respect to the externally mounted payload alignment cube. Similarly, the spacecraft also has an alignment cube, known as the master reference cube (MRC), that carries information about the spacecraft roll, pitch, and yaw axes. Spacecraft level alignment measurements are carried out to establish the relationship between the payload (camera) alignment cube and the MRC. Thus, the camera viewing axes are known with respect to the MRC.

6.6.4.2 Image Format Matching and BBR

For extraction of information by multispectral data analysis, ideally all the bands should view the same area on a pixel-by-pixel basis at any given instant of time (or there should be a fixed known offset, which is corrected during data product generation). This interband registration among the different spectral bands of a multispectral camera is referred to as "band-to-band registration" (BBR). For achieving the required interband registration, the first step is that the image formats of all the bands should be same; that is, all the bands on the ground should have exactly the same swath. The variation in swath between the bands can happen due to the focal length difference and/or relative distortion variations among the lenses of the four bands and the spread in the length of the active elements of CCDs. As we discussed in Chapter 3, the focal length of lenses pertaining to the different spectral bands of one camera are first matched within the specified limits by design, fabrication, and assembly processes at the lens level. The overall length variation of CCDs is only a few microns. However, as a first step CCDs with longer length can be used for lenses with relatively longer focal length. Further adjustments are possible by adjusting the CCD location within the depth of focus.

The basic technique to evaluate the BBR is to illuminate all the bands with suitable target(s) and find out the location of its image for each band. Different types of targets could be conceived to assess the BBR. Here, we shall describe the use of an M-shaped target for BBR evaluation. Figure 6.17a shows the schematic of a typical test setup for measurement of image format and BBR adjustment for a multispectral camera. The setup essentially consists of an M-shaped target, consisting of vertical and slanted slits, placed at the focal plane of an off-axis parabolic mirror (scene simulator) and illuminated by a uniform white light source. The M-target is imaged at three different locations on the CCD to cover the center and extreme fields by rotating the camera (or the collimator), which is mounted on a controlled rotary stage.

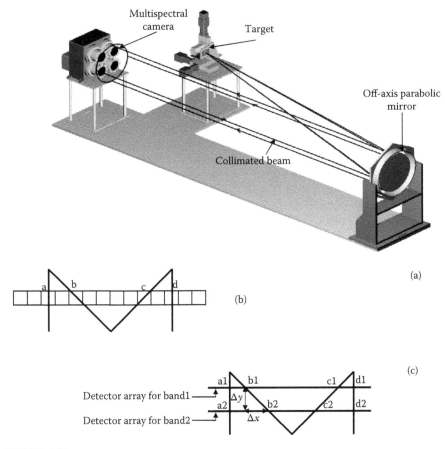

FIGURE 6.17
(a) Schematics of the format matching and band-to-band registration (BBR) bench. (b) Image of the target in the focal plane. (c) BBR procedure. Δx and Δy give the across-track (XT) and along-track (AT) missregistration.

The collimated beam aperture is chosen such that it comfortably covers all the spectral bands and the M-shaped target is imaged by them simultaneously. The image centroids corresponding to the vertical and slanted slits of the M-shaped target are computed for the respective bands. The image format is determined by taking the difference of the measured centroid values at different field points. It may be mentioned here that both focusing and the image format matching are parts of the performance optimization exercise for the multispectral camera system and may sometimes involve a trade-off between them. For example, to obtain a very high degree of image format matching between any two given bands, one of the bands (or both) may be required to be slightly defocused (within its depth of focus). Such a trade-off, however, is extremely small or negligible for a well-designed, correctly fabricated, and accurately integrated camera system.

Once the image format matching among the bands is accomplished, BBR is taken up. Here, the task is to align the detectors in all the bands so that their projection on the ground sees the same strip within the allowed error band. Thus, BBR is an interband alignment exercise wherein it is ensured that the overlap among the images of the same scene recorded in different spectral bands is maximum both in the XT and AT directions. The BBR is determined by comparing the centroids of the images formed by all the constituent bands simultaneously for a common target.

The test setup discussed earlier for format matching is used to evaluate BBR also. The "M" shaped target is imaged simultaneously by all the spectral bands. Thus, the image of the "M" target gives rise to four centroids—a, b, c, d—for each band under test (Figure 6.17b). If all the bands are exactly aligned, the centroids a, b, c, and d will be at the same pixel numbers for all bands. Next consider the case of two bands not aligned. Here, if the centroid falls at pixel no. a1, b1, c1, d1 in one band and a2, b2, c2, d2 in the other (Figure 6.17c), then the horizontal (along the pixel array) misregistration ΔH between the two bands is given by

$$\Delta H = \frac{[(a2 - a1) + (d2 - d1)]}{2}$$

and the vertical (across the pixel array) misregistration ΔV is given by

$$\Delta V = \frac{[(b2 - b1) + (c1 - c2)]}{2}$$

To compute the misregistration among the bands, one of the bands is selected as the reference band and the centroids of all remaining bands are compared with the centroids of the reference band. The misregistration among the bands is minimized by aligning the constituent bands with the reference band by adjusting the detector position.

6.6.4.3 *Flat Field Correction*

Ideally when a fixed, uniform light falls on the pixels of a linear or area array CCD, each pixel should give the same output voltage. However, due to a variety of factors this does not happen and for a uniform light incident onto the CCD the pixels give slightly different voltages (Figure 6.18). This difference in response to a uniform light source is referred to as PRNU. This nonuniform response leads to striping in the image data over uniform areas. The PRNU can vary from 1 to 3% for the highest grade CCD and can be 6–15% for midgrade devices (Weaver 2013). This PRNU needs to be corrected. In addition to PRNU, the illumination at the focal plane will also get affected by the response variation of the lens, vignetting, and so on. Therefore, a normalization procedure is applied such that when the camera sees an extended uniform source all the CCD pixels are normalized to give the same DN value.

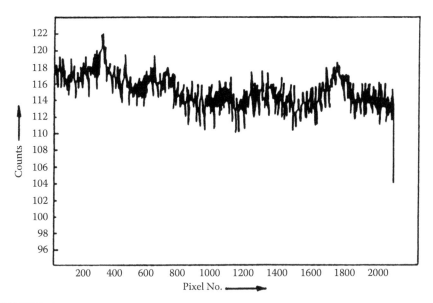

FIGURE 6.18

The response variation of Band 4 of the IRS LISS-2 camera. (Reproduced from Joseph, G., *Fundamentals of Remote Sensing*, 2nd edition, Universities Press (India) Pvt Ltd, 288, 2005. With permission.)

This is carried out by projecting a uniform light source onto the camera entrance aperture. The obvious next issue is how to generate a uniform source of light. The at-sensor nonuniformity in the light input should be much less than the tolerable mismatch between the pixels after correction. The light source should fill the total FOV of the camera under test. The test is usually carried out using a uniform light source, such as an integrating sphere. Since the integrating sphere performance primarily decides the final radiometric accuracy, we shall briefly describe the principles of operation of an integrating sphere.

An integrating sphere is a hollow spherical cavity with a diffusely reflecting internal surface, with a small opening (port) as an output source (Figure 6.19). Ideally, the coating on the inner side of the integrating sphere should be a diffuse reflector having very high reflectivity independent of the wavelength range for which the sphere is used. Light rays incident on any point on the inner surface produce multiple scattering/reflections, and the output from a small hole on the sphere produces a diffuse light, independent of the original direction of the illumination, giving a uniform light source, which is lambertian. Tungsten halogen lamps operated from a current-regulated power supply are most commonly used as integrating sphere sources. These lamps provide a continuous spectrum, free of emission lines. The exit port should not view the lamp or its initial few reflections. For this purpose, baffles coated with the same material as the integrating sphere wall are suitably located. The output radiance has to be controlled judicially. It is recommended not to control the output level by adjusting the current to

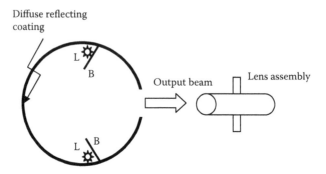

Diffuse reflecting coating

L B

Output beam   Lens assembly

L B

FIGURE 6.19
Schematics showing integrating sphere setup. L, tungsten lamp; B, baffle.

the lamps, since this may give rise to spectral shifts. The simplest way is to switch on (or off) the illuminating lamps is to vary the output radiance. However, the number of light levels that can be obtained will be limited. To have a controlled output in desired steps ND filters can be used in front of the imaging lens of the system under test.

Bigger-sized spheres produce a better integration of light and consequently better uniformity. However, signal level reduces as the square of sphere diameter when all other parameters are held constant and obviously larger spheres cost more. As a general rule of thumb, the integrating sphere diameter should be three times larger than the exit port diameter (McKee et al. 2007). The port size in turn depends on the aperture and FOV of the camera that will image the sphere's exit port. Integrating spheres are commercially available from various companies.

To correct for PRNU and other nonuniformities, a uniform source of light from the integrating sphere is used to illuminate different bands and DN values from all the pixels are measured. As mentioned earlier, the light source should fill the total FOV of the camera. The radiance from the integrating sphere is varied by known steps covering the whole dynamic range—from near saturation radiance to dark level. Thus, the light transfer characteristics (i.e., pixel output to light input) are measured for all the pixels. Since the CCD output varies linearly with the light input by applying linear regression to the digital counts and corresponding input radiance, each pixel's response to input radiance is characterized in terms of bias "d" and gain "g" such that

$$DN_i = d_i + g_i L$$

6.5

where "i" represents the ith pixel and "L" represents spectral radiance illuminating the CCD array in a given spectral band. The slope and intercept of the above function gives the gain (g) and offset for each pixel, respectively. The offset d in the above equation represents the dark level including any offset present in the electronics. In practice, a large amount of data samples is collected for each spectral band at every illumination level. The DN value for the same input radiance is distributed around a mean value. The spread depends on the noise. The mean value is used to generate the transfer equation. Since each pixel has slightly different gain and offset, the transfer curve will be different for each pixel.

The data so collected are used to generate a radiometric calibration (slope and offset) look-up table—(RADLUT). Using this RADLUT, the nonuniform response of individual detector elements are corrected by mapping from the input gray value to an output gray value for each detector and over the entire range of the input gray levels. To do so, all the pixel outputs may be normalized to one of the pixels having the least gain so that the corrected value is not saturated. If g_r is the gain of the reference pixel, then the corrected digital number $DN_{(corr)i}$ of the ith pixel is given by

$$DN_{(corr)i} = \frac{DN_{(raw)i} - d_i}{g_i} \times g_r$$

This normalization process transforms the raw digital number $DN_{(raw)}$ to a corrected digital number $DN_{(corr)}$ equal to the average detector response of the uniform scene. The result of this exercise is that all pixels in a band give the same output (subject to noise spread) when irradiated with a uniform radiance. The accuracy of normalization primarily depends on the number of illumination levels taken so as to estimate the gain and offset coefficients, the SNR, and the source stability. The accuracy can be judged by comparing the DN distribution (histogram) before and after normalization (Figure 6.20).

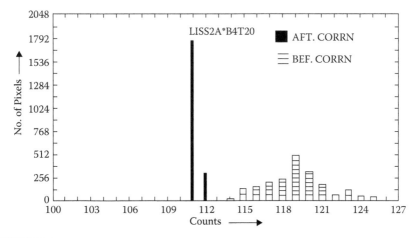

FIGURVE 6.20
Histogram of CCD camera output before and after radiometric normalization correction. Note after correction most of the pixels show 111 counts. Spillover to 112 counts is due to quantization. (Reproduced from Joseph, G., *Fundamentals of Remote Sensing*, 2nd edition, Universities Press (India) Pvt Ltd, 289, 2005. With permission.)

Figure 6.21 shows the raw image and image after radiometric normalization. The same setup can be used to get the relationship between absolute radiance (expressed in $mw/cm^2/sr/micron$) and DN value. For this purpose, a calibrated radiometer should be used to monitor the radiance emanating from the integrating sphere port. The setup can also be used to calculate the SNR of the integrated camera by calculating the root mean square deviation for a fixed radiance input.

(a)

(b)

FIGURE 6.21
Effect of flat field correction on image quality: (a) raw image; (b) image after flat field correction.

6.6.5 Qualification

After fully characterizing the camera for its electrical and electro-optical performance, it is subjected to various environmental tests as required by the satellite project. These include storing at different temperature regimes, a thermovacuum test, shock, vibration, and so on (refer to Chapter 10 for details). After every test, some of the critical parameters are tested to confirm proper functioning of the system. Specialized test and evaluation test benches are designed to carry out the integrated testing of the payload at various stages.

A reliable test and evaluation system is as important as the payload itself, since one has to depend on the quality of the payload based on the final test carried out with this setup. The checkout system has evolved over the years since the IRS-1A project. The current configuration of the checkout system consists of (Figure 6.22):

Spacecraft interface simulator: Provides all the electrical stimuli to the payload for its operation as received from the spacecraft main bus.

Payload data acquisition system: Acquires the payload data and sends it to the mainframe computer.

Payload status indicator: Monitors the health of the payload by displaying various parameters like voltage command status.

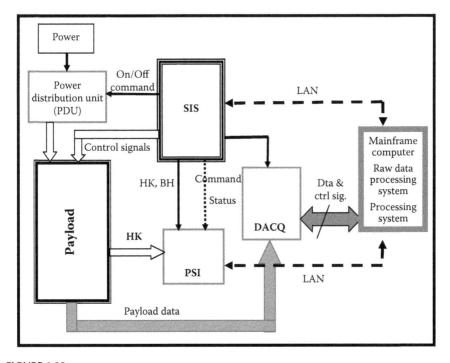

FIGURE 6.22
Functional block diagram of payload evaluation system. SIS, satellite interface simulator; DACQ, data acquisition system; PSI, payload status indicator; HK, housekeeping data.

Power distribution system: Provides the power to the payload as received from the satellite when the camera is operated without an onboard power package.

These are embedded microcontroller based systems, using FPGA for digital logic blocks. All the above systems are controlled by a computer to carry out various types of tests. The computer processes the raw data, displays the results, records the data, and logs the payload operations. The parameters of the payload that are evaluated during the payload testing include SNR, SWR, BBR, LTC, and radiometric calibration.

Once the camera has successfully undergone all mandatory tests, it is integrated with the spacecraft for further activities.

The IRS-1A satellite carrying LISS-1 and two LISS-2 cameras was launched on March 17, 1988 by a Russian rocket from the Baikonur Cosmodrome, Kazakhstan. IRS-1A/B operated in Sun-synchronous near-polar orbit at an inclination of 99° and an altitude of 904 km. One orbit around the Earth takes about 103 min and the satellite makes 14 orbits per day with 22 day repetivity. The IRS-1B satellite carrying similar payloads was launched on August 29, 1991. The equatorial crossing time for IRS-1A/B in the descending node was 9:40 am. As is the case with all Earth observation cameras from space, the raw data received has a number of artifacts produced by the camera, spacecraft, transmission/reception system, and so on. After applying necessary corrections, the data are suitably formatted and georeferenced before being supplied to the users. The quality of the data is periodically evaluated to ensure proper functioning of the system. Both IRS-1A and 1B outlived their designed life of 3 years and were extensively used for various applications.

6.7 IRS-1C/D Camera

The parameters of the next-generation remote sensing satellite IRS-1C were chosen to support the application needs arrived at based on various studies carried out using spaceborne data from IRS-1A/B and other remote sensing satellites. Three types of imaging systems have been identified:

1. A multispectral camera operating in the green (B2, 0.52–0.59 μm), red (B3, 0.62–0.68 μm), near-IR (B4, 0.77–0.86 μm), and SWIR (B5, 1.55–1.75 μm), with a nominal spatial resolution of 20 m—LISS-3.

2. A wide field sensor (WiFS) camera operating in red and near-IR, with a nominal spatial resolution of 160 m (8 × LISS-3 resolution) and a swath of about 700 km.

3. A camera operating in the panchromatic (PAN) band with a spatial resolution of 5 m (1/4 of LISS-3), with a capability of off-nadir viewing.

To make the camera optics less complex, an orbit lower than the IRS-1A/B orbit of 904 km is desirable. Therefore, initially the satellite was to be placed in a Sun-synchronous orbit around 700 km. However, considering various aspects such as orbit maintenance and coverage from the ground station, the satellite height was changed to 817 km. Since the cameras were designed as per the originally planned orbit height, the actual realized ground resolutions were correspondingly lower because of the change in the orbit height.

6.7.1 LISS-3 Design

The basic design of LISS-3 is similar to LISS-1, except for the addition of an SWIR band. For the VNIR bands, a 6000-element silicon CCD linear array having a pixel size of 10 μm (along the array) × 7 μm (across the array) with pitch of 10 μm was selected. Silicon CCD response does not extend beyond about 1 μm. Therefore, one has to look for a different technology for a SWIR detector. Indium gallium arsenide (InGaAs/InP) linear arrays, which can be operated near room temperature, were under development in the detector industry, specifically for imaging (Moy et al. 1986). Thomson-CSF, France, has successfully developed InGaAs photodiodes with a spectral response that extends from 0.9 to 1.70 μm, with readout CCD multiplexers. During the development of IRS-1C, the technology maturity was such that the smallest pixel size was 30 μm and each module had only 300 pixels with odd and even pixels staggered on two lines separated by 52 μm. To have a detector with the requisite number of pixels, the required numbers of modules are butted end to end. These modules could be butted with a very high geometrical accuracy, ±3 μm in focal plane directions and ±6 μm perpendicular to it (Hugon et al. 1995). Since the SWIR pixel size is three times that of VNIR, the focal length of the B5 channel has to be three times that of the VNIR channel. This would have necessitated a lens of more than 100 cm, which would make the camera design very complex. Therefore, it was decided to have SWIR resolution that is three times that of VNIR bands so that the focal length could be close to that of the VNIR bands. Thus, the LISS-3 camera from the operating altitude of 817 km has spatial resolution of 23.5 m for the B2, B3, and B4 and 70.5 m for the B5 band.

The optical configuration is similar to LISS-1/2, with one lens for each band. The basic electronics design is also similar to LISS-1/2 but tuned to cater to the electrical requirements of the devices. The data from all four bands are digitized to seven bits. The SWIR detector has to be cooled to reduce the dark noise. The detector is cooled by conducting the heat from the detector area through a copper braid to a passive radiative cooler. The temperature of the detector is monitored by a thermistor and controlled within ±0.1°C by a heater in a closed loop. For IRS-1C, there were two settings, −10°C and −12°C, which could be selected by a command. To avoid condensation during the laboratory test, the SWIR detector has to be hermetically sealed.

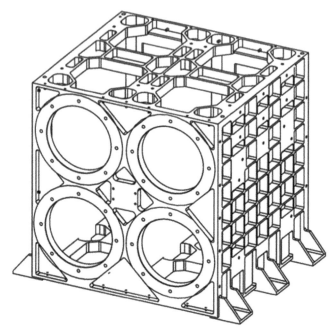

FIGURE 6.23
Mechanical structure of LISS-3 EO module.

Based on the experience of LISS-1/2, the mechanical structure of the electro-optics module structure of LISS-3 is redesigned to ensure that the alignment of the detector and the lens should not be disturbed beyond the permissible limits due to clamping loads and thermal environment. After studying various configurations, a cubical box type structure of Invar was chosen (Figure 6.23). The EO module is interfaced to the satellite platform by six bolts. The location where the EO module is interfaced with the satellite deck is maintained with a planarity better than 20 μm to reduce the mounting stresses. The EO module also carries some of the detector electronics (DE) boxes. The alignment, characterization, and calibration of LISS-3 is similar to LISS-11/2 as discussed earlier.

6.7.2 Wide Field Sensor

This is a unique sensor specific to the IRS constellation based on user feedback. Demand for a camera with higher temporal resolution came from agricultural scientists who were facing the challenge of providing national level wheat estimates (Navalgund and Singh 2010). The IRS LISS-1/2 based crop discrimination method used a single date classification approach. Looking at the limitations of single date classification from data acquired at a non-optimal biowindow, it was decided to increase the repeat coverage using a relatively coarser resolution camera. To prove the concept it was decided to have two bands operating in the red and near IR with a spatial resolution of about 188 m and swath of about 750 km.

To achieve the swath, the wide field sensor (WiFS) required FOV coverage of ±26°. We saw in Chapter 4 that for interference filters an increase in the angle of incidence means a shift toward a lower wavelength for λ_c. Therefore, if a single lens was used for WiFS to cover a full field at the extreme ends of the swath, the spectral response will be substantially different from that at the nadir. One could get over this situation by using a "telecentric" lens, as discussed in Chapter 3. Following much consideration, a simpler solution was adopted by covering the total swath using two optical heads mounted on the same mechanical structure and squinted with respect to nadir by ±13° (Figure 6.24). Thus, the impact of the angular dependence of the bandpass filter is considerably reduced.

The WiFS camera consists of two bands, B3 and B4, which are similar to Band 3 and Band 4 of LISS-1/2/3, and uses refractive collecting optics consisting of eight lens elements with an interference filter and a ND filter in the front. A 2048-element linear array CCD with a pixel size of 13 μm by 13 μm similar to the ones used in LISS-1/2 is used in WiFS. It should be noted that due to the availability of a large integration time, four times oversampling is used, which improves the AT MTF performance. Four independently selectable gains are provided for each band. The electronics, characterization, calibration, and so on are similar to LISS-1 and hence are not further elaborated.

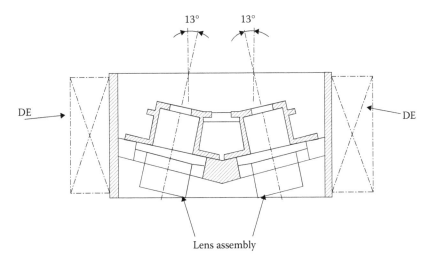

FIGURE 6.24
Mechanical layout of wide field sensor (WiFS) optical head showing the two lenses of each band tilted on either side of nadir by 13°. DE, detector electronics box.

6.7.3   PAN Camera

The PAN camera is designed to provide a spatial resolution of 5.8 m at nadir (1/4th of LISS-3) and operates in a single PAN (0.5–0.75 µm) spectral band. The camera is required to have a swath of about 70 km, which is steerable in the XT direction from nadir up to 26°. This off-nadir viewing provides the capability to acquire stereoscopic pairs from two different orbits and the ability to revisit any given site with a maximum delay of 5 days.

The required angular resolution can be obtained by a number of possible combinations of detector pixel size and optics focal length. For a compact system the optics focal length can be decreased by decreasing the detector pixel size. However, decreasing the pixel size has other issues, which we shall discuss in Chapter 7. The smallest pixel size available during the development phase was a 4 K CCD with a pixel size of 7×7 µm. To cater to the required spatial resolution from an orbit height of 817 km, an optical system with an effective focal length of about 980 mm and FOV of 2.5° is required. A 980 mm focal length refractive optics is very cumbersome for space use because of the long tube length requirement. From various system studies, the final chosen configuration to cover about 2.5° FOV is an unobscured TMA design, which we discussed in Section 3.2.2. The primary mirror is an off-axis hyperboloid and has an active circular aperture of 223 mm diameter. The beam reflected from the primary mirror converges toward the secondary mirror. Since the FOV in the AT direction is only about one-tenth of the XT direction, a secondary mirror of rectangular shape meets the full coverage of the field. The secondary mirror is a convex spheroid. After reflection at the secondary mirror, the beam diverges and falls on the tertiary, which is an off-axis oblate ellipsoid and is rectangular in shape. The beam reflected from the tertiary converges to the focus of the telescope (Joseph et al. 1996).

To cover the full swath of about 70 km requires a CCD with about 1200 elements. Since the CCD available during the development phase of the PAN camera had only 4096 elements, it was necessary to cascade three devices. As discussed in Chapter 3, by optically butting the CCDs it is possible to have them virtually appear as a continuous line in the focal plane. To maximize the radiance input to the CCDs, the PAN camera uses a different scheme. An isosceles prism is used to divide the focal plane and directs them to two zones as shown in Figure 6.25b. CCD1 and CCD3 are placed in one focal plane and CCD2 in the second focal plane. Each CCD covers a swath of about 23.9 km. The displacement of CCD2 with respect to CCD1 and CCD3 produces a shift of 8.6 km in the AT direction. Thus, the three CCDs are displaced both in AT and XT directions with respect to the satellite ground trace. However, since CCD2 has overlap with CCD1 and CCD3, a contiguous composite imagery from all three CCDs can be produced without gap. The composite image has a swath of 70 km. Each detector has separate interference filters. All three filters are made from the same batch so as to reduce spectral response variations. As discussed in Chapter 3, mounting of the mirrors onto the metering structure is one of the critical activities for any telescope development.

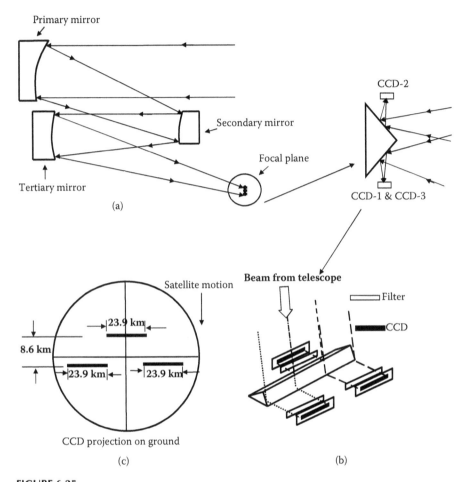

FIGURE 6.25
(a) Panchromatic (PAN) telescope ray diagram. (b) The focal plane layout. (c) The projection of CCDs on ground. (Reproduced from Joseph, G., *Fundamentals of Remote Sensing*, 2nd edition, Universities Press (India) Pvt Ltd, 195, 2005. With permission.)

Each mirror is first fixed to their respective "housing" with blade flexures using glue (Figure 6.26). (Later, cameras were fixed using different techniques to improve the isolation of the mirror from the structure.) Glue while curing due to shrinkage can exert stress onto the mirror. Therefore, the area over which the blade is attached and the thickness of glue (which depends on the gap between the blade and the mirror surface) have to be carefully chosen. For this purpose, initial trials are carried on a flat mirror of similar mechanical parameters. To have good dimensional stability under the operating temperature range, the metering structure is made of Invar. The structure is basically a cylinder of 630 mm inner diameter with an extended bracket for supporting the primary mirror. Ribs are provided in the radial direction to the bracket so as to increase the stiffness (Figure 6.27). The secondary and

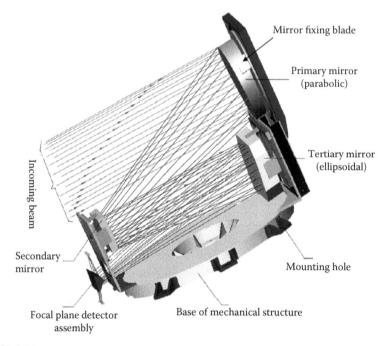

FIGURE 6.26
Mechanical layout of PAN mirror and focal plane assembly. (Reproduced from Joseph, G., *Fundamentals of Remote Sensing*, 2nd edition, Universities Press (India) Pvt Ltd, 195, 2005. With permission.)

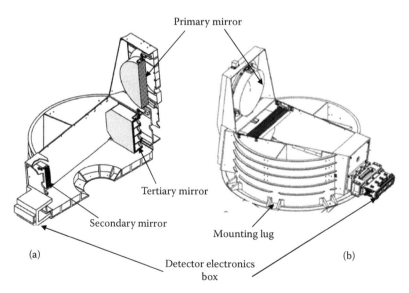

FIGURE 6.27
Schematics of PAN electro-optics module: (a) sectional view and (b) final assembled view.

territory mirrors are mounted on the circumference of the cylinder. Based on the ray tracing data, baffles are incorporated in the structure to ensure that no out of field radiation falls on the detector directly or after getting reflected by the mirrors, yet also ensure no vignetting takes place. The structure is fabricated to the designed values of separation and tilt between the mirrors within the fabrication tolerance and the final separation, and tilt is then realized by using appropriate shims based on the wavefront evaluation as explained in Section 3.4.4.

6.7.3.1 Payload Steering Mechanism

SPOT-1 is the first civilian satellite to provide off-nadir viewing capability. The off-nadir viewing enables the acquisition of stereoscopic imagery and provides a shorter revisit interval compared to temporal resolution, depending on the maximum off-nadir viewing geometry. This revisit capability facilitates change detection and is of immense value for monitoring the effect of disasters such as floods and cyclones. The SPOT satellite achieved this by using an elliptical plane mirror kept at 45° to the optical axis. The steering mirror optical performance is sensitive to the temperature gradient across the mirror. The degradation happens since the temperature gradient induces a curvature to the mirror surface due to uneven expansion of the front and back sides. Thus, the plane mirror becomes a powered element and hence the focal plane assembly (FPA) has to be repositioned to the new best focus. This activity calls for an onboard focusing mechanism, which adds to the complexity of the camera. The SPOT camera has an onboard focusing mechanism (Leger et al. 2003). IRS-1C has adopted a different technique for off-nadir viewing. To achieve the revisit capability, the telescope of the PAN camera is placed on a payload steering mechanism (PSM), which rotates the whole telescope so that the optical axis can be tilted to ±26° with respect to the nadir in the XT direction, which corresponds to an off-nadir coverage of ±398 km on the ground. The PSM can be rotated in steps of 0.09° and rigidly held in any position within a positional tolerance of 0.1° and with a stability of 0.1 arc s. The total angular movement of ±26° is covered in 15 min (Krishnaswamy et al. 1995). The mechanism also incorporates a pair of resolvers, one for coarse readout and the other for fine readout of the position of the camera with respect to the spacecraft roll axis. The PSM during launch is arrested by a hold-down mechanism and released in orbit by activating a pyro cutter by a command from the ground.

The electronics are similar to that discussed earlier for LISS-1/2 but tailored to the specific device and speed of operation. The CCD chosen has a loss of a fixed amount of charge during each transfer that is independent of the size of the signal charge. Therefore, it was necessary to have a fixed charge independent of the signal level, which is referred to as "fat zero." When operating with a fat zero, the fixed portion of the loss due to surface states is eliminated as well as any fixed loss from other sources. A fat zero can be created by an electrical bias or by an optical bias. The later method was used in the case of the PAN camera. Each CCD is illuminated by four LEDs using a cylindrical lens. Of these four LEDs, two are used for optical biasing and the other two for IFC. The IFC, as in the case of LISS-1/2, essentially evaluates the stability of the total system excluding the optics. The LEDs are operated in pulse mode and the duration to which the LED is on is varied to get different intensity values. Six nonzero exposure levels covering the full dynamic range are provided for each CCD. Four selectable gains are provided in the video chain, and the data are quantized to six bits. The parallel digital data from the PAN is formatted into two serial PCM streams, that is, PAN-I and PAN-Q, each with a data rate of 42.45 Mbps and transmitted via X-band.

The satellite has an onboard tape recorder with a recording capacity of 62 GB so that data can be collected over the areas that are not in the visibility of a ground station. The IRS-1C satellite, weighing about 1250 kg, was launched on December 28, 1995, from Baikonur Cosmodrome, Kazakhstan, into a polar Sun-synchronous orbit of 817 km, with the local time of equatorial crossing at 10:30 h. IRS-1D, a follow-on satellite to IRS-1C was launched on September 27, 1997 from the Indian Launch complex at SHAR by the Indian launch vehicle PSLV. Both the satellites performed satisfactorily for over a decade.

6.8   RESOURCESAT Series

To meet the additional needs of the remote sensing user community the next generation IRS satellites—RRESOURSESAT—are designed to have a set of cameras with improved spatial, spectral, and radiometric characteristics. The modifications are as follows:

1. Introduction of a high-resolution (5.8 m), multispectral camera. This will be useful in delineating individual fields as well as to carry out crop inventory for small administrative units, such as a village (LISS-4).

2. Improved LISS-3 with a SWIR resolution the same as the VNIR bands, that is, 23.5 m (LISS-3*). The improved resolution of the SWIR band will enhance the utilization of 23 m data for vegetation studies.

3. Improved WiFS with addition of green (B2) and SWIR (B5), in addition to the red and NIR bands in the IRS-1C/D WiFS and improved spatial resolution (56 m at nadir) designated as advanced WiFS (AWiFS). These improvements will enhance its application in many areas, especially for crop inventory.

Thus, RESOURCESAT provides a unique three-tier resolution imaging capability with the same spectral band characteristics (Figure 6.28). The configuration of the payloads is similar to those in IRS-1C/D and hence only the changes implemented will be presented here.

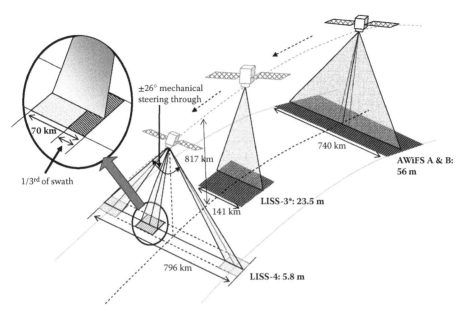

FIGURE 6.28
Schematics showing three-tier resolution capability of RESOURCESAT. Advanced WiFS gives 56 m resolution with swath of 740 km. LISS-3 has 23.5 m resolution and swath of 141 km, and LISS-4 has 5.8 m resolution. The 23-km width of LISS-4 can be steered anywhere within the AWiFS swath, thus providing three resolution data of any area within the LISS-3 swath.

6.8.1 RESOURCESAT LISS-3

The four spectral bands are realized using independent refractive optical assemblies and a linear array detector for each channel. The VNIR bands, that is, B2, B3, and B4, are configured as in IRS-1C/D in terms of detector, optics, and so on. The only major modification is the redesigning of the SWIR band to have the spatial resolution similar to the VNIR band, that is, 23.5 m. This is achieved by using a new detector device for the SWIR band. The device is an InGaAs photodiode array with 6000 pixels, where each pixel size is 13 × 13 µm. Six thousand elements are produced by butting 10 blocks of InGAs photodiodes each having 600 elements. The odd and even pixels are displaced by 26 µm across the array (AT while imaging). The detector is an active pixel type and provides two S&H video outputs. The SWIR detector temperature is regulated within ±0.1°C. The operating temperature can be selected (two values) through telecommand. As in the previous mission four gain settings are available for VNIR bands to cover the total dynamic range and the output is digitized to seven bits. For the SWIR band, there is only a single gain and it covers 100% albedo. SWIR output is digitized to 10 bits; however, only 7 selected bits are transmitted. Onboard calibration is carried out using LEDs as in the previous cameras (Dewan et al. 2006).

6.8.2   Advanced Wide Field Sensor (AWiFS)

The basic design concept of AWiFS is similar to WiFS but has improved performance in areas such as resolution (56 m vs. 188 m), dynamic range (10 bits vs. 7 bits), and additional spectral bands (B2, B3, B4, and B5). AWiFS also incorporates IFC, which was not available in WiFS. To cover the wide field imaging, the AWiFS camera is realized using two identical cameras, namely, AWiFS "A" and AWiFS "B." The four spectral bands for each camera are realized using independent refractive imaging optics having ±12.5° FOV. The two camera heads (electro-optical modules) are mounted on the spacecraft deck such that they are squinted with respect to nadir by ±11.84° so as to provide a combined field coverage of 47.94°. An overlap of 150 ± 20 pixels between camera A and camera B enables combining the data from two cameras to generate a wide swath imagery of about 740 km (Dave et al. 2006). A closed box type structure made out of stiffened Invar plates is used to house the lens detector assembly such that the lens is mounted on one side and FPA on the other side. It uses 6 K devices similar to those used for LISS-3. The camera covers the total dynamic range up to 100% albedo with a single gain setting and 12-bit digitization of which the 10 MSBs are transmitted. The camera saturation radiance of each band can also be selected through telecommanding (NRSA 2003). The camera provides an SNR of about 700 near saturation. As in the earlier LISS cameras LEDs are used for IFC.

6.8.3   LISS-IV Multispectral Camera

LISS-IV is a multispectral camera operating in three spectral bands, that is, in B2, B3, and B4, with IGFOV of 5.8 m and swath of 23 km at nadir. The camera retains the basic telescope configuration of IRS-1C/D PAN, that is, an off-axis TMA telescope consisting of an off-axis concave hyperboloid primary mirror, a convex spherical secondary mirror, and an off-axis concave oblate ellipsoidal tertiary mirror (Paul et al. 2006). To accommodate the three CCDs (which correspond to each spectral band) in the focal plane, an optical arrangement comprising an isosceles prism is used, which splits the beam into three imaging fields such that they are separated in the AT direction. The projection of this separation on the ground translates into a distance of 14.2 km between the B2 and B4 image lines. While Band-3 is looking at nadir, Band-2 will be looking ahead, and Band-4 will be looking behind in the direction of the velocity vector (Figure 6.29). Because of the different viewing geometry, the focal lengths for the three bands slightly differ. In addition, since there is a time delay when all the bands view the same strip, the spacecraft dynamics also affects the band-to-band registration. These aspects are taken care of during the processing on the ground to generate data products.

Each band has a linear CCD with 12,000 pixels with a pixel size of 7 μm × 7 μm (type THX31543A). Odd and even pixel rows are staggered by 5 scan lines (35 μm). To avoid any gap in the image due to this separation coupled

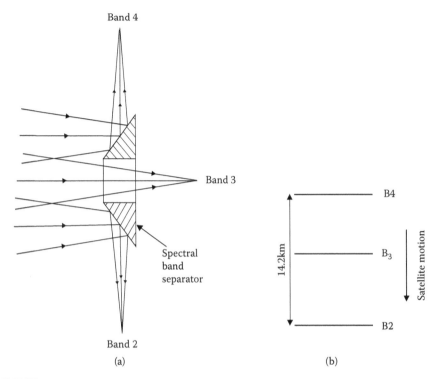

FIGURE 6.29
Schematics showing (a) LISS-IV focal plane beam splitter, (b) projection of detectors on the ground. (Reproduced from Joseph, G., *Fundamentals of Remote Sensing*, 2nd edition, Universities Press (India) Pvt Ltd, 197, 2005. With permission.)

with the Earth's rotation, the spacecraft is given a rotation about the yaw axis (NRSA 2003). To reduce the overall operating speed of electronics, the video output is available on eight ports. Eight video processors simultaneously process each port to achieve greater speed. The analog output signals from the CCDs are processed through independent port-wise electronic chains comprising amplifiers, DC restoration, and ADC. Though the CCD output is digitized to 10 bits, only 7 selected bits are transmitted (Paul et al. 2006). Dynamic range is adjusted to cover 100% albedo.

An important design consideration is to ensure the CCDs are maintained within a specified temperature range. Considering that each device is dissipating about 1.8 W and the dissipation of the circuitry nearby, special arrangements are made using a radiator and heat pipe to conduct away the heat. Another thermal issue is that the CCD temperature gradually increases when it is switched on and takes more than 10 min to stabilize. To minimize this effect, a heater kept near the CCD is switched on when CCD is off (thereby maintaining a temperature close to the condition when the

CCD is on) and switched off when the CCD is on. The CCD temperature is monitored by thermistors and an active thermal control is done using an on/ off temperature controller within the limits of 20 ± 2°C (Rao et al. 2006).

As in the case of the earlier LISS cameras, to monitor the long-term performance of the detector and processing electronics, an IFC scheme is implemented using LEDs. Eight LEDs are positioned in front to illuminate the CCD and are driven with a constant current, and the integration time is varied to get 16 exposure levels, covering the dynamic range.

The LISS-IV camera is kept on a PSM as in the IRS-1C/D PAN camera and can be tilted up to ±26° in the XT direction, thereby providing a revisit period of 5 days.

The LISS-IV sensor can be operated either in monochromatic (mono) mode or in multispectral mode (MX). In both cases, the resolution at nadir is 5.8 m. The mono mode has a 70 km swath corresponding to the full coverage of the 12,000 pixels of the CCD. However, only one single band is transmitted, which can be selected by ground command. Nominally, the red band (B3) is used because it is closest to nadir. To keep the data transmission rate the same as that of IRS-1C/D PAN, only one-third of the swath covering the same strip is transmitted in the multispectral mode, covering a swath of 23 km. To generate multispectral data 4000 pixels are collected from each of the three bands. Any pixel number from 1 to 8000 can be chosen as the start of the 4000-pixel subset, and hence the 23 km wide multispectral scene can fall anywhere within the mono 70 km footprint.

To collect data from regions not covered by any ground station, the satellite carries an Onboard Solid State Recorder with a capacity of 120 GB. Thus by advance planning data from any part of the globe can be collected.

The RESOURCESAT-1 was launched in October 2003 and RESOURCESAT-2, a follow-on mission to RESOURCESAT-1, was launched in 2011. Though RESOURCESAT-2 also carries three payloads as in the earlier mission, there is substantial improvement in radiometry in all the payloads (NRSC 2011). The LISS-IV camera uses 10-bit digitization both during RESOURCESAT-1 and RESOURCESAT-2. In RESOURCESAT-1 though, the data were digitized to 10 bits only 7 selected bits out of 10 were transmitted. However, in RESOURCESAT-2, the data are compressed so that all 10 bits are transmitted. This is achieved using a DPCM algorithm wherein 10 bits are mapped to 7 bits and reconstructed to 10 bits on ground. The same was incorporated in the LISS-III camera also. Therefore, the data are available without gain change covering up to 100% reflectance, even after taking into account the seasonal change in radiance. The AWiFS payload uses a 12-bit digitization scheme and provides 10 bits for data transmission so as to maintain the data rate. This has been achieved using multilinear gain (MLG). This scheme provides enhanced radiometric performance while viewing lower scene radiance.

In addition to the improvement in radiometry, the camera electronics hardware is miniaturized resulting in significant improvement in terms of size, weight, and power. Table 6.2 gives the major characteristics of the RESOURCESAT cameras.

TABLE 6.2

Performance Parameters of RESOURSESAT-2 Cameras

Parameters		LISS-3		AWiFS		LISS-4
		VNIR	SWIR	VNIR	SWIR	VNIR
Optics	Type	Refractive (DG)	Refractive (DG)	Refractive (DG)	Refractive (DG)	Reflective (TMA)
	EFL (mm)	347.5	451.75	139.5	181.35	980.0
	f/no.	4.5	4.7	5	5	4
Detector	Detection	Si	InGaAs	Si	InGaAs	Si
	Readout	Si CCD	Si CCD	Si CCD	Si CCD	Si CCD
	No. of pixels/size (µ)	6K/10×7	6K/13×13	6K/10×7	6K/13×13	12K/7×7
Bandwidth[a] (µm)		B2:69; B3:58 B4:95.1	B5:149.3	B2:69.5; B3:58.6; B4:101.6	B5:151.6	B2:72.2; B3:63.5; B4:87.9
IGFOV (m)		23.5	23.5	56 (nadir), 70 (off-nadir)	56 (nadir), 70 (off-nadir)	5.8
Swath (km)		141	141	740	740	70 (Mono) 23.5 (Mx)
Off-nadir view		nil	nil	nil	nil	±26° Roll
Mean saturation radiance (mw/cm²/sr/µ)		B2:54.4; B3:50.0; B4:32.7	B5:6.3	B2:52.4; B3:48.6; B5:30.5	B5:7.0	B2:50.4; B3:53.8; B4:34.6
S/N at saturation radiance		B2:>600; B3:>600; B4:>600	>500	B2:>600; B3:>600; B4:>600	>800	B2:>275; B3:>300; B4:>225
SWR		B2:>60; B3:>60; B4:>50	>35	B2:>50; B3:>50; B4:>40	>30	B2:>25; B3:>25; B4:>30
Quantization/transmitted (bits)		10/7	12/10	10/7	12/10	10/7
Compression ratio/type		1.42/DPCM	1.42/DPCM	1.2 MLG	1.2 MLG	1.42/DPCM

Source: Pandya, M. R., K. R. Murali, A. S. Kirankumar. 2013. *Remote Sensing Letters,* 4(3): 306–314.

Note: [a]Bandwidths based on the moments method.

The data from RESOURCESAT are being extensively used by the international scientific community. The AWiFS sensor with a better spatial resolution (56 m) than Landsat MSS and a wider swath (740 km) is found quite useful for many applications. In fact, AWiFS was the imagery of choice for the development by the USDA National Agricultural Statistics Service of their 2006 Cropland Data Layer (Bailey and Boryan 2010).

Currently, there are a number of Earth observation satellites acquiring imagery in the pushbroom mode. Giving an account of all such instruments is beyond the scope of the present book. However, to appreciate the engineering complexities of cameras other than IRS we shall now discuss the pushbroom camera systems of SPOT and Landsat.

6.9   SPOT Earth Observation Camera

The first operational space system to acquire imagery in the pushbroom mode was the Earth observation satellite SPOT (Satellite Pour l'Observation de la Terre) by Centre National d'études Spatiales of France in partnership with Belgium and Sweden. The first satellite, SPOT-1, was launched on February 22, 1986. The SPOT-1, -2, and -3 satellites are equipped with two identical high-resolution visible cameras—HRV (Haute Resolution Visible)—providing three-band multispectral image with 20 m resolution and a PAN band of 10 m resolution. The multispectral bands are B1 covering 0.50 to 0.59 μm (green), B2 covering 0.61 to 0.68 μm (red), and B3 covering 0.79 to 0.89 μm (near IR). The PAN band covers 0.51 to 0.73 μm. There are two such cameras on board, each providing a 60 km swath. SPOT is the first Earth observation civilian system providing off-nadir viewing capability. The camera view axis can be pointed in the cross-track direction within a range of ±27° from nadir. The off-nadir viewing enables the instrument to revisit a specific region within 5 days and can produce stereo pairs from two different orbits, which can provide height information. We shall now discuss the important design details of the SPOT camera system.

The HRV telescope has an f/3.5 aperture spherical primary mirror with 1.082 m focal length.

For extending the FOV, instead of using a single corrector plate as in the case of a classical Schmidt camera, a pair of refractive elements (lens doublet) is located at the center of curvature of the primary mirror. The incoming radiation is bent through 90° at the telescope entrance by a plane elliptical mirror (Midan 1986) driven by a stepper motor for off-nadir viewing. A fold mirror is used to reduce the overall dimension of the camera. After reflecting from the fold mirror, the radiation falls onto the primary mirror and the reflected beam passes through a hole in the fold mirror to form the image (Figure 6.30a). Before the focal plane an additional doublet lens is placed to

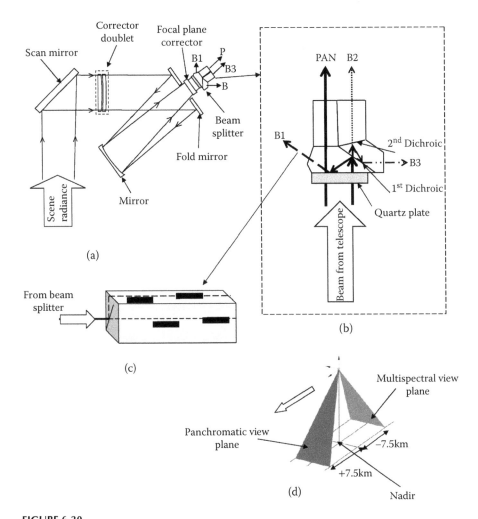

FIGURE 6.30
SPOT Haute Resolution Visible (HRV) camera. (a) Optical configuration used in SPOT. (b) Optical schematic of HRV beam splitter. (c) Optical butting scheme. (d) PAN and MXL projection on ground. ((a) and (c) adapted from Westin (1992); (b) from Henry, C., et al., *Acta Astronautica*, 17(5), 545–551, 1998. Reproduced with permission.)

provide further corrections to have a flat field. This optical scheme provides good image quality over a FOV of 4.3°.

The spectral separation and detection is carried out by a FPA, which consists of a beam splitter assembly and an optical butting unit with associated CCDs. The optical schematic of the beam splitter is shown in Figure 6.30b. The assembly essentially consists of a set of prisms with dichroic coatings at appropriate faces. The green band B1 is reflected by dichroic mirror-1. The red (B2) and IR (B3) are separated by dichroic mirror-2. The PAN band is directly transmitted through the prism. Spectral responses for each band are achieved by means of filters at the output ports. For each channel, the separator behaves as a plane parallel block of glass. To reduce the polarization sensitivity arising out of dielectric mirrors, a "multiwave" quartz plate is placed in front of the beam splitters.

The focal plane detector system is made of four Thomson-CSF Th7811 CCD detectors having 1728 detector elements of $13 \times 13\,\mu m$ size. To form a contiguous line, the four CCDs are optically butted by gluing to the face of an optical divider called a "Divoli." One thousand five hundred elements from each CCD are used to form a contiguous detector of 6000 elements. For the PAN channel, all 6000 detector elements are read out, whereas for multispectral channels, a ground resolution of 20 m is obtained by binning together two detector elements to form a line of 3000 detector elements.

For the SPOT-1/2/3 satellites, an XT line on the ground recorded by the PAN channel and the multispectral channel is not acquired at the same time due to the different locations of the CCDs in the focal plane of the instrument. The image line acquired at a given instant by any one of the HRV instruments is approximately +7.5 km in front of the subsatellite point in PAN mode and about −7.5 km behind the subsatellite point in multispectral mode (Spotimage 2002).

To improve the radiometric accuracy of the instrument and to correct for any changes in response that may occur during its lifetime in orbit due to aging of optical coatings, contamination of optical surfaces, degradation of CCD detectors, and so on, the HRV is equipped with a calibration system that can be operated by ground control. The relative calibration is performed by a lamp, which provides quasi-uniform illumination across the entire length of each CCD array. To carry out absolute calibration, a fiber optics unit projects solar irradiance onto some of the detectors in each band (Begni et al. 1986). The SPOT cameras are equipped with a refocusing mechanism to take care of any eventuality of defocus in orbit. On the ground, this mechanism is positioned at the best focus measured in vacuum (Meygret and Leger 1996).

The signals generated by the CCDs are processed by two separate image processing electronics for the PAN and MX channels. The amplifier gain for each of the CCDs can be changed by command. The data are digitized to eight bits. Due to the higher resolution, the data rate of PAN is one-third more than the MX data rate. To overcome this the PAN processing electronics includes a data compression function, which ensures that the PAN output bit rate is identical to MX. Each HRV delivers two 25 Mbps data streams corresponding to the PAN and MX modes.

The second-generation SPOT satellites, namely, SPOT-4/5, have an additional band in the middle infrared region (MIR) for multispectral imaging (HRVIR) and also carry a new five-band imaging system called a "vegetation instrument" with a resolution of about 1 km. The HRVIR focal plane arrangement is modified to accommodate the SWIR band. The modified focal plane arrangement is shown in Figure 6.31. As in the case of the HRV camera, each of the VNIR spectral bands is made of four Thomson-CSF Th7811 CCD detectors

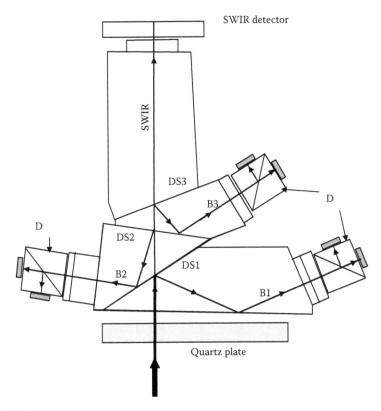

FIGURE 6.31
Schematics showing HRVIR focal plane arrangement for splitting the beam from the telescope to accommodate the CCDs for the four bands. DS, dichroic splitters. DS1 transmits radiation above about 0.6 μm; DS2 reflects in the red region, whereas DS3 separates NIR and SWIR. D, Divoli optically butting the four CCDs. (From Herve, D., et al., *Proc. SPIE*, 2552, 833–842, 1995. Reproduced with permission from SPIE.)

mounted on a linear optical divider (Divoli) so as to make up a line of 6000 elementary contiguous detectors. As in the case of the SPOT-1/2/3 HRV camera, ground resolution of 20 m is obtained by binning together detectors to form a line of 3000 points for channels B1, B2, and B3. The PAN band is replaced by band B2 with 10 m resolution, when all 6000 detectors are read out (Fratter et al. 1991). InGaAs photodiodes are used for the SWIR band. The SWIR detector is built as 10 contiguous blocks, with each block containing 300 pixels of $30 \times 30\,\mu m$ with an interval of $26\,\mu m$. The odd and even pixels are separated by $52\,\mu m$ and lie along two parallel lines. The look directions of SWIR detector odd pixels are coincident with the multispectral bands, whereas look directions for even SWIR detectors are displaced on the ground by about 40 m. This difference is corrected during ground processing.

The vegetation instrument is a separate and independent add-on payload of SPOT-4/5 satellites (Arnaud and Leroy 1991), and includes the camera, a solid-state recorder (which can record up to 97 min of imagery), an X- and L-band telemetry subsystem, and a computer. The instrument has four spectral bands of which three bands, namely, red (0.61–0.68 μm), NIR (0.78–0.89 μm), and SWIR (1.58–1.75 μm) are similar to the HRVIR bands. A fourth band in blue (0.43–0.47 μm) is added for oceanographic studies. Unlike the HRVIR, the vegetation instrument uses separate collecting optics covering the total FOV for each spectral band, as in the case of LISS-1/2/3. Since the vegetation cameras cover a FOV of 101°, the lenses have a telecentric design. The nominal resolution defined by pixels is 1.165 × 1.165 km and has a swath of about 2200 km with 10-bit digitization. A dedicated onboard calibration device monitors the radiometric performance of the cameras.

The SPOT-5 satellite launched in 2002 provides improved resolution with 5 m for the PAN band, which can be processed to 2.5 m (super mode), 10 m resolution for the three VNIR bands, and 20 m resolution for the SWIR band with its two HRG instruments. It also carries a new stereoscopic camera. We shall discuss the design aspects of these cameras in subsequent chapters.

The follow-on SPOT-6/7 satellites launched by the Indian Launch vehicle PSLV from India in 2012/2014 are substantially different from the earlier SPOT satellites in terms of satellite bus and the cameras. The camera is based on All-SiC telescope technology at EADS-Astrium. The telescope optics is a 200 mm aperture TMA design with three aspheric mirrors and two folding mirrors. The satellite simultaneously acquires five bands, namely, four multispectral bands in VNIR (blue: 0.450–0.520 μm, green: 0.530–0.590 μm, red: 0.625–0.695 μm, and near IR: 0.760–0.890 μm) with nadir spatial resolution of 6 m and a PAN band (0.450–0.745 μm) with 1.5 m resolution. A swath of 60 km is covered by two similar instruments. The focal plane detector array consists of CCDs operating in the time delay integration (TDI) mode. SPOT-6/7 data are digitized to 12 bits compared to the 8 bits of SPOT-5. The spacecraft can be maneuvered in all three axes. Using Control Moment Gyros as actuators, the satellites are highly agile, allowing stereo and tristereo acquisition of 60 km × 60 km scenes for the production of DEM. The high agility also allows a variety of acquisition scenarios in a single satellite pass.

6.10 Landsat 8: Landsat Data Continuity Mission

The Landsat program started with the launch of its first satellite in 1972 and is the longest running activity for acquisition of satellite imagery of the Earth. Landsat 8 (formerly called the Landsat Data Continuity Mission or LDCM) was launched on February 11, 2013 and is NASA's eighth satellite in the Landsat series. The imaging cameras—MSS/TM—on board Landsat 1 to 7 are optomechanical scanners. Landsat 8 (L8) is the first in the series operating in the pushbroom mode of scanning. L8 carries two Earth observing sensors, the operational land imager (OLI) and the thermal infrared sensor (TIRS). OLI and TIRS will coincidently collect multispectral digital images of the Earth. OLI and TIRS are the most advanced Earth observation system in the mid-spatial resolution regime. The basic improvement is the result of advancements in the focal plane detector system. Since the basic mode of operation is the same as LISS/HRVIR imagers, we shall here point out only some aspects that are specific to the L8 cameras. The OLI and TIRS data are stored on board on a solid-state data recorder with a capacity of 3.14 terabits for transmission to ground receiving stations.

6.10.1 Operational Land Imager

The OLI has nine spectral bands, of which eight work in the multispectral mode with a spatial resolution of 30 m and one band works in the PAN mode providing a spatial resolution of 15 m. Six of the multispectral bands are similar to the spectral bands collected by the enhanced thematic mapper-plus (ETM+) aboard Landsat 7, thereby giving continuity of data to ETM+ users. In addition, OLI will image in two new bands: a blue spectral band (Band 1) added mainly to provide ocean color data for coastal regions and a SWIR band (Band 9) added to facilitate the detection of cirrus clouds in OLI images. Table 6.3 gives the comparison between ETM+ and OLI bands and their application potential.

The OLI collecting optics is a four-mirror off-axis telecentric telescope with an aperture of 13.5 cm with aperture stop in the front (Figure 6.32a). The telescope has a cross-track (XT) FOV greater than 15° with excellent stray light rejection capability. The EFL in the AT direction is 862 mm, with an average EFL of 887 mm in the XT direction (Dittman et al. 2010). The mirrors are positioned inside a carbon composite optical bench. The FPA consists of 14 focal plane modules (FPMs) mounted on a single plate (Figure 6.32c). Each FPM consists of a sensor chip array (SCA) mounted on a motherboard. Each SCA consists of a HgCdTe detector array for the three SWIR bands, a Si PIN detector array for the six VNIR bands, and a CMOS silicon ROIC. Thus, each FPM contains nine detector arrays for the nine spectral bands arranged in the AT direction (Figure 6.32b). Each of the multispectral FPA has 494 active pixels with nominally 36 µm pitch. The PAN band FPA has 988 active pixels with nominally 18 µm pitch (Lindahl et al. 2011). In addition to these bands, there is a tenth band consisting of covered SWIR detectors, referred to as the "blind" band, that will be used to estimate variation in detector bias during nominal Earth image acquisitions. The spectral bands are defined by interference filters, which are critically aligned on the FPM such that each filter is directly over its corresponding detector. To take care of any detector failures during the operational life, there are redundant detectors at each pixel—two detectors for each of the VNIR pixels and three detectors per SWIR pixel.

TABLE 6.3

Comparison between ETM+ and OLI Bands and Their Application Potential;
Landsat 8 Thermal Band Is Generated by a Separate Instrument

Landsat 8-OLI				Landsat 7-ETM+	
Band No.	Spectral Range (µm)	Band Name	Application	Band No.	Spectral Range (µm)
1	0.433–0.453	Deep blue	Aerosol/ coastal zone	–	
2	0.450–0.515	Blue	Pigments/ scatter/ coastal	1	0.45–0.52
3	0.525–0.600	Green	Pigments/ coastal	2	0.53–0.61
4	0.630–0.680	Red	Pigments/ coastal	3	0.63–0.69
5	0.845–0.885	NIR	Foliage/ coastal	4	0.78–0.90
6	1.560–1.660	SWIR 2	Foliage	5	1.55–1.75
7	2.100–2.300	SWIR 3	Minerals/ litter/no scatter	7	2.09–2.35
8	0.500–0.680	PAN	Image sharpening	8	0.52–0.90
9	1.360–1.390	SWIR 1	Cirrus cloud detection	—	
		Thermal	Temperature	6	10.40–12.50

Source: Data consolidated from L8 EO Portal, https://directory.eoportal.org/web/eoportal/ satellite-missions/l/landsat-8-ldcm.

The desired detector for each pixel is selectable with the ROIC data words. Fourteen FPMs are organized in two rows of seven each such that the detectors have overlap so as to produce a contiguous detector array (after processing) to cover the desired swath of 185 km. For optimal SWIR performance, the FPA is passively cooled to 210 K. The AT spectral band separation leads to an approximately 0.96-s time delay between the first and the last band. Since the detectors are displaced AT, the yaw axis is steered to compensate for cross-track image motion due to the Earth's rotation. The satellite also has an off-nadir acquisition capability (up to one path off-nadir) for imaging high priority targets.

There is an elaborate calibration system to monitor the radiometric performance (Markham et al. 2008). This includes the following:

- Tungsten lamp assemblies that illuminate the OLI detectors through the full optical system. The lamps are operated at constant current and are monitored by a silicon photodiode.

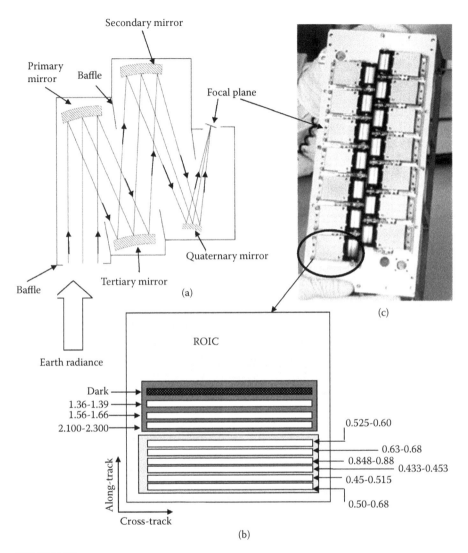

FIGURE 6.32
(a) Schematics showing operational land imager (OLI) telescope optical path. (b) Detector layout in one focal plane module. The numbers are the wavelength region in microns. (c) Photograph of OLI focal plane assembly. ((a) and (b) Adapted from L8 EO Portal, directory. eoportal.org/web/eoportal/satellite-missions/l/landsat-8-ldcm. (c) From Ball Aerospace Technologies brochure.)

- Solar diffuser-based calibration. The spacecraft is maneuvered to point the OLI solar calibration aperture toward the Sun so that the Sun enters the solar lightshade and the diffusers reflect light diffusely into the instruments aperture thereby providing full aperture calibration.
- A shutter that when closed provides a dark reference.

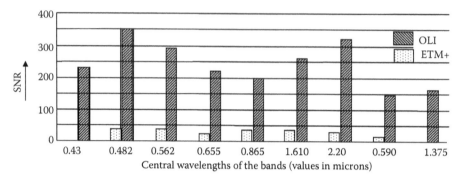

FIGURE 6.33

Comparison of ETM+ and OLI signal to noise ratio at typical values of radiance. (Adapted from Knight, E. J., et al., The Operational Land Imager: Overview and Performance. http://calval.cr.usgs.gov/JACIE_files/JACIE11/Presentations/TuePM/325_Knight_JACIE_11.070.pdf, 2011.)

As a part of the calibration scheme, the spacecraft is maneuvered every lunar cycle to view the Moon for tracking the radiometric stability over the mission.

The data are digitized to 12 bits and merged with TIRS data stream for transmission. The OLI data have a provision for lossless compression. The compression can be enabled or bypassed on an image-by-image basis.

OLI is a practical example of the decisive advantage of the pushbroom scan technique over whiskbroom scanning. What the ETM+—a whiskbroom scanner—achieved with a telescope of 1020 cm² could be achieved by OLI with only 143 cm² aperture with additional spectral bands, yet much better radiometric performance (Figure 6.33).

6.10.2 Thermal Infrared Sensor

TIRS is designed to produce imagery in two spectral bands in the thermal IR region with the central wavelength at 10.8 and 12 µm. The instrument produces 100 m spatial resolution from 705 km and generates a swath of 185 km. This is the first spaceborne instrument in the thermal IR region to operate in pushbroom mode.

The imaging telescope is a four-element refractive lens system made of three Ge elements and one ZnSe element. The lens has 178 mm focal length and f/no. 1.64. The telescope will focus the incident thermal energy onto the focal plane while providing a 15° FOV. To reduce the contribution of background thermal emission, the optics is radiatively cooled to a nominal temperature of 185 K and stabilized to ~0.1 K, using heaters. There is an ingenious nonmechanical way to adjust the focus if need arises. There is a fairly strong thermal dependence of the index of refraction of Ge. This property is used to adjust the focus by providing a provision to change the Ge lens temperature by ±5°K (Reuter 2009).

The focal plane consists of three 640 × 512 quantum well infrared photodetector (QWIP) GaAs arrays. Each pixel is 25 µm × 25 µm producing an

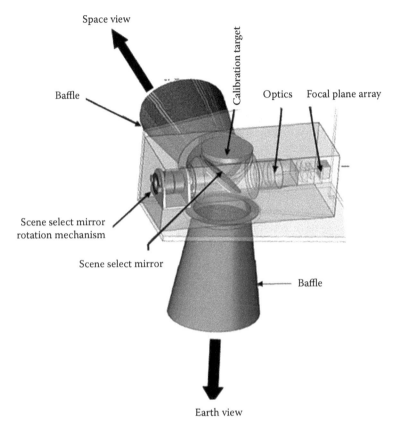

Space view

Calibration target

Baffle

Optics   Focal plane array

Scene select mirror
rotation mechanism

Scene select mirror

Baffle

Earth view

FIGURE 6.34
Schematics showing the thermal infrared sensor (TIRS) electro-optics module configuration. (Adapted from L8 EO Portal, directory.eoportal.org/web/eoportal/satellite-missions/l/landsat-8-ldcm.)

IFOV of 142 µrad, which from the satellite orbit gives 100 m spatial resolution. The 640 pixel lines are organized XT and arranged in a staggered configuration as depicted in Figure 6.34. The three arrays are precisely aligned to each other in the horizontal and vertical directions. Though the combined pixels XT for all three detectors are 1920, only 1850 detectors per band are used for generating an image line XT by discarding approximately 8 columns on either end of the line and allowing for approximately 27 columns of overlap between adjacent arrays. Thus, with each pixel having a spatial resolution of 100 m, the 1850 pixels project a line of 185 km on ground. The spectral selection is carried out by two interference filters mounted onto each QWIP.

Thirty-two rows are available under each filter band separated by 76 rows of covered pixels used for dark current estimation. For normal imaging, every line will include two rows each from the two spectral bands and from the dark strip area. But only one row of each spectral band is required to reconstruct the ground scene. However, since the mission requires that 100% of all the pixels in each band meet certain quality specifications, there is an option to combine two adjacent rows of a given band to construct an equivalent fully operational single row (Jhabvala et al. 2011; Arvidson et al. 2013). The focal plane is cooled by a mechanical cryocooler to 43° and controlled to 0.01 K.

For on-orbit radiometric calibration at the entrance to the lens a scene select mechanism flips a flat mirror to change the look angle from nadir (i.e., Earth view) to either an onboard blackbody calibrator or a deep space view (Figure 6.35). The blackbody is a full aperture calibrator whose temperature may be varied from 270 to 330 K. It is a curved plate whose surface is machined with V-grooves to improve the overall emissivity of the plate and its temperature is actively controlled to better than 0.1 K to meet the 2% absolute radiometric accuracy requirement (Thome et al. 2011).

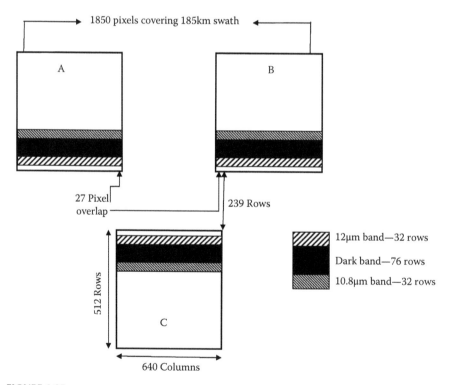

FIGURE 6.35
Schematic diagram showing the layout of detectors in TIRS focal plane. The three quantum well infrared photodetector QWIP detectors are staggered XT (a & b) and AT (c) with overlap. Of the total of 1920 pixels in a row (3 × 640), by including array overlap and eliminating the first and last 8 columns, 1850 pixels forms one image line AT. (Adapted from Arvidson, T., et al., Landsat and Thermal Infrared Imaging. ntrs.nasa.gov/archive/nasa/casi.ntrs.nasa. gov/20120015404_2012015279.pdf, 2012; Jhabvala (personal communication)).

6.11 Hybrid Scanner

Both whiskbroom and pushbroom imaging systems, on a low Earth orbiting satellite, generates images in one direction (AT) due to spacecraft motion. We have seen in Chapter 5 that from a geostationary orbit (GEO) the camera system has to produce scanning in two directions (E-W & N-S) to form the image. It is possible to incorporate a linear CCD in one direction and mechanical scanning in the other direction to form a two-dimensional image from a GEO. That is a hybrid of optomechanical scanning and pushbroom scanning.

Although imaging from GEO was started for meteorological applications, there were NASA-sponsored studies carried out by Perkin-Elmer Corporation to design a system for an Earth resources survey from GEO: "Synchronous Earth Observation Satellite" (Oberheuser 1975; Young 1975). The projected requirements for the camera were 13 spectral bands within 0.6 to 13 μ with spatial resolution requirements ranging from 100 m in the VNIR spectrum to 800 m in the long wavelength IR. Based on various trade-off studies for the optics and scanning technique, a telescope with a FOV of 0.6° × 1.2° was found to be optimum for satisfying the requirements projected. The system conceived uses pushbroom detector arrays subtending 1.2° in one direction and scanning in the other direction by slewing the entire spacecraft. However, the concept was not pursued further for a practical system. Now there is a renewed interest to build an Earth observation system from GEO that can give performance comparable to LEO in terms of spatial and spectral resolution (Puschell et al. 2008).

The first CCD-based imaging system from GEO was launched in 1999 onboard the ISRO satellite INSAT-2E. The camera has three spectral bands (0.62–0.68 μm, 0.77–0.86 μm, and 1.55–1.69 μm), and each of the bands produces 1 km IGFOV at the subsatellite point. The basic optical assembly consists of an RC telescope, scan mechanism, and dichroic beam splitters to separate the bands (Figure 6.36). Consistent with the FOV of the telescope, a CCD array was chosen to cover 300 km at nadir. A silicon linear array is used for the VIS and VNIR bands, while an InGaAs array is used for the SWIR band. The camera provides a nadir spatial resolution of 1 km in all three bands (Iyengar et al. 1999). The CCDs are aligned in the north-south direction, producing one scan line for each band. The motion in the east-west direction is achieved by sweeping the scan mirror. Each scan generates a three-band image strip of 300 km wide (N-S) by electronic scanning and 6300 km long (W-E) by mechanical scanning for a duration of 1 min. Thus, both mechanical scanning and electronic scanning are used to generate imagery. Successive image strips are produced by stepping the scan mirror by about 0.4° southward after a west to east scan. The imaging field can be positioned to generate imagery of any part of visible imaging disc. The instrument still needs two-axix scanning since only 300 km is covered in the N-S direction.

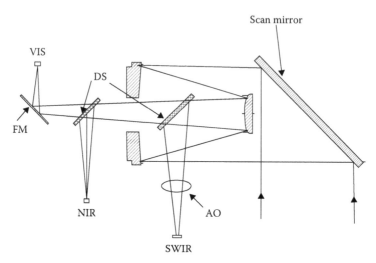

FIGURE 6.36
Optical schematics of INSAT CCD camera. The separation of the bands in the focal plane is achieved by dichroic beam splitters (DS). The SWIR channel has auxiliary optics (AO) to match the IFOV with the VIS/NIR channels. A fold mirror (FM) is used to reduce system size.

With the advancement in focal plane detector technology now it should be possible to cover the full Earth disc in N-S direction thereby requiring only one axis scan in the W-E direction.

Though the NASA study related to a mission for Earth resources survey from GEO was conducted in 1974, no further work was carried out and GEO platforms were primarily designed for meteorological applications. This could be due to the technology gap. However, now there is a renewed interest in having a multispectral imaging system from GEO giving spatial resolution close to what is being achieved from LEO. The primary advantage of GEO imaging is fast responsiveness and short revisit times, of the order of minutes, not available from LEO missions. That is, a GEO system can reach a target region almost immediately, compared to current LEO systems where it could be hours to days depending on where the satellite is. The higher temporal resolution substantially improves the data availability in periods of scattered clouds. These capabilities enhance monitoring of Earth's dynamic features and also can be effectively used for various types of disaster monitoring and needed relief operations. This requires systematic monitoring of the area to be serviced preferably continuously or at frequent intervals. Another unique capability is rapid imaging of a spot so as to generate "video" type images, enabling rapid detection and characterization of moving targets. This has a number of applications in maritime security. A GEO system should have wide FOV for high temporal resolution for the whole region to be serviced and high spatial resolution to discern details. The baseline design, ideally, should cater to land and ocean applications. The first instrument specifically designed for ocean color monitoring satellites from GEO is the Korean Geostationary Ocean Color Imager (GOCI) launched in 2010 (Cho and Youn 2006). ESA has conducted studies aiming at identifying the most relevant applications and principal technical concepts for a high-resolution GEO imaging system. ISRO has an approved program for a GEO satellite with multiple data collection capability—from multispectral to hyperspectral—covering wavelengths from visible to thermal IR. Before we discuss these systems, let us find out what capabilities the existing LEO systems have when they are placed on a GEO platform.

Let us do some "thought" experiments! What happens if we place the Landsat 8 OLI payload in the GEO? From 705 km orbit it was providing 30 m in multispectral imaging, which will become about 1.5 km from GEO. With 15° FOV the camera can cover about 60° N and 60° S. Thus, by attaching a scan mirror with a single scan the camera can take imagery with nadir resolution of 1.5 km in nine spectral bands and 750 m in PAN covering about ±60° latitude. In low Earth orbiting satellite missions, the integration time depends on the subsatellite velocity, which is fixed. In cameras on board GEO, the designer can choose the integration time, subject to frequency of Earth coverage, which is a great advantage. If we keep the Landsat 8 OLI dwell time of about 4.5 ms, then 15° (~6 × 10³ IFOVs) can be covered in less than 30 s. Since the Earth subtends only about 17° from GEO, it is possible to design an optical system covering the full disc and by re-engineering FPA it is possible to achieve 1 km resolution at the subsatellite point. This will provide unparallel capability compared to the present 1 km resolution imaging systems such as SPOT vegetation and NOAA AVHRR.

The current highest spatial resolution Earth observation camera is WorldView-2 (WV-2), providing 0.46 m in PAN (IFOV 0.6 μrad) and 1.84 m (IFOV 2.4 μrad) in 9 multispectral bands, with a swath of 16.4 km, from an altitude of 770 km. If placed in GEO, the instrument can give about 21.5 and 86 m IGFOV at nadir for the PAN and multispectral channels (MXL), respectively, with a swath of about 760 km. The number of IFOVs per degree for MXL is about 7.27 × 10³ (0.175/2.4 × 10⁻⁶). Thus, if we tilt the line of sight (LOS) by 1° from nadir, the camera will cover an MXL strip of about 625 km with a 760 km swath.

The WV-2 dwell time per pixel for PAN is about 0.07 ms and for MXL is about 0.286 ms. (WV-2 has a maximum of 64 stages of TDI and the required number of stages can be selected by command; therefore, the integration time will be the number of stages used multiplied by pixel dwell time.) At GEO since the subsatellite point is stationary the scan speed can be adjusted to get the required radiance and one need not resort to TDI mode. Assuming nominal operation of WV-2 is with 32 stages of TDI, the effective integration time is 9.2 ms (0.286 × 32). To achieve this, the LOS has to scan at a rate of 67 s per degree [(7.27 × 10³) × (9.2 × 10⁻³)]. This is well within the satellite's agility, which is 2.5°/s. The satellite can view any specific area of interest on demand with very little delay. However, the current spacecraft agility is only for positioning the camera view within the field of regard and the jitter is not important. But if the same mechanism is used for imaging, the jitter needs to be controlled within a fraction of IFOV. Except for this, WV-2 can be used as it is from GEO to get a medium resolution image. From the applications point of view, lack of SWIR is a drawback for vegetative studies. IF PAN can be replaced by SWIR, the application potential will substantially improve.

Currently, AWiFS of the IRS constellation is extensively used by many agencies for crop area estimation. Therefore, if WV-2 optics can be redesigned to get 50 m resolution, say by increasing the focal length by 70%, it will be an invaluable source of multispectral data for crop monitoring. Since the presence of clouds is transitory, imaging from GEO has the added advantage of choosing cloud-free areas in association with weather maps.

6.11.1 GEO High-Resolution Imaging Systems: Technology Challenges

Before we discuss some of the studies/projects carried out to achieve "LEO Earth observation capability from GEO," let us look at some of the technological challenges.

Collecting optics: The aperture size of the collecting optics is decided by the energy to be collected and the fundamental resolution limit due to diffraction. Let us consider, based on Rayleigh's resolution criteria, the minimum size of the mirror to achieve 1 m resolution from GEO. The IFOV required is 2.79×10^{-8} rad. A Rayleigh resolution criterion is the generally accepted measure for resolving two point targets. As we discussed in Chapter 3, the angular separation of two point targets that can be resolved as per the Rayleigh criterion for a circular aperture is given by $1.22\dfrac{\lambda}{D}$, where D is the aperture diameter. Using this criterion, the minimum diameter of the imaging system to resolve 1 m from GEO at 0.5 μm works out to be 22 m. A monolithic mirror of this dimension cannot be accommodated in the envelope of any of the present rockets. This will necessitate segmented mirrors as in the case of the James Webb Space Telescope (JWST), which is being fabricated for space-based IR observation. The JWST primary mirror is made up of 18 hexagonal-shaped mirror segments of 1.32 m in diameter, giving an effective diameter of 6.5 m (Lightsey et al. 2004). If the JWST telescope is used from GEO, by the same rationale discussed earlier it can give a resolvable spatial resolution of 3.4 m at 0.5 μm. The largest monolithic optical telescope in space is the Hubble Space Telescope with primary aperture of 2.4 m, which can provide a diffraction-limited resolution of about 9 m from a GEO. These resolution numbers assume the system has no other aberrations, which is not realistic.

Spacecraft drift/jitter: Image motion essentially reduces the MTF. This is discussed in detail in Chapter 7. The MTF degradation due to drift and jitter depends on their magnitude in relation to the IFOV of the system. Since from

GEO the IFOV is much smaller than the LEO system for the same spatial resolution, the satellite should have attitude stability commensurate with the IFOV. This is a challenging task.

FPA: The complexity of the FPA depends on the services expected from the imaging system. There are some specific advantages in imaging from GEO. We have seen that for AWiFS to cover about 750 km from LEO, we had to use two detector heads to reduce the angle of incidence onto the interference filter. However, to get a 750 km swath from GEO requires an angular coverage less than ±0.6°. Therefore, a single optical head with a filter at the input of the lens as in the case of the LISS cameras is adequate. From GEO either a linear array operating in pushbroom mode by scanning or an area array can be used for imaging. Though an area array requires more pixels, since the camera can be held stationary with respect to the scene imaged, there is no MTF degradation as in the case of a linear array with image smear in the AT direction. One of the requirements of high-resolution imaging from GEO orbit is to achieve rapid detection and characterization of moving targets to cater to the needs of various security agencies. This requires imaging the same area a few times in quick succession so as to have "video" type imaging. For a linear array in pushbroom mode to cover "n" lines requires "n" times the integration time, whereas an area array with $n \times n$ pixels can acquire the image of the whole area in just one integration time, all other things remaining the same. Thus, images in quick succession of a specific area is possible only with an area array operating in framing mode. Frame imaging is also not susceptible to line-wander (a low-frequency phenomenon) as in the case of pushbroom imaging. However, multispectral imaging with an area array is more complex. Multiple bands using a linear array is possible by laying appropriate filter strips over multiple lines and detectors requiring cooling can be separated in the focal plane by standard techniques. But area arrays require more elaborate arrangements for multispectral imaging, which will be discussed later.

We shall now discuss some of the operational and planned GEO missions for resource monitoring.

6.11.2 GEO-Resource Survey Systems

6.11.2.1 ESA GEO-HR System

Based on detailed interaction with the user community, ESA has identified a set of potential applications that require a combination of fast response, high-revisit, near-real-time, and high-resolution observations, which is possible only from GEO. Astrium GmbH conducted a feasibility study to have a multipurpose GEO-based observation facility (Geo-Oculus) offering relatively high spatial resolution (10 to 300 m), fast access time (minutes), and fast revisit with a specific focus on the area of the European Union. Though the system proposed is for two important application domains—maritime security/

surveillance and disaster management, and monitoring and mitigation—the combination of spectral bands, spatial and spectral resolutions, and revisit capabilities can be fruitfully used for many other applications. We shall briefly discuss the Geo-Oculus imaging capabilities and instrument concepts.

The instrument provides simultaneous imaging of Earth scenes on four multispectral focal planes (UV-blue, red-NIR, MWIR, and TIR) with a ground FOV of 300 km × 300 km (0.48 × 0.48°). In addition, a PAN focal plane provides a reduced FOV (157 km × 157 km, corresponding to 0.25 × 0.25°), but with higher resolution (10.5 m nadir, 21 m over Europe). Table 6.4 gives the spectral channels proposed for the Geo-Oculus. The ground sampling distance (GSD) for the VNIR channel is 40 m (at 52.5°N, i.e., over Europe). For marine applications to improve SNR, 2 × 2 pixel binning is used and hence the ground resolution is 80 m. The SWIR and MWIR have a GSD of 300 m and TIR of 750 m.

The collecting telescope is 1.5 m diameter all-SiC monolithic construction. The PAN channel requires almost twice the focal length compared to multispectral bands. This is achieved in an ingenious way by incorporating a third mirror in the Korsch configuration, while the main Cassegrain telescope beam is suitably split into four focal planes to accommodate the multispectral bands (Figure 6.37). The focal plane separation is carried out by a set of dichroic beam splitters.

The S/N requirement and fast image generation ruled out a pushbroom integration mode of operation. If pushbroom technology is used a very slow scan speed (about 1 h per image) is required to meet the S/N requirement. Using TDI, the imaging duration can be reduced but at the cost of further

TABLE 6.4

Proposed Spectral Channels for Geo-Oculus

UV–Blue $\lambda_c (\Delta\lambda)$ nm	Red–NIR $\lambda_c (\Delta\lambda)$ nm	SWIR–MIR $\lambda_c (\Delta\lambda)$ nm	TIR $\lambda_c (\Delta\lambda)$ nm	PAN $\lambda_c (\Delta\lambda)$ nm
318 (10)[a]	620 (10)	1375 (50)[a]	10,850 (900)	655 (155)
350 (10)[a]	665 (10)	3700 (390)	1200 (1000)	
412 (10)	681 (8)			
443 (10)	709 (10)			
490 (10)	753 (8)			
510 (10)	779 (15)			
555 (10)	865 (20)			
	885 (10)			
	900 (10)			
	1040 (40)[a]			

Source: ESA, Contract No. 21096/07/NL/HE, Geo-Oculus: A Mission for Real-Time Monitoring through High-Resolution Imaging from Geostationary Orbit. http://emits.sso.esa.int/emits-doc/ESTEC/AO6598-RD2-Geo-Oculus-FinalReport.pdf, 2009.

Note: [a]Optional channels.

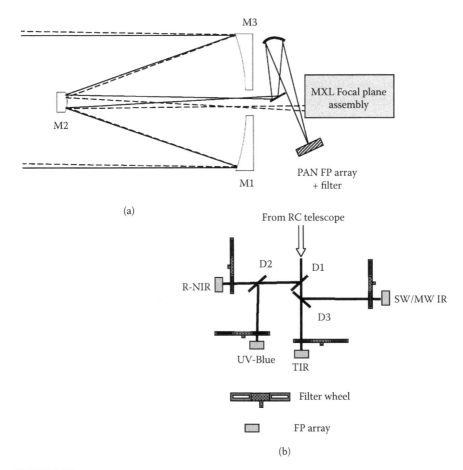

FIGURE 6.37
(a) Proposed optical schematics of Geo-Oculus camera. Mirrors M1 & M2 form the RC telescope. Part of the field is directed to a concave tertiary mirror. M1, M2, and M3 form a Korsch telescope that is used for PAN detection. (b) Schematics of focal plane arrangement for multispectral imaging. Ds, dichroic beam splitters. Auxiliary optics not shown. (Adapted from Vaillon, L., Geo-Oculus: High Resolution Multi-spectral Earth Imaging Mission from Geostationary Orbit. *Presented at ICSO 2010 8th International Conference on Space Optics Rhodes.* http://www.congrexprojects.com/custom/icso/Presentations%20Done/Session%209b/04_ICSO2010_GeoOculus.pdf, 2010.)

constraints on platform stability. Therefore, area array sensors with a filter wheel are found to be optimal.

For UV and VNIR spectral bands, a monolithic CMOS array is preferred to CCD for its good maturity to build large arrays and its better immunity to the harsh radiation environment in GEO. To optimize the sensitivity two kinds of detectors are used, one optimized for UV and short visible wavelengths ("UV-blue" detector) and the other for red and NIR wavelength ("Red-NIR" detector). For SWIR and MWIR detection, an HgCdTe detector with a response from 1.3–3.7 μ with a CMOS ROIC was chosen. QWIP is the preferred candidate for the TIR channel. Current technology does not allow a single area array of the required number of pixels. Therefore, a number of pixels are "stitched" together to get the required size detector. To reduce the dark current, the SWIR/MIR detector and the thermal detectors are cooled by mechanical cryocoolers (ESA 2009).

LOS pointing stability is an important factor in generating good quality imagery. The requirement for the spacecraft is dictated by the PAN requirement. The requirements projected are relative pointing error (stability over the integration time) 0.15–0.2 μrad peak-to-peak for high-frequency jitter (>10 Hz), relative maximum measurement error (over image acquisition time) 0.1 μrad over 5 s acquisition time, and pointing drift error (drift over integration time) 5 μrad/s. This is a challenging task.

Needless to say the ground system should match to ensure quick delivery to the end user.

6.11.2.2 Geostationary Ocean Color Imager

The first Korean multifunctional geostationary satellite—Communication, Ocean, and Meteorological Satellite—launched in 2010 and carried an ocean color monitoring instrument—GOCI—as one of the three payloads on board. GOCI takes imagery in 8 spectral bands covering 400–900 nm with a GSD of 500 m covering a target area of 2500 km × 2500 km centered on 130E, 36N. The GOCI optical system is a TMA design with 140 mm aperture and 1171 mm focal length. The mirrors and the structure are made of silicon carbide.

The image generation is carried out by a CMOS/APD area array having rectangular pixel size to compensate for the Earth projection. The detector is passively cooled and maintained at a temperature of 10°C. The detector has 1415 × 1432 pixels producing one frame of imagery at a time. Sixteen frames (as a 4×4 matrix) are required to cover the service area of 2500 km × 2500 km. Successive frames are generated by a pointing mirror, located close to the entrance of the instrument, providing a bidirectional scan. Thus, by successively pointing the line-of-sight in 16 directions, the detector array is moved in the FOV so as to image the complete area under surveillance. Spectral band selection is carried out by a nine-position filter wheel, of which eight positions carry an interference filter corresponding to the eight spectral bands while the ninth one blocks the input for the detector dark current measurement (Faure et al.). Spectral selection with a filter wheel has the advantage that the integration time can be adjusted for each band independently to meet the SNR requirement. One frame takes about 100 s to acquire the 8 bands and dark signal and the full image covering 18 frames takes less than 30 minutes, which includes image integration and read-out time and the time taken for filter wheel motion. The signal generated is amplified with an amplifier that has bilinear gain to accommodate the signal from ocean and land, digitized to 12 bits and transmitted via L-band to the ground station.

The follow-on system, GOCI 2, is expected to have better spatial resolution, full disk coverage, and five more bands compared to GOCI 1.

6.11.2.3 ISRO Geostationary Imaging Satellite (GISAT)

The ISRO has an approved program for a high-resolution imaging system from GEO. The imaging covers a broad spectral region extending from visible to thermal IR with different spectral and spatial resolution and is capable of imaging the full disk or any user specified region. GISAT will provide four types of data. They are multispectral VNIR (MX-VNIR), hyperspectral imager VNIR (HySI-VNIR), hyperspectral imager SWIR (HySI-SWIR), and multispectral LWIR (MX-LWIR).

MX-VNIR provides spatial resolution of 50 m in six spectral bands. The bands chosen are 0.45–0.52, 0.52–0.59, 0.62–0.68, 0.77–0.86, 0.71–0.74, and 0.845–0.875 (all values in micrometers); the first four bands are similar to the B1–B4 bands of the LISS cameras, thereby giving continuity to LISS data users. AWiFS data are widely used globally for vegetation studies and MX-VNIR, though it does not have the SWIR channel, can replace this data, with the added advantage of frequent revisit.

The HySI-VNIR instrument will have hyperspectral imaging capability with about 60 bands covering about the 0.4 to 1.0 μm spectral range with a spectral resolution less than 10 nm, with nadir IGFOV of around 500 m. The data, among other uses, are of immense value for ocean color monitoring. HySI-SWIR will provide hyperspectral data in more than 150 bands with a spectral resolution of less than 10 nm in the SWIR bands covering about the 0.9 to 2.5 μm spectral range with a footprint at nadir of about 500 m. Both the hyperspectral imagers are the first of their kind from GEO and will generate spectral signatures for Earth's surface features such as agriculture, forestry, minerals, and ocean. The multispectral TIR imager will collect data in six bands between about 7 and 14 μm, with nadir IGFOV of 1.5 km × 1.5 km. The system covers 100% albedo in the VNIR/SWIR range and scene temperature range of 100 to 340 K in LWIR.

The basic optical design is similar to CARTOSAT2, which we shall discuss in detail in Chapter 7. The 700 mm telescope operates in RC configuration with appropriate field correcting optics to generate the required FOV. The major challenge is configuring the image plane to accommodate four types of detector system. This is achieved by suitably separating the telescope field followed by appropriate reimaging optics. The VNIR and thermal multispectral channels use appropriate linear arrays with suitable filters overlaid. The hyperspectral imagers use grating as the dispersing element with appropriate area arrays. The scanning is accomplished by controlled spacecraft motion. The scan rates can be varied depending on the SNR requirement. The data can be acquired in different modes such as full globe ($18 \times 18°$), subcontinent ($10 \times 12°$), and user-defined area (Kumar 2013).

6.11.2.4 Geostationary Hyperspectral Imaging Radiometer (NASA)

To explore and understand the biological and biogeochemical processes of oceans, NASA's Ocean Biology and Biogeochemistry Working Group (OBBWG) has proposed a long-term integrated plan covering systematic observational strategies, research, analysis, and modeling.

OBBWG recommended that to better assess, understand, and predict natural and anthropogenic-driven variability in coastal zones, in addition to polar orbiting missions, there is a need for dedicated regional/coastal missions with high temporal coverage using GEO platform.

The major requirements for a Geostationary Hyperspectral Imaging Radiometer proposed by the working group are summarized below (NASA 2006).

1. Spectral coverage from ~350 to 1050 nm (1300 nm goal) with spectral sampling of 2 to 4 nm.
2. Spatial footprint 50 to 200 m nadir with more than 1000 pixel swath.
3. Minimum SNR of 500–1500, with image summing improving to greater than 3000 and high dynamic range (14-bit digitization).
4. For proper interpretation of the data, the instrument should have polarization sensitivity/change less than 0.2%.

One of the major advantages of a GEO system is the temporal resolution. The projected need is to provide complete regional coastal coverage (e.g., CONUS) a minimum of 4 times per day, with regional repeats more than 10 times per 6 hours, and event coverage at 15 minute intervals.

The end user requirements of high spatial, spectral, and radiometric resolution, and varying temporal resolution make realization of Earth observation sensors from GEO a challenge.

References

Arnaud, M. and M. Leroy. 1991. SPOT 4: A new generation of SPOT satellites. *ISPRS Journal of Photogrammetry and Remote Sensing* 46(4): 205–215.

Arvidson, T., J. Barsi, M. Jhabvala, and D. Reuter. 2012. Landsat and Thermal Infrared Imaging. ntrs.nasa.gov/archive/nasa/casi.ntrs.nasa.gov/20120015404_2012015 279.pdf (accessed on July 12, 2013).

Arvidson, T., J. Barsi, M. Jhabvala, and D. Reuter. 2013. Landsat and thermal infrared imaging. *Thermal Infrared Remote Sensing: Sensors, Methods, Applications, Remote Sensing and Digital Processing.* ed. Kuenzer C. and Dech S. Springer.

Bailey, J. T. and C. G. Boryan. 2010. Remote Sensing Applications in Agriculture at the USDA National Agricultural Statistics Service. http://www.fao.org/fileadmin/templates/ess/documents/meetings_and_workshops/ICAS5/PDF/ICASV_2.1_048_Paper_Bailey.pdf (accessed on May 14, 2014).

Begni, G., M. C. Dinguirard, R. D. Jackson, and P. N. Slater. 1986. Absolute calibration of the SPOT-1 HRV cameras. *Proceedings of SPIE* 660: 66–76.

Brodersen, R. W., D. D. Buss, and A. F. Tasch Jr. 1975. Experimental characterization of transfer efficiency in charge-coupled devices. *IEEE Transactions on Electron Devices* 22(2): 40–46.

Cho, Y. and H. Youn. 2006. Characteristics of COMS meteorological imager. *Proceedings of SPIE* 6361: 63611G1–63611G8.

Dave, H., C. Dewan, S. Paul, et al. 2006. AWiFS camera for Resourcesat. Proceedings of SPIE 6405: 6405X.1–6405X.11.

Dewan, C., S. Paul, H. Dave, et al. 2006. LISS-3* camera for Resourcesat. *Proceedings of SPIE.* 6405: 64050Y.1–64050Y.7.

Dittman, B., G. Michael, and O. Firth. 2010. OLI telescope post-alignment optical performance. *Proceedings of SPIE* 7807: 780705.1–780705.5.

ESA. 2009. Contract No. 21096/07/NL/HE, Geo-Oculus: A Mission for Real-Time Monitoring through High-Resolution Imaging from Geostationary Orbit. http://emits.sso.esa.int/emits-doc/ESTEC/AO6598-RD2-Geo-Oculus-FinalReport.pdf (accessed on May 14, 2014).

Faure, F., P. Coste, and G. Kang. The GOCI instrument on COMS mission—the first Geostationary Ocean Color Imager. http://www.ioccg.org/sensors/GOCI-Faure.pdf (accessed on April 24, 2014).

Fratter, C., J.-F. Reulet, and J. Jouan. 1991. SPOT 4 HRVIR instrument and future high-resolution stereo instruments. *Proceedings of SPIE* 1490: 59–73.

Gamal, El A. and H. Eltoukhy. 2005. CMOS image sensors. *IEEE Circuits and Devices Magazine* 21(3): 6–20.

Grindel, M. W. Collimation testers: Testing collimation using shearing interferometry. http://www.oceanoptics.com/Products/ctcollimationtesterarticle.asp (accessed on June 18, 2014).

Guntupalli, R. and R. Allen. 2006. Evaluation of InGaAs camera for scientific near infrared imaging applications. *Proceedings of SPIE* 6294:629401.1–629401.7.

Henry, C., A. Juvigny, and R. Serradeil. 1998. High resolution detection sub-assembly of the SPOT camera: On-orbit results and future developments. *Acta Astronautica* 17(5): 545–551.

Herve, D., G. Coste, G. Corlay, et al. 1995. SPOT 4's HRVIR and vegetation SWIR cameras. *Proceedings of SPIE* 2552: 833–842.

Hugon, X., O. Amore, S. Cortial, C. Lenoble, and M. Villard. 1995. Near-room operating temperature SWIR InGaAs detectors. *Proceedings of SPIE* 2552: 738–747.

Iyengar, V. S., C. M. Nagrani, R. K Dave, B. V. Aradhye, K. Nagachenchaiah, and A. S. K. Kumar. 1999. Meteorological imaging instruments on-board INSAT-2E. *Current Science* 76: 1436–1443.

Janesick, R. J. 2001. Scientific Charge-Coupled Devices. ebooks.spiedigitallibrary.org (accessed on May 12, 2014).

Jhabvala, M., K. Choi, A. Waczynski, et al. 2011. Performance of the QWIP focal plane arrays for NASA's Landsat Data Continuity Mission. *Proceedings of SPIE* 8012: 80120Q.1–80120Q14.

Joseph, G. 2005. *Fundamentals of Remote Sensing*, 2nd edition, Universities Press (India) Pvt Ltd, Hyderabad, India.

Joseph, G., V. S. Iyengar, R. Rattan, et al. 1996. Cameras for Indian Remote Sensing satellite IRS 1C. *Current Science* 70(7): 510–515.

Kevin Ng. Technology review of charge-coupled device and CMOS based electronic imagers. http://educypedia.karadimov.info/library/ng_CCD.pdf (accessed on April 24, 2014).

Knight, E. J., B. Canova, E. Donley, G. Kvaran, and K. Lee. 2011. The Operational Land Imager: Overview and Performance. http://calval.cr.usgs.gov/JACIE_files/JACIE11/Presentations/TuePM/325_Knight_JACIE_11.070.pdf (accessed on May 14, 2014).

Krishnaswamy, M., P. M. Varghese, M. Y. S. Prasad, S. K. Sam, and P. Pandian. 1995. Payload steering mechanism for IRS-IC. *Journal of Spacecraft Technology* 5(2): 142–145.

Kumar, K. 2013. Indian payload capabilities for space missions. *Proceedings of the International ASTROD Symposium*, July 11–13, Bangalore, India. http://www.rri.res.in/ASTROD/ASTROD5-Wed/Kirankumar_Indian-payload.pdf (accessed on June 22, 2014).

L8 EO Portal. directory.eoportal.org/web/eoportal/satellite-missions/l/landsat-8-ldcm (accessed on April 8, 2014).

Leger, D., F. Viallefont, E. Hillairet, and A. Meygret. 2003. In-flight refocusing and MTF assessment of SPOT-5 HRG and HRS cameras. *Proceedings of SPIE* 4881: 224–231.

Lightsey, P. A., A. A. Barto, and J. Contreras. 2004. Optical performance for the James Webb Space Telescope. *Proceedings of SPIE* 5487: 825–832.

Lindahl, A., W. Burmester, K. Malone, et al. 2011. Summary of the operational land imager focal plane array for the Landsat data continuity mission. *Proceedings of SPIE* 8155: 81550Y.1–81550Y.14.

Majumder, K. L., R. Ramakrishnan, I. C. Matieda, G. Sharma, A. K. S. Gopalan, and D. S. Kamat. 1983. Selection of spectral bands for Indian remote sensing satellite (IRS). *Advances in Space Research* 3(2): 283–286.

Markham, B. L., P. W. Dabney, J. C. Storey, et al. 2008. Landsat data continuity mission calibration and validation. www.asprs.org/a/publications/proceedings/pecora17/0023.pdf (accessed on May 14, 2014).

McKee, G., S. Pal, H. Seth, A. Bhardwaj, and H. S. Sahoo. 2007. Design and characterization of a large area uniform radiance source for calibration of a remote sensing imaging system. *Proceedings of SPIE* 6677: 667706.1–667706.9.

Mehta S., K. Bera, V. D. Patel, A. R. Chowdhury, and D. R. M. Samudraiah. 2006. Low noise high speed camera electronics for Cartosat-1 imaging system. *Journal of Spacecraft Technology* 16(2): 35–46.

Meygret, A. and D. Leger. 1996. In-flight refocusing of the SPOT1 HRV cameras. *Proceedings of SPIE* 2758: 298–307.

Midan, J. P. 1986. The SPOT-HRV instrument: An overview of design and performance. *Earth-Oriented Applications of Space Technology* 6(2): 163–172.

Moy, J. P., J. J. Chabbal, S. Chaussat, J. Veyrier, and M. Villard. 1986. Buttable arrays of 300 multiplexed InGaAs photodiodes for SWIR imaging. *Proceedings of SPIE* 686: 93–95.

NASA. 2006. Earth's Living Ocean: The Unseen World. An advanced plan for NASA's Ocean Biology and Biogeochemistry Research. http://oceancolor.gsfc.nasa.gov/DOCS/OBB_Report_5.12.2008.pdf (accessed on May 14, 2014).

Navalgund, R. R. and R. P. Singh. 2010. The evolution of the Earth observation system in India. *Journal of the Indian Institute of Science* 90(4): 471–488.

NRSA. 2003. *IRS-P6 Data User's Handbook*. IRS-P6/NRSA/NDC/HB-10/03.

NRSC. 2011. *Resourcesat-2 Data Users' Handbook*. NRSC:SDAPSA:NDC:DEC11-364.

Oberheuser, J. H. 1975. Optical concept generation for the synchronous Earth observatory satellite. *OpticalEngineering* 14(4): 295–304.

Pandya, M. R., K. R. Murali, A. S. Kirankumar. 2013. Quantification and comparison of spectral characteristics of sensors on board Resourcesat-1 and Resourcesat-2 satellites. *Remote Sensing Letters* 4(3): 306–314.

Patel, V. D., S. Bhati, S. Paul, et al. 2012. 3D packaged camera head for space use, *Proceedings of the IEEE First International Symposium on Physics and Technology of Sensors (ISPTS)*: 63–66.

Paul, S., H. Dave, C. Dewan, et al. 2006. LISS-4 camera for Resourcesat. *Proceedings of SPIE* 6405:640510.1–645010.8.

Puschell, J. J., L. Cook, Y. J. Shaham, M. D. Makowski, J. F. Silny. 2008. System engineering studies for advanced geosynchronous remote sensors: Some initial thoughts on the 4th generation. *Proceedings of SPIE* 7087: 70870G1–70870G18.

Rao, V. M., J. P. Gupta, R. Rattan, and K. Thyagarajan. 2006. RESOURCESAT-2: A mission for Earth resources management. *Proceedings of SPIE* 6407: 64070L.1–64070L.8.

Reuter, D. 2009. Thermal Infrared Sensor: TIRS: Design and Status. Landsat Science Team Meeting. June 22–24, 2009, http://landsat.usgs.gov/documents/7_Reuter_TIRS_Status.pdf (accessed on May 14, 2014).

Spotimage. 2002. *Spot Satellite Geometry Handbook*. Spotimage document. S-NT-73-12-SI. Edition 1, Revision 0.

Tamilarasan, V., S. K. Sharma, and S. R. Nagabhushana. 1983. Optimum spectral bands for land cover discrimination. *Advances in Space Research* 3(2): 287–290.

Texas Instruments. 1997. CMOS Power Consumption and CPD Calculation. www.ti.com/lit/an/scaa035b/scaa035b.pdf (accessed on May 14, 2014).

Thome, K., A. Lunsfordb, M. Montanaroc, et al. 2011. Calibration plan for the thermal infrared sensor on the landsat data continuity mission. *Proceedings of SPIE* 8048: 804813.1–804813.9.

Vaillon, L. 2010. Geo-Oculus: High Resolution Multi-spectral Earth Imaging Mission from Geostationary Orbit. *Presented at ICSO 2010 8th International Conference on Space Optics Rhodes* http://www.congrexprojects.com/custom/icso/Presentations%20Done/Session%209b/04_ICSO2010_GeoOculus.pdf (accessed on April, 24, 2014).

Weaver, W. L. 2013. *A decade of innovation. Technology from the first decade of the 21st century*. ISBN: 978-1-4689-2200-4 (eBook).

Weimer, P. K., G. Sadasiv, J. E. Meyer, Jr., L. Meray-Horvath, and W. S. Pike. 1967. A self-scanned solid-state image sensor. *Proceedings of the IEEE* 55(9): 1599–1602.

Westin, T. 1992. Interior orientation of spot imagery. http://www.isprs.org/proceedings/xxix/congress/part1/193_XXIX-part1.pdf (accessed on April 25, 2014).

Wey, H. M. and W. Guggenbuhl. 1990. An improved correlated double sampling circuit for low noise charge-coupled devices. *IEEE Transactions on Circuits and Systems* 37(12): 1559–1565.

Young, P. J. 1975. Scanning system tradeoffs for remote optical sensing from geosynchronous orbit. *Optical Engineering* 14(4): 289–294.

7

Submeter Imaging

7.1 Introduction

The first multispectral imagery of Earth from satellite, specifically designed for Earth resources surveys, started with Landsat 1, which had a spatial resolution of approximately 40 m from the Return Beam Vidicon (RBV) camera and of about 80 m from the optomechanical scanner multi-spectral scanner system (MSS). Since then, to cater to the increasing needs of the user community, efforts have been made to provide space imagery with better and better performance capabilities. One area where consistent efforts are being put in is to generate images with improved spatial resolution. The Thematic Mapper launched in 1982 had a resolution of 30 m. The launch of SPOT in 1986 gave panchromatic (PAN) imagery with 10-m spatial resolution, whereas the Indian Remote Sensing Satellite (IRS) 1C launched in 1995 produced PAN imagery with about 6-m resolution. However, during this period, highly classified spy satellites have been operating with spatial resolution of a few tens of centimeters resolution. These Earth observation satellites were funded by their respective governments. The U.S. policy of allowing private entrepreneurs to launch and operate satellites with resolution up to 1 m opened a new era in Earth observation. The first such class of satellite was IKONOS launched in 1999. The IKONOS could provide PAN imagery with a spatial resolution of 1 m and 4-band multispectral (B, G, R, near infrared [NIR]) imagery with 4-m spatial resolution. Thus the gap between civilian and military remote sensing capabilities started narrowing. The major improvement has been for the PAN band. Figure 7.1 gives the improvement of spatial resolution of the PAN channel since the launch of SPOT-1. In this chapter, we shall discuss the design considerations and challenges for realizing spaceborne cameras with a resolution of 1 m or better, which we shall refer to as a *high-resolution* system.

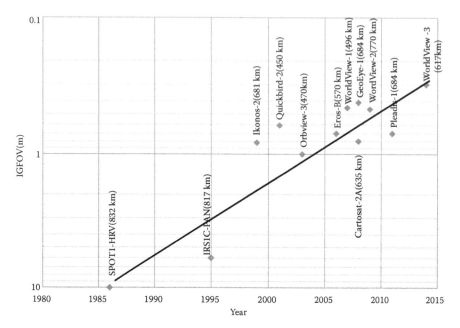

FIGURE 7.1
Improvement of spatial resolution of panchromatic band since the launch of SPOT-1. The figures in the parentheses give the altitude of the satellite. The list is not exhaustive.

7.2   Considerations for Realizing a High-Resolution Imaging System

The basic question is, as resolution increases how do we realize an instrument that collects adequate radiance to provide the desired signal-to-noise ratio (SNR)? For continuity, we shall repeat what we discussed in Chapter 2, Section 2.8. If we observe a target for a time τ seconds (integration time/dwell time), then the on-axis energy received by the detector is given by

$$Q_d = \frac{\pi}{4} O_e \, \Delta\lambda \, L_\lambda \, \beta^2 \, D^2 \, \tau \text{ joules} \tag{7.1}$$

where L_λ is the target radiance (W·m^{-2}·sr^{-1}·µm^{-1}); $\Delta\lambda$ is the spectral bandwidth of the radiation to be measured (µm); O_e is the optical efficiency—transmittance of the optical system including atmosphere (<1); if β radian is the instantaneous field of view (IFOV) then the solid angle subtended by the detector at the lens—β^2 steradian—can be expressed as A_d/f^2, where A_d is the detector area and f is the effective focal length (EFL) of the imaging optics; and D is the effective aperture diameter of the collecting optics.

Substituting for β^2 in terms of focal length, we can rewrite Equation 7.1 as

$$Q_d = \frac{\pi}{4} O_e \, \Delta\lambda \, L_\lambda \frac{A_d}{f^2} \, D^2 \tau \text{ joules}$$

$$Q_d = \frac{\pi}{4} O_e \, \Delta\lambda \, L_\lambda A_d \frac{D^2}{f^2} \tau \text{ joules}$$

As discussed in Chapter 2, Section 2.2, f/D is f-number (F/#). Equation 7.1 can be rewritten as

$$Q_d = \frac{\pi}{4} O_e \, \Delta\lambda \, L_\lambda A_d \frac{1}{(F/\#)^2} \tau \text{ joules} \tag{7.2}$$

The energy Q_d received by the detector can be increased by using optics with lower F/# and increasing the integration time τ. That is, we can increase the light-gathering capacity of a telescope by using a faster optics or increasing the amount of time the telescope is pointed at an object.

7.3 Increasing the Integration Time

The submeter class imaging systems generally operate from an orbital height of around 650 km. From this altitude, the subsatellite ground speed is about 7 km/sec, which gives a dwell time of about 0.14 millisecond for 1-m pixel resolution. Even with a reasonable increase in size of the optical system, such a short integration time cannot generate an adequate signal to produce imagery with the required radiometric quality. Therefore, it is essential to resort to techniques that can increase the dwell time. This can be done in two ways. One method is to observe the same pixel a number of times and add the signal generated in each observation to produce the final signal. This technique is called *Time Delay and Integration* (TDI). This technique has been used in IKONOS. In the second technique, the camera is forced to *stare* at the same pixel for a longer duration by appropriately tilting the camera's optical axis. This reduces the effective ground velocity as seen by the imaging system. This technique is used in the IRS Technology Experiment Satellite (TES). Another technique, though not very common, is to use two charge-coupled device (CCD) linear arrays displaced in the along-track and across-track directions by 0.5 pixel. In such an arrangement, it is possible by interpolation to generate imagery that has a factor of 2 higher resolution (instantaneous geometric field-of-view [IGFOV] halved) than that generated during the integration time by each array. We shall now briefly discuss these techniques and their merits and limitations.

7.3.1 Time Delay and Integration

As mentioned earlier, one of the techniques used to increase the effective dwell time for high-resolution imaging is called TDI (Barbe 1976). The TDI mode uses a two-dimensional array instead of the single linear array used in conventional pushbroom cameras. Consider an array consisting of M pixels (across track) and N rows (along track). The charge accumulated in each row is transferred to the successive lines at a rate exactly compensating the image motion, thereby increasing the effective dwell time N times. The TDI technique thus enables the integration of charges for the same objects through multiple lines positioned along track (Figure 7.2). The signal increases linearly, while, in general, the noise adds incoherently, thereby producing an improvement of SNR by a factor of $N^{1/2}$. IKONOS is the first civilian remote sensing satellite system to use TDI for generating high-resolution imagery. The number of CCD stages depends on the *well* capacity of the CCD, that is, how many electrons can be stored in one photo site without saturation. QuickBird, OrbView-3, and Formosat-2 are some of the other satellites that employ this technique for

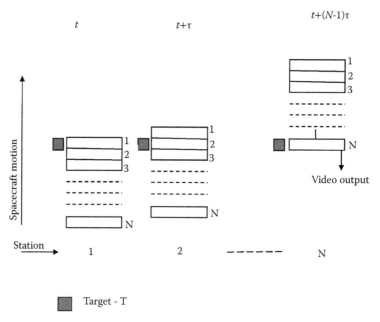

FIGURE 7.2
Schematic showing concept of Time Delay and Integration. At time t (station 1), target "T" is seen by the first array. When the sensor has moved through a distance corresponding to one integration time τ, at station 2 the target is seen by the second array (2) and the information in array "1" is transferred to array "2." This process continues till the last array N is reached after $(N-1)$ transfers and data is read out serially from the Nth array. (Reproduced with permission from Joseph, G., *Fundamentals of Remote Sensing*, 2005, Universities Press (India) Pvt. Ltd., Hyderabad, Telangana, India.)

imaging (Jacobsen 2005). While TDI mode effectively increases the dwell time, the full performance advantage can be achieved only if the charge transfer from each line to the next is in exact synchronization with image motion, that is, all the TDI stages should see the same ground resolution elements during one stage of operation. However, in practice, all the TDI stages cannot see the same ground scenes due to the spacecraft attitude perturbation, drift, and so on, which essentially reduces the overall modulation transfer function (MTF) of the image. The effect becomes more serious as the number of TDI stages increases. Therefore, increasing the number of TDI stages to improve SNR, without the required attitude stability, will be counterproductive.

7.3.2 Asynchronous Imaging

In the conventional pushbroom technique, when the image is taken along track, the optical axis is along the radius vector of the satellite orbit (Figure 7.3a). The signal is usually integrated for a duration equal to the time taken by the subsatellite point to move through one resolution element, that is, the dwell time. We may name this mode of operation *synchronous mode*, because the data is collected synchronously with the satellite motion (or the optical axis moves synchronously with subsatellite velocity). If we can reduce the relative velocity between the optical axis and the ground, we can increase the dwell time. In practice, this can be achieved by continuously tilting the optical axis, so that its projection on the ground moves in the opposite direction with respect to the subsatellite velocity vector, such that the camera *stares* at a pixel for a longer time, that is, the ground scanning velocity is different than the satellite's ground velocity. Because this mode is asynchronous to the subsatellite velocity, it is referred to as the *asynchronous* mode or *step and stare* mode of operation. In Figure 7.3b, initially the subsatellite point is at A. The optical axis is tilted such that its projection on the Earth is at C. Now, as the satellite moves from A to B, the optical axis is continuously moved, opposite to the subsatellite velocity vector such that as the subsatellite point travels through the arc length AB, the optical axis projection only travels through CD. Thus a relative velocity reduction of CD/AB is achieved. The increase in dwell time not only allows a better SNR, but also produces a lower data rate. However, this mode of operation has its own limitations. During a satellite pass, we can collect data only for a shorter strip compared to the classical mode of pushbroom operation. This mode produces an additional geometric distortion because the viewing angle of the camera continuously changes during imaging. At the beginning of the imaging, the IGFOV is large, becoming minimum as the distance between the satellite and ground becomes minimum (usually around the center of the imaging area), and then increases again toward the end of the imaging process. In those cases, when the scanning starts and finishes at the same scan angle, the IGFOV size varies as "$x \sec\Phi$" across track and "$x \sec^2\Phi$" along track (Purdue University 2014), where x is

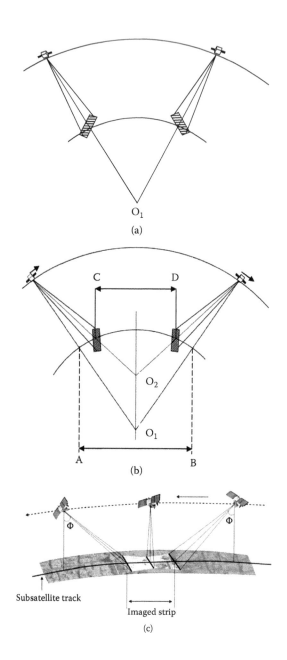

FIGURE 7.3
Schematics showing synchronous (a) and asynchronous (b) modes of imaging for a push-broom sensor. In the synchronous operation, the optical axis continuously *looks* at the center of the Earth. In the asynchronous mode, the optical axis is continuously tilted such that it always points toward a point that is at a certain distance above the center of the Earth. While the sub-satellite point moves through a distance AB, the optical axis traverses only through a distance CD. AB/CD gives the improvement in the dwell time. (c) Geometric distortion produced by asynchronous scanning. (Adapted from https://engineering.purdue.edu/~bethel/eros_orbit3. pdf, Purdue University, accessed on May 15, 2014.)

the nadir IGFOV and Φ the view angle (Figure 7.3c). The geometric distortion produced is corrected during data product generation. The EROSA1 launched in December 2000 successfully demonstrated this concept.

7.3.3 Staggered Array Configuration

A staggered line array consists of two identical CCD linear arrays with one array shifted half a pixel with respect to the other in the along-track and across-track directions. Using the data from the two CCDs, in principle, it is possible to interpolate and produce imagery as good as for a nonstaggered line with half pixel size (Jahn and Reulke 2000). Thus, this technique effectively reduces the pixel size without affecting the limitations of a smaller size pixel array. The basic assumption behind this technique is that the data in a CCD pushbroom camera is not sampled at a rate satisfying Shannon's criteria. According to Shannon's sampling criteria, to recover the signal from its samples exactly, it is necessary to sample at a rate at least twice the highest frequency present in the signal. Because of diffraction, all optical systems act as a low-pass filter. The highest frequency of a diffraction-limited system is given by

$$\frac{1}{\lambda(F\,/\,\#)} \text{ cycles/millimeter}$$

where λ is the wavelength expressed in millimeters and F/# is the f-number of the optics. We shall explain this by taking the example of the SPOT 5 imaging system where the staggered array technique was first applied (Latry and Rougé 2003). The SPOT telescope has an F/# of 3.5, and the corresponding cutoff frequency at 0.6 μm wavelength is ≈500 cycles/mm. That is, if we assume that the optics is diffraction limited, the focal plane can have a frequency component extending up to 500 cycles/mm (in practice the value is lower depending on the aberrations). The imagery is generated by sampling the focal plane by the detector array. Sampling is the process of measuring the value of a continuous function at regular intervals. For a CCD with a pitch of x mm, the sampling is done at intervals of $1/x$ times/mm or we may say that the sampling frequency is $1/2x$ line pairs/mm. For the 6.5 μm pitch of the CCD used in SPOT HRG, this works out to be 77 line pairs/mm. Thus the information content at the focal plane is not fully extracted. In a pushbroom system, the data in the along-track direction can be sampled at any rate consistent with the sensor sensitivity required to achieve the desired SNR. However, as the CCD is a *discrete* sampling device, in the across-track direction the sampling rate is defined by the detector size and cannot be increased with a single device. To improve the resolution, two linear arrays are used displaced with each other by 0.5 pixel in the along-track and across-track directions and the data is interpolated to provide a resolution a factor of two better than that produced by each CCD strip. Theoretically, the resolution is as good as for a nonstaggered line with half the pixel size. However, the slightly different viewing angle of both the lines of a staggered array can lead to deteriorations because of platform motion and attitude fluctuations and nonflat terrain (Reulke et al. 2004).

A new concept called *supermode*, patented by the French space agency CNES, is adopted in SPOT 5. Here two identical CCD arrays are displaced 0.5 pixel in the across-track direction and 3.5 pixels (corrected to 0.5 pixel during ground processing) in the along-track direction (Figure 7.4). The two 5-m images generated by the two CCDs are processed independently by the on-board system and transmitted. Processing these shifted images to obtain the final 2.5-m image involves specialized image processing operations on the ground. When the shifted data from the two CCDs are interleaved, a quincunx sampling grid is generated. (A quincunx is a geometric pattern consisting of five coplanar points, four of them forming a square or rectangle, and a fifth at its center.) From this a new image with a 2.5-m sampling grid is created. With the new sampling scheme, higher frequencies are now available. However, the MTF of the instrument for these frequencies is low. The next operation, called

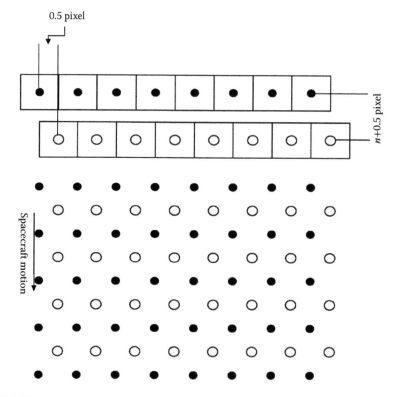

FIGURE 7.4

Schematics showing the supermode operation to improve spatial resolution. Two charge-coupled devices (CCDs) A and B are displaced 0.5 pixels in the across track and $n + 0.5$ pixels along track. As the spacecraft moves, each CCD produces independent image strips. (From Latry, C. and B. Rougé, Super resolution: quincunx sampling and fusion processing, *Proceedings of SPIE*, 4881, 189–199. Reproduced with permission from SPIE.)

deconvolution, applies a filter representing the instrument's inverse transfer function. As deconvolution amplifies noise at high frequencies in the image, in the final step a noise-removal algorithm is applied (Latry and Rouge 2003). It should be noted that to have a quincunx sampling scheme, the offset between the two shifted images should be half the sampling interval. As mentioned previously, the offset depends not only on the CCD itself, but also on the spacecraft attitude perturbations because there is a time delay due to the physical separation between the two linear arrays. Therefore, suitable registration has to be performed first so as to maintain the offset close to the 0.5 sampling interval.

This technique was first used in the SPOT 5 launched in 2002 and later by the OrbView-3 launched in 2003, wherein 1-m sampled data was produced from two staggered arrays, which have an inherent resolution of 2 m (Topan et al. 2007). In all these cases, the actual quality of images is slightly poorer than the theoretical value. The basic advantage is that one can use an optics with half the focal length compared to a sensor specifically designed to produce higher resolution, thereby making the instrument less complex and cost effective. The data to be transmitted (if compression is not used) is also halved. However, does the imagery so produced in the *supermode* will have the same quality from a sensor designed for the same resolution with a single array?

7.4 Choosing Faster Optics

A faster optics (lower F/#) can be realized by increasing diameter D of the collecting optics or by reducing the focal length. However, as the IFOV is the required sensor parameter, if we reduce the focal length, the detector pixel area also needs to be proportionally reduced to keep IFOV the same. From a system design perspective, the question is what trade-offs are available in choosing a CCD with lower pixel size.

7.4.1 Choosing Charge-Coupled Device Pixel Dimension

During IRS-1A/B, the CCD device used had a pixel size of 13×13 μm. The pixel sizes of successive generations of CCDs have decreased. Linear arrays of 2.5 μm pixels have been reported in the literature (Tatani et al. 2006; Nixon et al. 2007). We shall briefly review the implications of reducing the pixel dimension. As explained in Chapter 6, in each pixel, photons are converted into electrons. The limit to the number of electrons a pixel can hold, called the well capacity (expressed in units of electrons), among other things depends on the pixel size. When electrons in excess of the well capacity are accumulated, the excess electrons spill over to the neighboring pixels and cause a blooming artifact. However, there are techniques developed to store the accumulated electrons either in a storage gate adjacent to the photosensitive area or in an elongated pinned photodiode (Nixon et al. 2007). In general, larger pixels have higher well capacity. The well capacity decides the dynamic range of the camera. The dynamic range is the ratio of the full well capacity expressed in terms of number of electrons to total noise expressed in electrons. Though the CCD and associated circuit design tries to minimize the noise, those devices with larger pixels, which can collect more photons without saturation, will have a higher dynamic range. Dynamic range is an important parameter in an imaging system, as it decides the ability of the system to record faithfully very dim and very bright parts of a single image.

Another important consideration is the MTF degradation due to charge diffusion. Nixon et al. (2007) calculated the MTF for 650-nm illumination for 5, 3.5, and 2.5 μm pinned photodiode pixels using a theoretical model. Their investigation shows that as the pixel size is reduced, the MTF also decreases (Figure 7.5). Thus the spatial resolution benefit of a smaller pixel is negated by the degradation in MTF due to charge diffusion.

A smaller pixel size also requires the optics to operate at a higher spatial frequency. For example, a device with 13-μm pixel size as in LISS-1 has a Nyquist frequency of 38.5 cycles/mm, while for a 5-μm device it is 100 cycles per millimeter. Realizing telescope optics at higher frequency has its own fabrication complexities. The major advantage of using a device having a

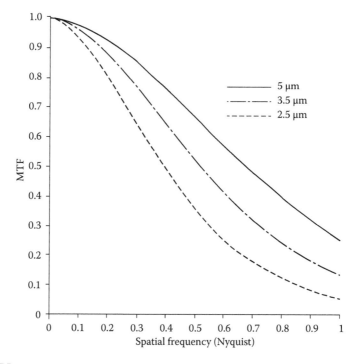

FIGURE 7.5
Charge-coupled device (CCD) horizontal modulation transfer function as a function of pixel size (From Nixon et al., 2.5 μm Pixel Linear CCD. http://www.imagesensors.org/Past%20 Workshops/2007%20Workshop/2007%20Papers/085%20O%20et%20al.pdf, 2007, accessed on May 15, 2014).

smaller pixel size is that the telescope focal length can be smaller and hence to achieve same f-number a smaller aperture can be used thereby reducing the telescope realization time and cost. The system designer has to weigh all the pros and cons and of course availability of space-quality devices while choosing the pixel size of the CCD. While choosing a CCD, one should also make sure that the speed of operation of the device is commensurate with the integration time. The charge collected at the photo site is transferred to *transport shift registers* and the data in the transport shift register has to be read out serially during one integration time. The maximum speed of the readout clock should be commensurate with this requirement. As discussed in Chapter 6, to achieve higher speed, CCDs generally have parallel transport shift registers on both sides of the photodiode array; even-numbered pixels are transferred to one register while odd pixels are connected to the other. In such a scheme, each shift register has to clock out only half the data during one integration time and the charges are transferred to two separate amplifiers. When a higher data rate is required, the transport registers on both sides can be further subdivided. For example, the 12K device used in LISS-4 has four separate shift registers on each side of the photosensitive array.

7.4.2 Increasing Collecting Optics Diameter

Irrespective of the type of telescope used, the first element—the collecting optics—has the maximum size. As seen from Table 7.1, a submeter telescope has a primary mirror size of 50 cm and above. A larger aperture telescope brings about a number of design challenges for realizing the mirror and associated mechanical system. Realizing an optical telescope requires a synergy of expertise from diverse engineering and scientific disciplines such as optics, mechanical, thermal, material, metrology, the fabricator, and so on. We have already discussed in Chapter 3 the critical issues in realizing mirrors for a spaceborne optical system as the mirror size increases and hence that is not repeated here.

7.5   Data Transmission

Demand for better spatial resolution images from remote sensing satellites increases the amount of data generated. Let us try to understand how the data generated is dependent on other sensor parameters. If there are n pixels in the linear CCD array used for imaging, the number of bits generated by one cross-track line is $n \times b$, where b is the quantization level (number of bits) of the video data. If there are m bands, the number of bits to be transmitted for the multispectral data is $m(n \times b)$. This has to be transmitted during the integration time τ. Referring to Equation 6.1, the integration time for a pushbroom scanner is given by

$$\tau = \frac{\beta}{(v / h)} \tag{7.3}$$

TABLE 7.1

Major Specifications of Some Representative Cameras Operating in 1 m and Below

		Ikonos-2 (USA)	QuickBird-2 (USA)	Orbview-3 (USA)	Eros-B (Israel)	Resurs-DK1 (Russia)	Kompsat-2 (Korea)	World View-1 (USA)	Cartosat-2A (India)	Geoeye-1 (USA)	World View-2 (USA)	Pleiades-1 (France)
Launch		1999	2001	2003	2006	2006	2006	2007	2008	2008	2009	2011
Orbit (km)		681	450	470	510	350–610	685	496	635	684	770	694
Optics	Type	TMA (Korsch)	Off-axis TMA	TMA	Cassegrain	?	R-C	?	RC+FCO	TMA	TMA	TMA (Korsch)
	Aperture (cm)	70	60	45	50	50	60	60	70	110	110	65
	EFL (m)	10	8.8	2.77	5	4	9	8.8	5.6	13.3	13.3	12.9
Focal plane	Type	TDI-32 stages	TDI-32 stages	Linear CCD	TDI-96 stages	TDI-128 stages	TDI-32 stages	TDI-64 stages	Linear 12kCCD	TDI-64 stages	TDI-64 stages	TDI-20 stages
	PAN pixel size (micron)	12	12	6×5.4	7	9	?	8	7	8	8	13
	No. of MXL channels	4	4	4	Nil	3	4	Nil	Nil	4	8	4
IGFOV PAN/MXL (meters)		0.82 /3.82	0.61 /2.44	1/4	0.7/ Nil	1/2.5 -3.5	1/4	0.45/ Nil	0.8/ Nil	0.41/ 1.65	0.46/ 1.84	0.7/2.8
Swath (km)		11	16.5	8	14	28@350 km	15	17.6	9.6	15.2	16.4	20
Quantization (bits)		11	11	11	11	10	10	11	10	11	11	12
Imaging scheme		TDI	AS	AS	S or AS	?	?	TDI	AS	TDI	TDI	TDI

S, synchronous scanning; AS, asynchronous scanning.

where β is the IFOV, v is the ground velocity of the platform, and h is the platform height.

The data rate to be transmitted is given by

$$\frac{m(n \times b)}{\tau} = \frac{m \times n \times b(\frac{v}{h})}{\beta} \qquad (7.4)$$

If Ω radians is the angular FOV, which defines the swath, then the number of pixels n can be expressed as $n = \Omega/\beta$.

The data rate to be transmitted from a pushbroom camera as given in Equation 7.4 can be expressed as

$$\frac{m \times \Omega \times b \times (v/h)}{\beta^2} \qquad (7.5)$$

This is the minimum data rate because parity and some auxiliary data are also transmitted with the data. To understand how the data rate increases as the resolution is improved, consider the following example. LISS-1 with 72.5-m resolution and 148-km swath has a data rate of 5.2 megabits per second (Mbps). Using the Equation 7.5, we find that to produce 1-m resolution with 148-km swath, the data rate will be about 27,000 Mbps. You can now appreciate why meter-class imageries generally have a limited swath.

Even *if* you are able to generate such a large volume of data during an integration period, the task is how to transmit them in real time. Space communication links are both power and bandwidth limited. Therefore, a wide variety of techniques, such as use of higher frequencies, frequency reuse, spectrally efficient modulation techniques, and so on have been investigated and adopted for efficient use of transmitted power and channel bandwidth. The details of the various techniques adopted are beyond the scope of the present book. However, it is worth understanding the frequencies used for data transmission. The use of the electromagnetic frequency spectrum for all telecommunications applications is done under the auspices of the International Telecommunication Union (ITU), which is a specialized agency of the United Nations (UN). The ITU has developed rules and guidelines called radio regulations based on extensive consultations with various stakeholders. For space-to-Earth data transmission, the frequency bands allocated by ITU are in the S (2200–2290 MHz), X (8025–8400 MHz), and Ka (25.5–27 GHz and 37.5–40.5 GHz) bands. The bandwidths available in these bands are about 90 MHz in the S band, 375 MHz in the X band, and 4500 MHz in the Ka band. At present, data transmission from remote sensing satellites is confined to the S and X bands as Ka band signals suffer from significant rain attenuation. To reduce the complexity of the data transmission system and the storage space required on board (when data collected is stored), the data is suitably compressed and transmitted/stored. On the ground, it is decompressed to get back the original data.

7.5.1 Data Compression

Images can be compressed because there is some sort of redundancy within the image. For example, when the neighboring pixels are correlated (which is the case when imaging extensive uniform terrain) instead of transmitting all the bits from both the pixels, the information could be suitably coded to reduce the number of bits. That is, we make use of spatial redundancy for compression. There could also be spectral redundancy wherein correlation happens between different spectral bands, which also could be exploited for compression. Thus in image compression, the number of bits required to represent an image is reduced by removing the redundancy present in the data. The reduction in data rate achieved through compression is specified as the compression ratio (CR), which is the ratio of compressed data rate to uncompressed data rate. There are two methods of data compression: lossless image compression and lossy image compression. In lossless compression, when decompressed, the original data can be reproduced exactly. That is, if we compare the original (before compression) digital number (DN) values and the DN values of the decompressed data, they will be identical on a pixel-by-pixel basis. However, lossless compression can provide only a modest reduction in data rate. The CR achievable for lossless compression is approximately 1.5–2 for single-band images, but can be as high as 3 or 4 for multiband images when band-to-band correlation is present (Tate 1994). In lossy image compression, a high CR can be achieved but the decompressed data will not be an exact replica of the original image. Different methods of achieving compression are available in literature (Rabbani and Jones 1991). We shall now discuss how lossy compression affects the end utilization of the remote sensing data.

Lossy compression is becoming increasingly used in remote sensing for high spatial resolution and hyperspectral data though its effect on the processing results has not yet been fully understood. The accuracy of information extraction from lossy decompressed data depends on many factors such as the type of data analysis, the algorithm used, and nature of the terrain, apart from the extent of loss of data due to compression. For example, if one is trying to extract automatic linear feature extraction from PAN data, one may not find any significant effects of lossy compression. A number of investigators have studied the effects of lossy compression on the accuracy of information retrieval. Mittal et al. (1999) investigated how compression affects the achievable precision of automated procedures like automated conjugate points matching and digital elevation model (DEM) extraction from satellite stereo pairs. IRS-1C stereo pair data having a high relief was used to study the effects of various CRs starting from 1:2 to 1:25 for a standard JPEG and wavelet-based set partitioning in hierarchical trees (SPIHT) coding on the accuracy of DEM generation. Their study concluded that DEMs derived from the space imageries are not affected much by compression even at CRs of

1:25. There are a number of similar studies on the effect of compression on DEM extraction (Lam et al. 2001; Shih and Liu 2005; Liang et al. 2008).

However, lossy compression affects the accuracy of classification using multispectral imagery. In lossy compression, for decompressed multi-spectral data, the deviation of DN value need not be at the same location for each band. Thus if multispectral classification is carried out using per-pixel classifiers, the accuracy of feature extraction decreases. Paola and Schowengerdt (1995) have compressed four remotely sensed multispectral images to varying degrees and investigated the resulting supervised clas-sifications obtained by the minimum-distance (MD), maximum-likelihood (ML), and three-layer back propagation neural network classifiers. Figure 7.6 gives the performance of the classifiers when both training and classification are carried out on the images that have undergone compression. The devia-tion from the classification of the original data, in general, decreases as the CR increases and the effect is greatest for the maximum-likelihood classifier. Though the impact of lossy compression on the images themselves as well as on the obtained classification has been studied by a number of investigators (Lau et al. 2003; Zabala and Pons 2011), one does not find a unique solution that can be applied for all data. This is an area where the system engineer should make a careful assessment based on the available literature or even carry out his own studies to arrive at the best compression algorithm that meets the objectives of the applications for which the camera is utilized. In summary, one has to assess for a particular application what the highest CR is beyond which a lossy compression negates the advantage of going for a higher resolution.

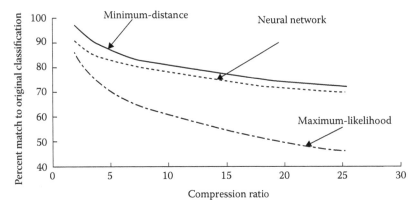

FIGURE 7.6
Performance of different types of classifier as compression ratio increases. Classification is carried out on the images that underwent compression when trained on the data that has also undergone similar compression. (From Paola, J. D. and R. A. Schowengerdt, The Effect of Lossy Image Compression on Image Classification, *NASA-CR-199550*, 1995.)

7.6   Constraints on the Satellite

The best image product from an Earth observation camera can be generated only if there is *teamwork* between the camera, the satellite, and the processing algorithm. We should know the location of each pixel on the surface of the Earth within a certain acceptable error. Therefore, the satellite trajectory (orbit) and its attitude must be known with a certain amount of precession. For higher spatial resolution imaging systems, the knowledge of the absolute value of these parameters becomes more demanding. Though the imaging payload itself can serve as a precession attitude sensor with ground control points (GCPs), this necessitates identifying a large number of GCPs and accurately determining their location. The current high-resolution satellites carry a number of sensors, to determine the orbit and attitude, such as a star sensor, precision gyros, global positioning system (GPS), and so on. Taking these inputs, the WorldView-2 (WV-2) can locate each pixel on the ground within an error circle of 5 m. Of course, higher accuracies can be obtained by using GCPs (Thomassie 2011).

In an ideal case, when the data is collected, there should not be any relative motion between the line of sight (LOS) of the camera and the scene. In classical photographic frame cameras placed in aircraft/spacecraft, this in practice is achieved by a forward motion compensation technique, wherein the film is moved in the opposite direction to exactly compensate for the forward motion of the vehicle (Kraus 2007). However, in the pushbroom mode of operation, the basic functioning is such that the camera is *exposed* during the dwell time when the spacecraft moves through the IGFOV. This relative motion reduces the MTF in the along-track direction. The effect can be reduced by using an exposure time shorter than the dwell time, or by using a detector having a dimension in the along track direction less than that required to cover the IGFOV. Both these techniques reduce the signal collected.

Apart from the relative motion we discussed previously, for any satellite in orbit, there are factors producing undesirable perturbations of the LOS of the camera. A satellite in orbit is subjected to a number of external and internal forces, which disturb the stability of the orbit and the attitude of the satellite. Some of the forces affecting the attitude stability include the changing speed of the momentum wheel, solar array/solar sail flutter, any moving mechanisms within the satellite, and so on. The camera LOS movement during the integration time effectively reduces the MTF. The LOS disturbance can be considered under two categories—the first is a low frequency component that can be considered linear motion during the integration period, producing smear. The second contribution comes from high frequency random disturbance of the LOS-jitter. Both these disturbances reduce the MTF of the system. However, they have different effects on image degradation. Figure 7.7

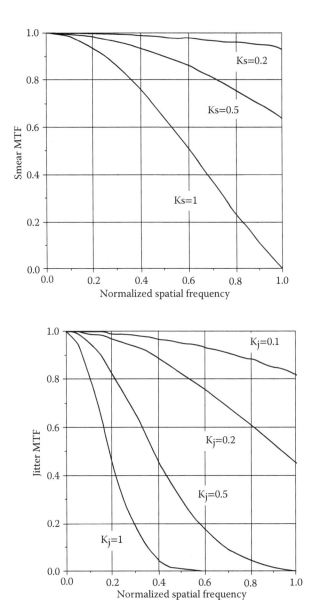

FIGURE 7.7
Smear and jitter modulation transfer functions. The relative frequency is normalized to 1/instantaneous field of view (IFOV). The relative frequency 0.5 corresponds to the Nyquist frequency. K_s is the angular change in line of sight direction during the integration interval normalized to IFOV (i.e., as a fraction of IFOV). K_j is the RMS jitter amplitude relative to IFOV. (From Auelmann, R. R., 2012 Image Quality Metrics. http://www.techarchive.org/wp-content/themes/boilerplate/largerdocs/Image%20Quality%20Metrics.pdf, accessed on May 15, 2014.)

gives the MTF degradation due to smear and jitter. As seen from the figure, the MTF is more sensitive to jitter than to smear. However, for high-resolution imaging since the exposure time is generally short, most of the disturbances appear as smear rather than jitter (Auelmann 2012). As the magnitude of the impact depends on the amplitude of the disturbance relative to IFOV, for a high-resolution system the satellite attitude stability has to be much better than required for a low-resolution system. Most of the submeter imaging systems use TDI to improve image collection. In TDI, at every stage the data is supposed to be collected by the detector pixels from the same IGFOVs as in the previous stage. However, if the LOS of camera has perturbations, this will not happen, resulting in reduction in the MTF. The reduction in the MTF increases as the number of TDI stages increases (Holst 2008).

Apart from the normal requirements of a spacecraft carrying imaging payloads, the satellites carrying high-resolution cameras need to have some additional capabilities. The high-resolution imagery usually has a swath of 10–15 km. This is primarily because as the swath increases, the number of pixels of the CCD also linearly increases and there is a technological limitation of increasing the length of the CCD device. However, this can be tackled to a certain extent by butting a number of detector assemblies. Another issue is the increase in the data rate, though this could be tackled by appropriate data compression. With a camera having a 15-km swath, in the conventional mode of operation, in one pass the satellite generates a strip of 15 km around the subsatellite track. Such a narrow swath is inadequate for many applications. In addition, there may be a number of regions of interest that may not lie on the subsatellite track necessitating *spot* images. To take care of these issues, the high-resolution satellites are designed to slew their optical axes in different off-nadir directions (Figure 7.8). The off-nadir viewing can produce imageries of a specific area of interest to the users, either as a strip image of width that of the camera swath anywhere within the field of regard or a contiguous imagery of the required width—a mode of operation that is usually referred to as the *paintbrush* mode of operation. The paintbrush mode provides a wider combined swath by imaging adjacent strips from same orbit. The slewing capability of the satellite also helps in generating same pass stereo pairs. The slewing time directly influences the imaging capacity. To maximize the possibility of selecting the required ground scenes in a pass as well as improving the stereo-imaging capacity, the satellite should have fast slew rate and quick settling time. A higher orbit also helps to get quickly to an area, because for the same off-nadir distance, the angle through which the spacecraft axis has to be moved is less at a higher altitude. The design of the attitude control system of the satellite should take these aspects into consideration for maximum utilization of the camera.

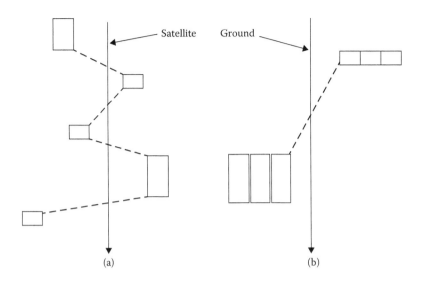

FIGURE 7.8
Imaging modes with agile satellite. (a) Any area of interest to users within the field of regard with a width of the swath of the camera; (b) Paintbrush mode with a width larger than the swath of the camera within the field of regard. The number of possible image strips in a pass depends primarily on the agility of the satellite.

7.7 Imaging Cameras with Submeter Resolution

After having deliberated on the various considerations to accomplish high-resolution cameras for Earth observation, we shall now present the technical details of some of the sensors that have been realized and successfully operated in space. What is presented does not cover all submeter imaging missions, but representative systems to understand and appreciate the technology behind achieving such instruments.

7.7.1 IKONOS

Until the launch of IKONOS in 1999, submeter resolution satellite imagery was in the domain of defense/national security applications. IKONOS was also the first commercial Earth observation satellite. The satellite's camera, designed and built by Kodak Co., could generate PAN imagery of 0.82 IGFOV and four-band 3.28 IGFOV multispectral images, with a swath of 11.3 km. The satellite has off-nadir viewing capability up to ±30° in any direction for better revisit and stereo capabilities.

The optical assembly consists of a three-mirror anastigmatic (TMA) (Korsch) telescope with a focal length of 10 m and f-number 14.3. The telescope consists of three powered mirrors with two additional flat mirrors to *fold* the optical beam to reduce the overall size of the telescope (Figure 7.9). The 70-cm-aperture primary mirror mass was reduced by cutting a honeycomb pattern into its core using abrasive water-jet technology and fusing thin mirror plates to each face (Eo Portal Directory 2014b). The focal plane consists of two TDI CCD arrays for the PAN channel, one for forward and the other for reverse-mode scanning (Baltsavias et al. 2005). Two CCD arrays with forward and reverse help to scan more images within a given time by reducing the time needed to rotate the satellite body when acquiring multiple neighboring strips (Jacobson et al. 2008). The PAN CCD has 12-μm pixel pitch and 13,500 pixels. The TDI has a maximum of 32 stages with a choice of selecting by command 10, 13, 18, 24, or 32 stages depending on the expected scene radiance. There are four multispectral arrays without TDI, each with 3375 effective pixels. In the present case, the multispectral CCDs are identical

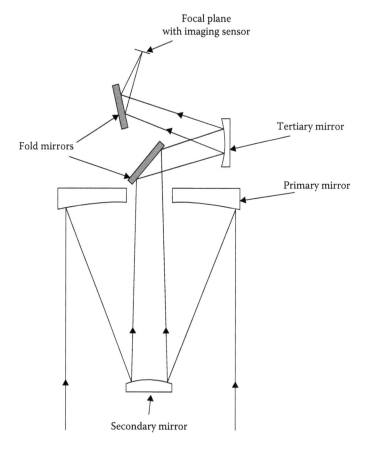

FIGURE 7.9
Schematics of the optical layout of IKONOS telescope. The primary, secondary, and tertiary mirrors are curved surfaces. The fold mirrors are used to reduce the size of telescope (not to scale). Based on a photograph at https://engineering.purdue.edu/~bethel/cams.pdf.

to the PAN CCDs with 12-µm pixels with very thin optical filters overlaid over the array. To have four times coarser resolution for the multispectral channels, a larger effective pixel size of 48 µm is achieved in the along-track direction by increasing the integration time four times and in the across-track direction by averaging (binning) four pixels (Jacobson et al. 2008) This mode of generation leads to better image quality than producing images with real 48 µm, because the along-track smear effect is reduced. Each of the two PAN and four multispectral arrays consists of three separate CCDs that are mechanically butted to form a virtual contiguous long line. The central of the three CCDs is offset from the other two in the along-track direction, while in the across-track direction, the central CCD overlaps with the other two. The overlap allows radiometric and geometric corrections when stitching the three CCD line images to produce a contiguous one line image. The data is digitized to 11 bits and compressed to 2.6 bits per pixel using adaptive differential pulse-code modulation (ADPCM) to reduce the transmission data rate. The IKONOS spacecraft is agile to orient the optical axis in the pitch and roll axis up to $\pm 30°$, which enables the satellite to have various modes of data collection such as spot image, wide swath monoscopic coverage, and same pass stereo collection.

7.7.2 QuickBird-2

The QuickBird-2 of DigitalGlobe Inc. (formerly EarthWatch) launched in 2001 with a spatial resolution of 0.61 m in the PAN band, and took the number one position at the time of launch in providing the highest spatial resolution imagery on a commercial basis. Ball Aerospace was responsible for the entire space segment including the imaging instrument.

The QuickBird camera telescope is a 60-cm aperture, wide field of view, diffraction-limited unobscured TMA design. To reduce the overall size of the telescope, a fourth mirror is used to fold the light rays. The telescope is an F/14.7 system with a focal length of 8.8 m. The primary and the secondary mirrors are off-axis, while the tertiary mirror is rotationally symmetric. The mirrors are fabricated using Zerodur. The secondary, tertiary, and fold mirrors are solid, while the primary is lightweighted by 50% (Figoski 1999). The metering structure (optical bench) to mount the optics is a lightweighted rectangular box made of graphite–cyanate ester laminate with stiffening. The bench is designed to have a first-mode frequency of 70 Hz, which is stiff enough to support the mirror weight with no need to compensate for gravity deflections. The mirrors are mounted on to the optical bench in a stress-free kinematic configuration by bonding specially designed flexures to the outside edge of each of the mirrors in three locations (Figoski 1999).

The focal plane consists of six staggered linear CCD arrays each with 32 TDI stages. TDI levels of 10, 13, 18, 24, or 32 can be selected depending on the scene radiance. The multispectral imager also uses six CCD blocks staggered in the across-track direction with each CCD block having four CCD linear arrays corresponding to each band. Spectral selection is carried out by interference filter strips aligned over each CCD block. The overlaps between the CCD blocks enable generation of a contiguous image across track. Both PAN and multispectral data is digitized to 11 bits per pixel. At the system level, the typical MTF at the Nyquist frequency is 0.17 along track and 0.21 across track (Liedtke 2002).

The QuickBird satellite was launched into a 450-km 98° Sun-synchronous orbit. The spacecraft can be maneuvered ±30° into the along-track and cross-track directions to generate imageries in various modes such as snapshot (16.5 × 16.5 km, single scene), strip map mode (16.5 × 225 km), area mosaic patterns (typically, 32 × 32 km), and same pass along-track stereo (typically, 16.5 × 16.5 km) (Eo Portal Directory 2014c).

7.7.3 GeoEye-1

When GeoEye-1 (formerly known as OrbView-5) was launched in 2008, it became the world's highest resolution commercial Earth-imaging satellite, providing imagery with a ground resolution of 0.41 m in PAN mode and 1.65 m for multispectral mode with a swath of 15.2 km. As per U.S. government requirements, PAN imagery sold to commercial customers (non-U.S. government) is resampled to 0.5-m resolution. The multispectral channel operates in blue (450–510 nm), green (510–580 nm), red (655–690 nm), and NIR (780–920 nm).

The imaging system was designed and developed by ITT Space Systems Division. The telescope optics consists of a TMA design. Two additional mirrors in the optical path reduce the overall size of the telescope. The primary mirror has a clear aperture of 1.1-m diameter and the telescope f-number is 12.1, providing a nominal focal length of 13.3 m. To provide the required swath of 15.2 km, the telescope is designed with a total field of view of 1.2° (Eo Portal Directory 2014a). The focal plane assembly consists of linear array CCD TDI detectors having 8-μm pixel size for the PAN channel and 32-μm pixel size for the multispectral channel. The data is digitized to 11 bits per pixel. The digitized data is compressed and stored on solid-state onboard recorders of 1.2 Tb capacity. The data downlink in the X band is at 740 Mb/s (or at 150 Mb/s). The data can also be provided in real time.

The satellite was launched into a Sun-synchronous orbit of 681 km altitude with an equatorial crossing time of 10:30 am. The spacecraft is highly agile, and can point along and across track up to 60° at a rate of 2.4°/s with acceleration to 0.16°/s², thereby enabling image collection in various modes including same pass stereoscopy.

The follow-on GeoEye-2 when launched will collect images with a ground resolution of 34 cm in the PAN and 1.36 m in multispectral.

7.7.4 WorldView Imaging Systems

WorldView-1, owned by DigitalGlobe and launched in 2007, carries a PAN (0.4–0.9 μm) imaging system with a spatial resolution of 0.5 m and swath of

17.7 km from an altitude of 496 km. The satellite can be pointed ±40° about nadir, thereby providing a field of regard of 775 km in cross track, giving a revisit capability of 1.7 days at 1 m ground sample distance (GSD) at 40°N latitude.

The satellite's high agility enables rapid targeting and efficient in-track stereo collection and can collect over 1 million km² per day. The follow-on to WorldView-1, the WV-2 was launched in 2009 to an altitude of 770 km, with a descending nodal crossing time of approximately 10:30 am. We shall briefly describe the design aspects of WV-2.

The WV-2 satellite launched in 2009 is the first commercial imaging satellite to collect very high spatial resolution data in nine spectral bands—eight multispectral (MXL) bands in the spectral range from 400 nm to 1050 nm at a spatial resolution of 1.84 m and a PAN band (450–800 nm) with a spatial resolution of 0.46 m. Both the PAN and MXL channels provide a swath of 16.4 km. The MXL channel contains in addition to the normal bands (red, blue, green, NIR) four new bands (coastal blue, yellow, red edge, longer wave NIR—NIR2), thereby enhancing the application potential of the WV-2 data. Table 7.2 gives the spectral bands and representative applications (not exhaustive). The imaging optics is a 1.1-m aperture TMA telescope with 13.3-m EFL. The PAN channel uses Si CCDs with 8-μm pixel size having 64 TDI stages of which 8, 16, 32, 48, 56, or 64 stages can be chosen by command. To generate the required swath, the PAN focal plane is made of 50 staggered detector

TABLE 7.2

WorldView-2 Spectral Channels and Their Application Potential

Bands	Designation	Spectral Coverage (nm)	Major Application Potential
PAN	Panchromatic	447–808	Mapping
MS1	NIR1	765–901	Vegetation types, vigor and biomass survey, delineating water bodies, soil moisture discrimination
MS2	Red	630–690	Vegetation analysis
MS3	Green	506–586	Green reflectance of vegetation, cultural feature identification
MS4	Blue	442–515	Coastal water mapping, soil/vegetation discrimination
MS5	Red edge	699–749	Vegetative condition, plant health revealed through chlorophyll production
MS6	Yellow	584–632	Turbidity analysis
MS7	Coastal blue	396–458	Bathymetric studies, atmospheric correction techniques.
MS8	NIR2	856–1043	Vegetation analysis and biomass studies. Less affected by atmospheric influence

subarrays (DSAs). The PAN array uses two separate readout registers for each of its 50 DSAs. Each readout register has its own analog-to-digital converter. The multispectral channels use 10 staggered Si CCDs with 32-μm pixel size also operating in the TDI mode with selectable levels of 3, 6, 10, 14, 18, 21, or 24 stages. To maximize the radiometric resolution while minimizing the number of saturated pixels in an image, the TDI setting is selected using a look-up table based on the estimated solar elevation angle of the image being acquired. In the focal plane, the eight multispectral channels are organized in two rows of four bands each positioned on either side of the PAN array. Each DSA contains four parallel rows of detectors, each with a different color filter. For each DSA, the individual bands are collected by a separate readout register. The data is digitized to 11 bits per pixel and transmitted at a rate of 800 Mb/s. The nominal operating configuration for WV-2 is with 2.4 CR for PAN and 3.2 CR for the multispectral data (Updike and Comp 2010). Onboard storage capacity of 2.2 Tb enables data to be recorded for future transmission.

The off-nadir viewing is accomplished by pointing the entire spacecraft. The satellite, using the Control Moment Gyro (CMG) assembly as actuators, is highly agile with a slew rate of 3.5°/s and takes only 10 seconds to slew 200 km. The maximum off-nadir viewing angle is ±45°, covering a swath of 1355 km (Satellite Imaging Corporation 2014). To enhance data collection efficiency, there is provision for bidirectional scanning. Using only data from the WV-2 star trackers, gyros, and GPS receivers, it is possible to locate each pixel in an image to less than 5 m CE90 (Anderson and Marchisio 2012). Better accuracy is possible if GCPs are used. These capabilities of the satellite increase image collection opportunities, enhance change detection applications, and enable accurate map updates.

The follow on to WV-2, WorldView-3 (WV-3) was launched in August 2014. WV-3 provides 31 cm resolution for panchromatic (PAN) channel and 1.24 m resolution for the multispectral (MXL) bands (these numbers refer to ground sampling distance [GSD] at nadir). The imaging optics is 1.1 m aperture telescope similar to WV-2 and the improvement in resolution is achieved primarily by bringing down the altitude of the satellite to 617 km compared to WV-2 altitude of 770 km. Correspondingly the swath is also reduced from 16.4 km of WV-2 to 13.1 km for WV-3. The spectral coverage of the panchromatic and multispectral channels is similar to that of the WV-2 sensor. WV-3 has additional 8 bands in the short wave IR (SWIR) covering 1.195–2.365 μm, with a GSD of 3.7 m and swath 10.8 km at nadir. The SWIR focal plane is separate from the Pan/MXL focal plane, with dedicated processing electronics. The data are digitised to 11-bits per pixel for Pan and MXL and 14-bits per pixel for the SWIR bands. WorldView-3 also carries a 12 band atmospheric sounder called CAVIS (Clouds, Aerosols, Vapours, Ice, and Snow), covering a spectral range of 0.405 μm–2.245 μm, with a spatial resolution of 30 m and swath 14.8 km at nadir. The CAVIS telescope is an all aluminum construction reflective triplet. There are two channels with the same spectral coverage (2.105–2.245 μm), to produce a parallax for height measurement. The CAVIS imager data will be used for correction of high-resolution imagery data when WV-3 images Earth objects through haze, soot, dust etc. With unique combination of spectral channels WorldView-3 offers newer opportunities for remote sensing analysis of vegetation, coastal environments, agriculture, geology, and many others.

7.7.5 Indian Remote Sensing Satellite High-Resolution Imaging Systems

The TES was the first meter-class imaging system realized by the Indian Space Research Organization (ISRO) and can be considered the forerunner of the next-generation high-resolution Earth observation system of the IRS constellation. The TES was an experimental satellite to demonstrate and validate in orbit some of the critical technologies so that they could be used in future operational satellites. The technologies demonstrated in TES included an attitude and orbit control system, high-torque reaction wheels, new reaction control system, lightweight spacecraft structure, solid-state recorder, X-band phased-array antenna, improved satellite positioning system, and miniaturized telemetry and telecommand (TTC) and power systems. The satellite carried a PAN camera operating in the 0.5–0.85 µm spectral region having 1-m IGFOV and about 15-km swath. The TES, weighing 1108 kg, was launched on October 22, 2001 from Sriharikota, India, in a 572-km Sun-synchronous orbit.

Submeter imaging systems started with the Cartosat-2 series. We shall briefly describe the Carto2A system. The camera operates in the PAN spectral band 0.5–0.85 µm providing an IGFOV of 0.8 m with a swath of 9.6 km. The optical system is a catadioptric configuration. It is essentially a Ritchey–Chretien (RC) telescope with field correcting optics (FCO) to obtain the desired flat field coverage. The primary mirror is a concave hyperboloid made of Zerodur having a clear aperture of 700 mm. The secondary mirror is a convex hyperboloid also made of Zerodur. Both the primary and secondary are weight relieved to optimize the weight with the required stiffness. The FCO is a multielement refractive system kept near the focal plane. It is designed to provide a flat field over ±0.5° field. The telescope has an EFL of 5600 mm. Specially designed mirror-fixation devices attach the mirrors to the telescope structure. Baffles strategically located at the primary and secondary mirrors and the hood at the entrance of the telescope avoid any stray light reaching the detector. The mechanical layout of the telescope is given in Figure 7.10. The camera structure is made of lightweight carbon-fiber-reinforced plastic (CFRP), with the desired dimensional stability under the required environmental condition. The detector is a 12 K linear CCD with 7-µm pixel similar to that used in LISS-4 as discussed in Chapter 6. A band-pass interference filter kept in front of the CCD decides the spectral region covered. Two CCDs are placed in the along-track direction, each having its own electronics for redundancy. The electronics are similar to that of LISS cameras (Chapter 6), but tuned to the Carto2 data requirements. The data is digitized to 10 bits and compressed to 3.2 bits per pixel for transmission. The Cartosat-2 series

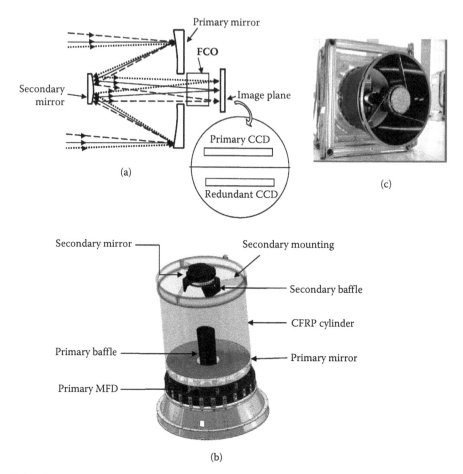

FIGURE 7.10
(a) Optical layout of Carto2 telescope. (b) Mechanical layout of Cartosat-2 telescope.
(c) Photograph of telescope assembly showing the secondary and baffle (credits SAC/ISRO).

satellites are agile to point the optical axis along and across track up to ±26° to
provide multiple spot/wide swath imageries from the same orbit. To achieve
an agile satellite, the complete spacecraft has been built around the telescope.
The major performance parameters of Cartosat-2 are given in Table 7.3.

7.8　What Limits the Spatial Resolution?

When one designs an Earth observation camera, there is one optical component on which the designer has no control! That is the intervening media—
the atmosphere—between the camera and the target. The scattering by the

TABLE 7.3

Major Performance Parameters of Cartosat-2

Parameters		Value
Optics	Type	RC
	EFL (mm)	5580.5
	f/#	F/8
Detector	Detection	Si CCD
	No. of Pixels/size (μ)	12K/7
Bandwidth (μm)		0.5–0.85
IGFOV (m)		0.8 m
Swath (km)		9.6 km
Off-nadir view		±26° (across track)
		±45° (along track)
Mean saturation radiance (mw/cm²/sr/μ)		58.9
Mean SNR at saturation radiance		233
SWR @ Nyquist		>10
Quantization (bits)		10
CR1/type		3.2:1/JPEG like

SWR, square-wave response.

aerosols and the turbulence degrades the image quality. We can attribute an atmospheric MTF associated with these processes. Turbulence is identified with fluctuations in air temperature, which in turn causes fluctuations in the refractive index of air. The random refractive index variation in the atmosphere associated with turbulence causes the refracted wave to produce directional fluctuation. Over a long exposure, image irradiance over many such small angles of arrival is integrated to form the recorded image, thus giving rise to image blur (Dror and Kopeika 1995). For astronomical observations, this increases the spot size of the image of the observed star.

The atmospheric degradation near the receiver has a much greater effect than the degradation occurring near the object. This is called the *shower curtain effect* (i.e., the effect of phase front distortions of an optical wave near the telescope is more severe than the wave front distortions of the same amplitude farther away from the telescope). Therefore the magnitude of image degradation depends on how far the observer is from turbulence; the closer to the turbulence, the larger will be the degradation. As turbulence strength is maximum near the surface, and approaches zero at very high altitude, turbulence blur can be ignored in imagery from satellites and very high elevation aircraft (N. S. Kopeika, pers. comm.).

The scattering by the atmospheric molecules and aerosols change the spatial distribution of both incident and upward-going radiation. The molecular scattering is dominant only at a lower wavelength—blue region. Therefore the limitation of attainable spatial resolution of a space camera should be based on the MTF degradation due to scattering by aerosols. This is called the *adjacency effect*. If the scattering by aerosols takes place near the camera, the light scattered at small angles can still enter the field of view and can contribute to the image formation. But if the scattering takes place far away from the camera, only small angle scattered rays enter the field of view. As the aerosol density decreases as one goes away from the Earth, the effect of aerosol MTF, as in the case of turbulence MTF, is less serious for imaging Earth from space compared to imaging upwards as in the case of astronomical observation (Kopeika 1998). Thus the atmosphere affects an observer looking up from ground to observe a star and looking at ground from space differently, the effect being less serious for the latter. Nevertheless, the aerosol scattering reduces the contrast seen through the atmosphere in most atmospheric conditions (Dor et al. 1997) thereby affecting the quality of the image taken.

To summarize, the intervening atmosphere reduces the contrast of the targets observed by an Earth observation camera. Various processes such as turbulence, aerosol scattering, absorption, and the atmospheric background radiation contribute to atmospheric MTF, which has an influence on the quality of the image generated.

Though there are a number of theoretical and experimental investigations on atmospheric MTF (Kopeika et al. 1998), they do not provide a direct answer to our problem, that is, what is the limiting resolution when imaging from space, looking down through atmosphere? Fried (1966) gives a partial answer to the problem by considering a point source on the ground and investigating the resolution with which it can be imaged under typical atmospheric turbulence. As one observes the ground from the sky, the size of the smallest detail resolvable increases with increasing altitude until about 10 km; beyond that the value reaches an asymptotic limit. Using a large-enough optics and no other limiting factors, this asymptotic value has been shown to be about 5 cm. More recent estimates indicate significantly lower optical strength of turbulence and correspondingly smaller values for the limiting resolution (D. L. Fried, pers. comm.). The exact value depends on what model for the strength of turbulence distribution one considers is applicable. That is, the impact of turbulence will be felt only when one goes to resolution below about 5 cm. This could be considered as the limiting resolution when optical depth is very low. However, as the optical depth increases, the aerosol scattering takes place and could further degrade the limiting resolution. As discussed previously, the effect of aerosol on the limiting resolution depends on the probability of a photon being scattered as it passes through the atmosphere before reaching the detector. The scattering reduces the contrast between two regions, thereby reducing the contrast in the imagery.

With increasing demand for getting better and better spatial resolution from space imagery, it is hoped that further work will be carried out to establish the limiting spatial resolution from space imagery due to intervening atmosphere.

References

Anderson, N. T. and G. B. Marchisio. 2012. WorldView-2 and the Evolution of the DigitalGlobe Remote Sensing Satellite Constellation. *Proceedings of SPIE*.8390, May 8, 2012: L1–L15.

Auelmann, R. R. 2012. Image Quality Metrics. http://www.techarchive.org/wp-content/themes/boilerplate/largerdocs/Image%20Quality%20Metrics.pdf (accessed on May 15, 2014).

Baltsavias E., Z. Li, and H. Eisenbeiss. 2005. DSM Generation and Interior Orientation Determination of IKONOS Images Using a Testfield in Switzerland. *Proceedings of ISPRS Workshop High-Resolution Earth Imaging Geospatial Inf.*, Hannover, Germany. http://www.isprs.org/publications/related/hannover05/paper/112-baltsavias.pdf (accessed on May 15, 2014).

Barbe, F. D. 1976. Time delay and integration image sensors. *Solid State Imaging*. eds. Jespers, P. et al. Noordhoff International Publishing, Leyden, MA.

Dor B. B., A. D. Devir, G. Shaviv, P. Bruscaglioni, P. Donelli, and A. Ismaelli. 1997. Atmospheric scattering effect on spatial resolution of imaging systems. *Journal of the Optical Society of America A* 14(6): 1329–1337.

Dror, I. and N. S. Kopeika. 1995. Experimental comparison of turbulence modulation transfer function and aerosol modulation transfer function through the open atmosphere. *Journal of the Optical Society of America A* 12 (5): 970–980.

Eo Portal Directory. 2014a. GeoEye-1. www.eoportal.org/directory/pres_GeoEye1-OrbView5.html (accessed on May 14, 2014).

Eo Portal Directory. 2014b. IKONOS. https://directory.eoportal.org/web/eoportal/satellite-missions/i/ikonos-2 (accessed on May 15, 2014).

Eo Portal Directory. 2014c. QuickBird. https://directory.eoportal.org/web/eoportal/satellite-missions/q/quickbird-2 (accessed on May 15, 2014).

Figoski, J. W. 1999. The QuickBird telescope: The reality of large, high-quality commercial space optics. *SPIE* 3779: 22–30.

Fried, D. L. 1966. Limiting resolution looking down through the atmosphere. *Journal of the Optical Society of America* 56(10): 1380–1384.

Holst, G. C. 2008. *Electro-Optical Imaging System Performance*, 5th Edition. SPIE Publications, Bellingham, WA.

Jacobsen, K. 2005. High resolution satellite imaging systems–overview. http://www.ipi.uni-hannover.de/uploads/tx_tkpublikationen/038-jacobsen.pdf (accessed on September 15, 2005).

Jacobsen K., E. Baltsavias, and D. Holland. 2008. Information extraction from high resolution optical satellite sensors. XXIst ISPRS Congress, Beijing, Tutorial-10. http://www.ipi.uni-hannover.de/fileadmin/institut/pdf/Tutorial10_1.pdf (Accessed on May 15, 2014).

Jahn, H. and Reulke, R. 2000. Staggered line arrays in pushbroom cameras: Theory and application. *International Archives of Photogrammetry and Remote Sensing*. Vol. XXXIII, Part B1. Amsterdam. http://www.isprs.org/proceedings/xxxiii/congress/part1/164_XXXIII-part1.pdf (accessed on May 15, 2014).

Joseph, G. 2005. *Fundamentals of Remote Sensing*. 2nd Edition. Universities Press (India) Pvt Ltd., Hyderabad, Telangana, India.

Kopeika, N. S. 1998. *A System Engineering Approach to Imaging*. SPIE Press Book, Bellingham, WA. ISBN: 9780819423771.

Kopeika N. S., D. Sadot, and I. Dror. 1998. Aerosol light scatter vs turbulence effects in image blur. *SPIE* 3219: 3097–3106.

Kraus K. 2007. *Photogrammetry: Geometry from Images and Laser Scans*. Vol. 1. Walter de Gruyter GmbH & Co., Berlin, Germany.

Lam, W. K. K., Z. L. Li, and X. X. Yuan. 2001. Effects of JPEG compression on the accuracy of digital terrain models automatically derived from digital aerial images. *The Photogrammetric Record* 17 (98): 331–342.

Latry, C. and B. Rougé. 2003. Super resolution: Quincunx sampling and fusion processing. *Proceedings of SPIE* 4881: 189–199.

Lau, W. L., Z. L. Li, and W. K. Lam. 2003. Effects of JPG compression on image classification. *International Journal of Remote Sensing* 24: 1535–1544.

Liang Z., X. Tang, and G. Zhang. 2008. Mapping oriented geometric quality assessment for remote sensing image compression. *Proceedings of SPIE*. 7146: 714610.1–714610.9.

Liedtke, J. 2002. Quickbird-2 System Description and Product Overview. JACIE Workshop, Washington DC. http://calval.cr.usgs.gov/wordpress/wp-content/uploads/16Liedtk.pdf (accessed on May 15, 2014).

Mittal, M. L., V. K. Singh, and R. Krishnan. 1999. Proceedings of Joint Workshop of ISPRS WGI/1. http://pdf.aminer.org/000/232/914/towards_a_model_relating_dtm_accuracy_to_jpeg_compression_ratio.pdf (accessed on May 15, 2014).

Nixon, O., L. Wu, M. Ledgerwood, J. Nam, and J. Huras. 2007. 2.5 μm Pixel Linear CCD. http://www.imagesensors.org/Past%20Workshops/2007%20Workshop/2007%20Papers/085%20O%20et%20al.pdf (accessed on May 15, 2014).

Paola, J. D. and R. A. Schowengerdt. 1995. The Effect of Lossy Image Compression on Image Classification. NASA-CR-199550.

Purdue University. 2014. Asynchronous imaging mode. https://engineering.purdue.edu/~bethel/eros_orbit3.pdf (accessed on May 15, 2014).

Rabbani, M. and P. W. Jones. 1991. *Digital Image Compression Techniques*. SPIE Press, Bellingham, WA.

Reulke, R., U. Tempelmann, D. Stallmann, M. Cramer, and N. Haala. 2004. Improvement of spatial resolution with staggered arrays as used in the airborne optical sensor ADS40. *Proceedings ISPRS Congress*, Istanbul, Turkey. http://www.isprs.org/proceedings/XXXV/congress/comm1/papers/22.pdf (accessed on May 15, 2014).

Satellite Imaging Corporation. 2014. http://www.satimagingcorp.com/satellite-sensors/worldview-2/ (accessed on May 29, 2014).

Shih, T. Y. and J. K. Liu. 2005. Effects of JPEG 2000 compression on automated DSM extraction: Evidence from aerial photographs. *Photogrammetric Record* 20(112): 351–365.

Tatani, K., Y. Enomoto, A. Yamamoto, T. Goto, H. Abe, and T. Hirayama. 2006. High-sensitivity 2.5-μm pixel CMOS image sensor realized using Cu interconnect layers. Proceedings of SPIE 6068: 77–85.

Tate, S. R.1994. Band ordering in lossless compression of multispectral images, *Proceedings of Data Compression Conference*, 1994, DCC '94, Snowbird, UT. IEEE: 311–320.

Thomassie, B. P. 2011. DigitalGlobe Systems and Products Overview. *10th Annual JACIE (Joint Agency Commercial Imagery Evaluation) Workshop*, March 29–31, Boulder, CO. http://calval.cr.usgs.gov/JACIE_files/JACIE11/Presentations/WedPM/405_Thomassie_JACIE_11.143.pdf (accessed on May 15, 2014).

Topan, H., G. Büyüksalih, and D. Maktav. 2007. Mapping Potential of Orbview-3 Panchromatic Image in Mountainous Urban Areas: Results of Zonguldak Test-Field. *Urban Remote Sensing Joint Event.* http://geomatik.beun.edu.tr/topan/files/2012/05/20_urs2007_zong_tam2.pdf (accessed on May 15, 2014).

Updike, T. and C. Comp. 2010. Radiometric Use of WorldView-2 imagery. Technical Note. *Digital Globe.* http://www.digitalglobe.com/sites/default/files/Radiometric_Use_of_WorldView-2_Imagery%20%281%29.pdf (accessed on May 15, 2014).

Zabala, A. and Pons, X. 2011. Effects of lossy compression on remote sensing image classification of forest areas. *International Journal of Applied Earth Observation and Geoinformation*, 13(1): 43–51.

8

Hyperspectral Imaging

8.1 Introduction

The cameras we discussed so far have spectral selection either in one broad spectral band, usually covering the visible–near infrared (NIR), which is known as the panchromatic band, or in a number of narrow spectral bands at specific thematic locations in the visible–infrared (IR) region, which is referred to as multispectral imaging. Thus, the IRS PAN camera takes imagery in 0.5–0.75 µm, whereas LISS-3 collects data in four spectral bands (0.52–0.59, 0.62–0.68, 0.77–0.86, and 1.55–1.70 µm) within the visible to shortwave infrared (SWIR) spectral region. In multispectral imaging, we are sampling the reflectance spectra of the scene at specific wavelengths. This is adequate for many applications such as identifying the surface features and for generating thematic information. However, when there are certain narrow spectral characteristics in the detected spectra the coarse spectral bandwidth (which averages the spectral response within the band) and the sampling at a few points do not bring out these characteristics. Figure 8.1 gives the reflectance spectra of the mineral halloysite (from the U.S. Geological Survey spectral library) in the 0.3 to approximately 1.8 µm spectral region. Two specific absorption bands can be seen in the spectra. The dotted line in Figure 8.1 shows how the reflectance spectra will look in the 0.52–1.7 µm region, if measured using the four bands of LISS-3. The two absorption peaks are not brought out in this case. That is, the multispectral sensors like LISS, TM, and so on undersample the spectral information content available in the reflectance spectra and, hence, the fine spectral details are missed.

The ability to detect subtle spectral features in reflected and emitted radiation from the Earth's surface and atmosphere has provided useful information to a number of remote sensing applications, from natural resource monitoring to detection of military targets. For this, we have to take imagery in a number of contiguous narrow spectral bands, which is referred to as hyperspectral imaging. Geologists/geophysicists were the first to realize the potential of hyperspectral imaging to identify the Earth's surface mineralogy by monitoring the minerals' unique spectral lines across the NIR–long-wave IR (LWIR) spectral region. The depth of the absorption band is closely related

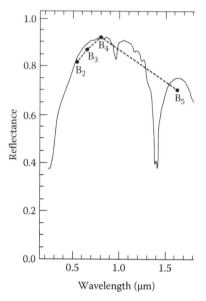

FIGURE 8.1

Reflectance spectra of the mineral halloysite. The solid curve gives the laboratory spectrometer data, and the dotted curve shows how the spectra will look like if the data are taken through IRS LISS-3. (Reproduced from Joseph, G., *Fundamentals of Remote Sensing*, Universities Press (India) Pvt Ltd., Hyderabad, India, 2005. With permission.)

to the amount of minerals in the rocks. Therefore, by locating the absorption bands and measuring their intensity the distribution of certain minerals can be identified from the imagery.

Multispectral sensors measure the radiation reflected (and/or emitted) from the Earth's surface at a few locations in the spectral region, with coarse spectral resolution, whereas hyperspectral sensors make *contiguous* measurements in the spectral region of interest in *narrow* spectral bands (Figure 8.2). The aforementioned definition of the hyperspectral sensor is still ambiguous, as it does not specify how narrow the spectral band should be. We may say that if a spectral feature has to be reproduced faithfully the sampling bandwidth should be less than one-fourth to one-sixth of the spectral width of the feature under investigation. That is, if a hyperspectral sensor has to resolve the spectral feature under investigation the spectral region should be sampled at close intervals. The actual design of a sensor takes a broader approach such that the spectral sampling exceeds that required for any particular application, so that the same sensor is useful for a wide suite of applications. However, engineering considerations limit the spectral resolution achievable, so that the data generated have a reasonable signal to noise ratio. For a spaceborne hyperspectral sensor with about a few tens of meters spatial resolution, a spectral resolution of approximately 10 nm seems reasonable and feasible. (For an aircraft platform, one talks of

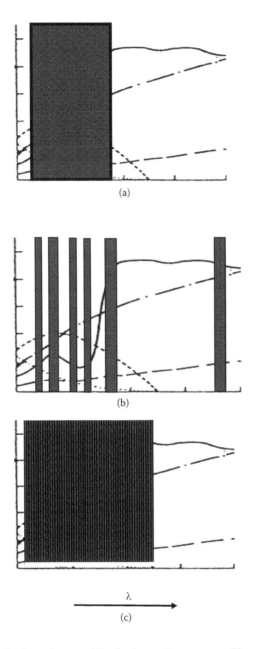

FIGURE 8.2
Different spectral selection schemes of Earth observation cameras. The spectral response of bands is shown superimposed on target scene spectra. (a) Panchromatic imaging, where the spectral band covered is a few hundreds of nanometers, usually covering part of the visible and near-infrared spectrum. (b) Multispectral imaging, where the spectral band covered is a few tens of nanometers. Location of each band depends on the thematic requirement. The width of the bands is not necessarily the same. (c) Hyperspectral imaging with hundreds of contiguous bands, each band covering a few fives of nanometers.

ultraspectral imaging [Meigs et al. 2008], where the spectral resolution is ~1 nm.) It is not the number of bands that qualifies a sensor as hyperspectral but rather the narrowness and contiguous nature of the measurements in the wavelength domain (Shippert 2004).

Hyperspectral imagers (HISs), also referred to as imaging spectrometers, used for remote sensing involve the convergence of two technologies, that is, *spectroscopy* and *imaging* from remote platforms there by providing, spatial and spectral information simultaneously. Thus, hyperspectral data give a three-dimensional organization: two spatial dimensions and one spectral dimension (Figure 8.3). In the preceding chapters, we discussed at length the principles and technology of realizing imaging systems from remote platforms. We shall now discuss how these imaging systems are modified to generate hyperspectral imagers.

8.2 Hyperspectral Imaging Configuration

In general, a hyperspectral imager consists of two parts: an imaging system and a spectrometer. In Sections 8.2.1 and 8.2.2, we give a general outline of how the optomechanical scanner and the pushbroom sensor are modified to generate hyperspectral imagery.

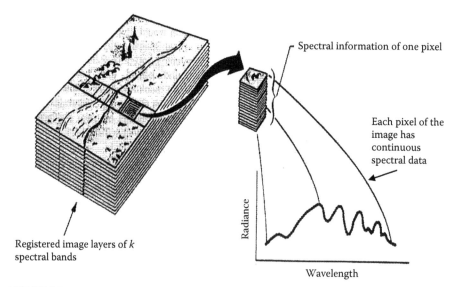

FIGURE 8.3
Schematics showing the concept of an image cube. Images taken simultaneously in 100–200 narrow spectral bands, inherently registered. Each pixel of the image cube has contiguous spectral information. (From NASA, *Earth Observation System—Instrument Panel Report-HIRIS*, Vol. 2c, 1987.)

8.2.1 Scanner Approach

Here, the image formation is similar to the optomechanical scanner we discussed in Chapter 5. An aperture at the focal plane of the imaging system defines a ground instantaneous geometric field of view (IGFOV). The collimated radiation from the slit falls onto a dispersing element like a prism or grating. The dispersed energy is focused onto a linear detector array wherein each detector element responds to different wavelength regions (Figure 8.4). That is, the detector output gives the spectral response of the pixel on the ground defined by the aperture and the number of spectral samples depends on the number of detector elements per array. A mirror scanning in the cross-track direction (like in the optomechanical scanner) provides successive pixel information in the across-track direction. In actual practice, to increase dwell time the entrance aperture at the focal plane may cover many pixels (say, n pixels)

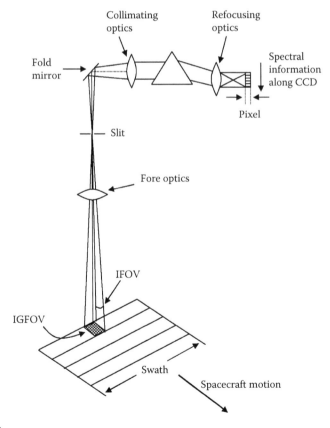

FIGURE 8.4

Concept of imaging spectrometer in optomechanical scanner configuration. The scan mirror in object space is omitted for clarity. (Reproduced from Joseph, G., *Fundamentals of Remote Sensing*, Universities Press (India) Pvt Ltd., Hyderabad, India, 2005. With permission.)

in the along-track direction and the detector will be an area array giving spatial information (of n pixels) in one direction (along track) and spectral information in the other direction.

8.2.2 Pushbroom Approach

Here, the slit at the focal plane of the imaging optics is aligned in the cross-track direction. The length of the slit defines the swath coverage (field of view [FOV]) of the sensor, whereas the width defines the along-track sampling interval, which is usually kept equal to the IGFOV. The collimated radiation from the slit falls onto a dispersing element like a prism or grating. The dispersed energy is focused onto an area array detector producing the spectral information in one direction and the spatial information in the other direction. The number of detector elements of the detector array that are aligned in the cross-track direction defines spatial resolution across track, and the number of elements in the other direction decides the number of spectral samples. The pushbroom scanning technique produces contiguous image coverage (Figure 8.5). Though no scan mirror is required to generate imagery, pointing mirrors may be used for along-track and/or across-track pointing.

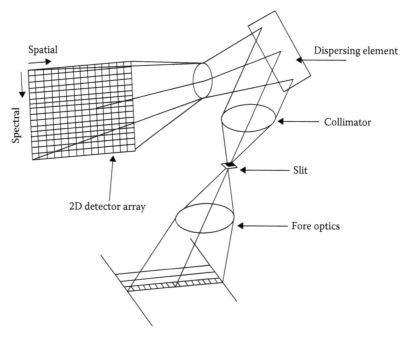

FIGURE 8.5
Imaging spectrometer concept using the pushbroom technique. (Reproduced from Joseph, G., *Fundamentals of Remote Sensing*, Universities Press (India) Pvt Ltd., Hyderabad, India, 2005. With permission.)

What distinguishes a hyperspectral imager from a multispectral camera is the addition of a spectrometer, which disperses the incoming radiation into the constituent spectrum. Therefore, an understanding of the principles of operation of spectrometers and their design considerations is important to have a full appreciation of a hyperspectral imager.

8.3 Spectrometers: An Overview

Spectroscopy is the science of studying the emission (or reflection) of the radiation emanating from a target or source to understand its properties. The beginning of spectroscopy may be attributed to Joseph von Fraunhofer who in the 1820s made careful measurements of more than 500 dark lines in the Sun's spectrum, which are now known as "Fraunhofer lines." Currently, spectroscopy is applied in various fields—from studies in planetary science to understanding molecular structure—covering different parts of the electromagnetic spectrum, each with numerous implementation techniques and applications. For the present discussion, we may consider spectroscopy as the study of the electromagnetic radiation from a target as a function of wavelength, and the instruments that carry out such measurements are spectrometers. The heart of the spectrometer is the spectral separation technique by which incoming electromagnetic radiation is divided into narrow, distinct spectral bands. One of the figures of merit of a spectrometer is the smallest difference in wavelength that can be separated, also referred to as the minimum resolvable wavelength difference—$\Delta\lambda$. As $\Delta\lambda$ can be wavelength dependent, a performance parameter that is usually used while describing a spectrometer is resolving power such that (Wolfe 1997)

$$RP = \frac{\lambda}{\Delta\lambda} \tag{8.1}$$

In spectroscopy, the spectral properties are also expressed in wave number $\tilde{v}$ (nue bar), which is the reciprocal of wavelength:

$$\tilde{v} = \frac{1}{\lambda} \tag{8.2}$$

Conceptually, wave number represents the number of waves in unit length and its SI unit is m^{-1}. The resolving power in terms of wave number is

$$RP = \frac{\tilde{v}}{\Delta\tilde{v}} \tag{8.3}$$

RP is a dimensionless quantity and has the same numerical value whether expressed in wavelength or in wave number.

The techniques of spectral separation can be broadly classified into

- Use of spectral dispersion devices
- Use of interferometers
- Use of interference filters

8.3.1 Dispersive Spectrometers

Prisms are the oldest component used in spectrometers for spectral selection. Other dispersive devices are gratings and sometimes a combination of a grating and prism (grism). The dispersion by a prism is the consequence of refraction. We saw in Chapter 2 that when electromagnetic radiation passes from one medium to another differing from the first in refractive index it undergoes angular deviation from its original direction according to Snell's law of refraction. As the refractive index varies with wavelength, for radiation falling on the prism the angle of refraction changes with the wavelengths at the entrance and exit faces and it leaves the prism with wavelength-dependent exit angle, thereby separating different wavelengths (Figure 8.6). The resolving power of a prism spectrometer is given by

$$RP = B\frac{\mathrm{d}n}{\mathrm{d}\lambda} \tag{8.4}$$

where B is the effective base of the prism, which is the maximum travel of the ray through the prism and $\frac{\mathrm{d}n}{\mathrm{d}\lambda}$ is the dispersion of the refractive index. Thus, the resolving power of the prism spectrometer is proportional to the dimension of the base of the prism and spectral dispersion of prism material.

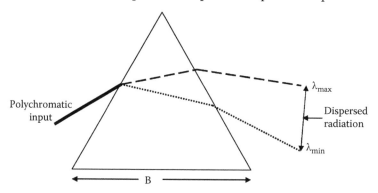

FIGURE 8.6
Schematics showing the path of polychromatic radiation through a prism.

The larger the prism base and the higher the dispersion of the material, the finer the spectral resolving power. However, in general, materials with high dispersion have high absorption, thereby reducing the throughput of the system. The wavelength region that can be covered by a prism spectrometer depends on the transmission property of the prism material. Thus, resolving power and wavelength region of operation depend on the prism material over which the designer has only limited choice. The conventional configuration of nonimaging spectrometers using a thick prism is not suited for an imaging spectrometer as the thick prism material in the optical path produces significant aberrations. However, there are a number of optical arrangements to circumvent this problem in an imaging spectrometer (Eismann 2012).

Grating is another device commonly used to separate polychromatic radiation into its constituent monochromatic components, though the principle involved is different from that of a prism. In the case of a prism the dispersion is caused by refraction, whereas in the case of grating the spectral separation is achieved by a combination of diffraction and interference. A diffraction grating consists of a number of transparent slits (apertures) on an opaque screen (for transmission grating) or reflecting grooves on a substrate (reflection grating). These periodic structures (slit/grooves) separated by a distance comparable to the wavelength of the radiation under study act as diffracting elements. The diffracted radiations from the different slits interfere in phase in certain directions where maxima of light intensity are observed. The radiation from adjacent grooves can constructively interfere if the path difference between them equals the wavelength λ or an integral multiple of λ. When the incident and diffracted rays lie in a plane that is perpendicular to the grooves, the diffraction equation is given by (Loewen and Popov 1997)

$$\sin\theta_d = \sin\theta_i + m\frac{\lambda}{s}; \; m \text{ can have values } 0, \; \pm 1, \; \pm 2, \; \pm\ldots \quad (8.5)$$

where s is the distance between the grooves, θ_i is the angle of incidence measured from the normal, θ_d is the angle of diffraction measured from the normal for a ray of wavelength λ (Figure 8.7a), and m is an integer number referred to as the order of the grating. From the aforementioned equation, it is clear that for a fixed incidence angle θ_i for the *same wavelength* there can be different angles of diffraction depending on the integer value assigned to m. These are called diffraction orders. For $m = 0$, the angle of refraction equals the angle of incidence for all wavelengths and, hence, the grating acts as a mirror and produces specular reflection without dispersion. The higher orders can happen on both sides of the normal, that is, we have both positive and negative orders.

The consequence of multiple orders is that at the focal plane the shorter wavelengths of higher order can overlap with the higher wavelength of the lower order. That is, the location of the first order of λ will also have $\frac{\lambda}{2}$ of second order ($m = 2$), $\frac{\lambda}{3}$ of third order, and so on, thereby producing spectral contamination (Figure 8.7b). Therefore, it is necessary to have suitable filters (referred to as "order sorting filters") located in the focal plane to filter out higher orders. The higher order takes away some of the radiation from the first order, which is normally used in a spectrograph.

The efficiency with which different wavelengths are diffracted is mainly determined by the groove facet angle, known as the blaze angle. The grating blaze angle can be designed to have maximum diffraction efficiency for a particular wavelength. The grating efficiency decreases at wavelengths other than the blazed wavelength. This can set a limit on the spectral coverage of the spectrometer.

The resolving power, RP, of a grating is given by the order and the number of lines in the grating (Wolfe, 1997):

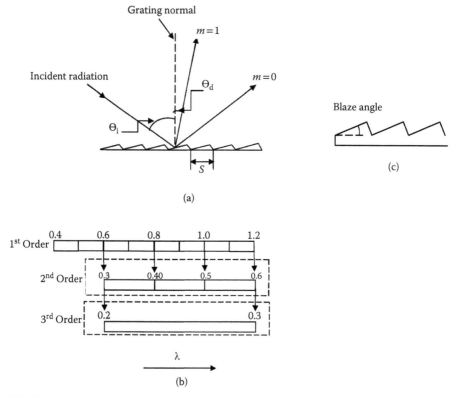

(a)

(c)

(b)

FIGURE 8.7
(a) Schematics showing dispersion by a plane grating. The grating grooves come out of the page (normal to the page), and the rays lie in the plane of the paper. For clarity, only one wavelength is shown for $m = 1$. (b) Schematics showing spectral overlap of different orders. Radiation of wavelength 0.3 µm of second order and wavelength 0.2 µm of third order are diffracted in the same direction of wavelength 0.6 µm of first order. (c) The blaze angle, the angle between the face of the groove and the plane of the grating.

$$RP = \frac{\Delta\lambda}{\lambda} = mN \qquad (8.6)$$

The resolving power depends on the order number, m, and the number of lines, N, on the grating (total number of grooves illuminated on the surface of the grating).

The gratings can be realized to operate in transmission or reflection mode. Plane reflection grating helps to reduce the system dimension by suitably steering the beam, whereas reflective grating with power (concave/convex grating) can be used as a powered component in optical design. Thus, grating can be used as an optical component to simplify overall optical design. In addition, grating has linear dispersion and broader spectral coverage compared to a prism. However, the light throughput in a grating spectrometer is much lower than that in a prism spectrometer because radiation is diffracted into different orders and the design has to take care of order overlap.

A typical prism or grating imaging spectrometer has input optics, which receives the input radiation from the target and focuses the radiation onto a slit. The slit at the focal plane defines the FOV. The slit is followed by a collimating optics and the collimated radiation falls on a dispersing element such as a grating or prism, which separates the radiation spatially into its constituent wavelength bands. The spectrum so generated is focused on a suitable two-dimensional (2D) detector array, which measures the energy of the spectrum (Figure 8.8a).

Figure 8.8b shows a practical hyperspectral imager operating in pushbroom mode. The system has a three-mirror anastigmatic (TMA) telescope as the fore optics followed by a grating spectrometer operating in Offner mode. As discussed in Chapter 3, a TMA telescope can be designed to have a large FOV in the cross-track direction. The basic Offner spectrometer has three spherical concentric reflective elements. The primary and tertiary are reflecting mirrors, whereas the secondary is a curved grating for dispersion and could be the limiting aperture of the system. This configuration has the advantage of being simple, compact, and both spatially and spectrally uniform. This combination is free from spherical aberration, coma, and distortion (Mouroulis and McKems 2000; Blanco et al. 2006). Offner convex-grating spectrometers have been used in a number of space missions such as Compact Reconnaissance Imaging Spectrometer for Mars on the Mars Reconnaissance Orbiter, ARTEMIS hyperspectral sensor for the Air Force Research Laboratory's TacSat-3 satellite, Earth Observation 1 (EO-1)/Hyperion, and so on (Silverglate and Fort 2004; Silny et al. 2010; Folkman et al. 2001).

8.3.2 Fourier Transform Spectrometers

The spectrometers using dispersive elements described in Section 8.3.1 are relatively simple in operation as they directly produce the spectral components; however, their achievable spectral resolution is not adequate for many applications. A class of instruments using principles of interferometers followed

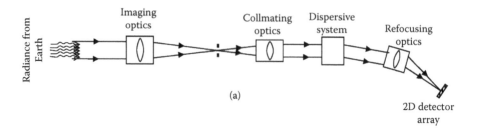

(a)

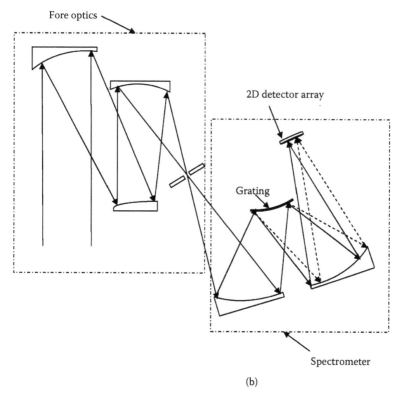

(b)

FIGURE 8.8
(a) Schematics showing arrangements of a spectrometer using prism/grating. (b) Schematics of the optical layout of a practical hyperspectral imager using grating as dispersive element operating in pushbroom mode.

by signal processing has been used as spectrometers for quite some time in the laboratory. As the primary data generated by the instruments have to be subjected to digital processing (Fourier transform [FT]) to generate spectral data, they are called Fourier transform spectrometers (FTSs). The basic principle behind all FTSs is to split the incoming beam into two coherent beams and introduce a variable optical path difference (OPD) between them. The two beams are then recombined, resulting in interference. The interference generates dark and bright patterns depending on the OPD. The resulting intensity variation with respect to OPD (referred to as an interferogram) is measured by a suitable detector system. The interferogram has information about the frequency components of the source. However, the interferogram gives information about the source spectrum in the spatial domain (path difference) and has to be converted into the frequency domain. FT is a mathematical operation that allows the interconversion of a function from one domain to another. Thus, by applying FT to the interferogram one obtains the intensity of each individual frequency (wavelength) in the original beam. The various types of interferometers primarily differ on how controlled OPD is generated between the two beams. When the variation in OPD occurs in time it is called time domain FTS (e.g., Michelson interferometer), and if the variation in OPD occurs in space it is called spatial domain FTS (e.g., Sagnac interferometer). We shall briefly describe their principles of operation.

8.3.2.1 Michelson Interferometers

The Michelson interferometer essentially consists of two plane mirrors located at right angles to each other with a beam splitter placed at a 45° angle relative to the two mirrors (Figure 8.9). The beam splitter is designed to reflect close to 50% of the incident radiation and transmit the other 50%, thereby splitting the incident radiation into two beams; one beam is reflected toward mirror M_1, and the other is transmitted toward mirror M_2. The Michelson interferometers usually incorporate a "compensator" in one of the arms using a plate optically identical to the beam splitter so that both beams have the same optical path when the mirrors are located at the same physical distance. In Figure 8.9,

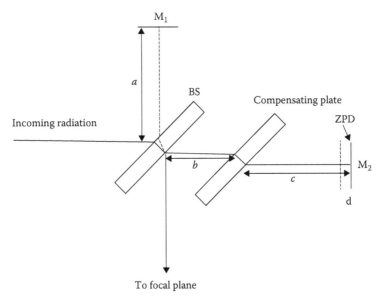

FIGURE 8.9
Optical layout of a Michelson interferometer.

we notice that after the beam is split the reflected beam to mirror M_1 travels two times through the beam splitter before combining with the transmitted beam reflected by mirror M_2. The compensating plate adds two passes for the transmitted beam, equivalent to the beam splitter traversal undergone by the reflected beam. The beams reflected back from M_1 and M_2 get recombined at the beam splitter. The two beams interfere to different degrees depending on the phase difference, Φ, between the beams produced due to the difference in the optical path between the beams reflected from M_1 and M_2. In Figure 8.9, as the reflection takes place at the bottom of the beam splitter, when $a = b + c$ the two beams have zero path difference (ZPD). This position is called the ZPD position. One of the mirrors can be moved parallel to the incident beam to produce a path difference. If mirror M_2 is moved through distance d, then the OPD produced between the reflected and transmitted beams is $2d$. The factor 2 comes as the ray travels twice through the distance d. The actual phase difference produced depends on d/λ and, hence, different wavelength components have different phase difference. When the phase difference is zero or an integral multiple of 2π there will be constructive interference, giving maximum intensity, and when the phase difference is an odd multiple of π the amplitudes of the waves cancel, that is, destructive interference. In practice, one of the mirrors is moved to produce varying OPD. We shall now present some of the performance parameters of the interferometer without getting into the actual derivation, which can be found in the work by Eismann (2012).

If the maximum displacement of the mirror from the ZPD in either direction is d, then resolving power is given by

$$RP = \frac{4d}{\lambda} \tag{8.7}$$

That is, the resolving power is directly proportional to the total OPD $4d$, from end to end over the double-sided interferogram. This is in contrast to the spectrometers using dispersive elements such as prisms and grating, wherein the optical properties of the dispersive component decide the resolving power.

In a Michelson FTS, data samples are acquired at regular intervals of mirror positions. The data sampling interval is dictated by the minimum value of wavelength to be measured. For a uniformly distributed set of samples with a differential mirror displacement δ between samples, to satisfy the Nyquist sampling criteria the minimum wavelength is given by

$$\lambda_{min} = 4\delta \tag{8.8}$$

The maximum operating wavelength is practically limited by optics transmission and detector response.

To convert the Michelson interferometer to a hyperspectral imager, a fore optics images the scene onto an intermediate image plane and a collimating optics relays the image to a Michelson interferometer. The resulting combined radiation from both the mirrors is directed to a focusing optics, which focuses the radiation onto a 2D focal plane array (FPA), like the area charge-coupled device (CCD), to capture the interferogram. As the mirror is moved, the FPA records a set of frames of the interferogram for each position of the mirror. The interferogram for a particular scene pixel has to be assembled from the series of frames. Thus, the recorded data can be considered a cube with spatial information in two dimensions and the interferogram in the third dimension (Figure 8.10). The content on each CCD element in the frame represents the interference data of a ground pixel at a particular position of the mirror. The trace of CCD element data over the set of frames represents an interferogram associated with a pixel assigned to a particular location on the ground. By Fourier transforming the signals for each pixel in a frame-to-frame direction, the spectrum can be generated for each pixel, thereby producing a hyperspectral image. One of the challenging tasks of realizing temporal FTS is the precise movement of the mirror and accurate estimation of its position.

The Thermal Emission Spectrometer (TES) and mini-TES for Mars observation are examples of Michelson-based imaging spectrometers successfully used in space missions (Christensen et al. 1992; Silverman et al. 1999). The TES interferometer carried on board Mars Global Surveyor, launched in 1966, has a 15.2-cm aperture afocal Cassegrain telescope as the fore optics. The output beam of 1.524 cm diameter after passing through the Michelson

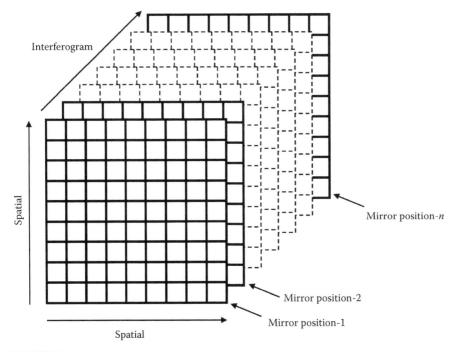

FIGURE 8.10
Interferogram generation by a temporal imaging Fourier transform spectrometer. At each mirror position, an interferogram is generated, which is recorded by the focal plane array.

interferometer is focused by an off-axis mirror onto a pyroelectric detector array. The spectrometer operates in the range of 6–50 μm with spectral sampling of 10 and 5 cm^{-1}.

8.3.2.2 Sagnac Interferometer

The Sagnac interferometer is a spatial domain FTS wherein the interference pattern is recorded simultaneously on a detector array, rather than temporally (i.e., frame by frame) as in the Michelson interferometer. The basic Sagnac interferometer also consists of two mirrors and a beam splitter like the Michelson interferometer. However, the mirrors are not orthogonal to the reflected/transmitted beam as in the case of the Michelson interferometer but have a fixed angle (<90°) between them. Such a configuration generates a triangular path for the reflected and transmitted beams from the beam splitter, and they travel in opposite directions (Figure 8.11). The reflected and transmitted rays follow exactly the same path but in opposite directions. Therefore, this type of interferometer is also referred to as a common path interferometer. If the two mirrors are placed symmetrically with respect to the beam splitter (Figure 8.11a), then the two rays, after reflection from the two mirrors, will exit the interferometer in the same direction and at the same position. If one of the mirrors is offset from the symmetric position (Figure 8.11b), then the two rays exit the interferometer in the same direction but at positions symmetrically offset from the optical axis. The path difference through the interferometer is a function of the angle of the ray entering the interferometer with respect to the optical axis but does not vary with the position of entry (Sellar et al. 2014). The rays emerging out from the beam splitter are then focused by a suitable optical system onto a 2D detector array, giving spatial

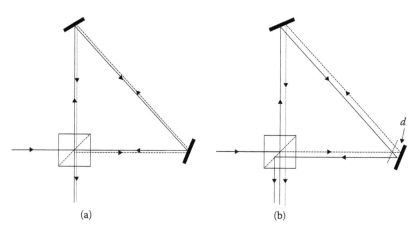

(a)   (b)

FIGURE 8.11
Optical path of a Sagnac interferometer. (a) Mirrors placed symmetrically. The reflected and transmitted rays overlap. However, in the diagram they are shown as displaced for clarity. (b) Optical path when one of the mirrors is offset by a distance d.

information in one direction and an interferogram corresponding to each pixel in the other direction (Figure 8.12). While what is presented is the basic concept, there are a number of configurations in which the counterpropagating rays within the interferometer can be realized (Sellar and Boreman 2003).

The aforementioned Sagnac interferometer uses beam splitters and air-spaced mirrors. The position of the beam at each pixel is very sensitive to the angle between the mirrors (Griffiths and Haseth 2007). For a space system, this is of concern as the instrument has to encounter various mechanical and thermal stresses. Monolithic design, incorporating the beam splitter and the reflecting mirrors in a single block, has been realized by a number of investigators (Dierking and Karim 1996; Rafert et al. 1995). In the design by Rafert et al., the interferometer is made up of two pieces (A and B) of glass that are bonded along the plane of the beam splitter coating (Figure 8.13). Such a design has very high immunity to vibration, shock, and thermal effects. The configuration also reduces the number of surfaces that could contribute to stray light, generating "ghost interferograms."

In a Sagnac imaging interferometer operating in pushbroom mode, the fore optics focuses the scene onto a slit, which defines the swath and instantaneous field of view (IFOV). The one-dimensional image is then passed through the

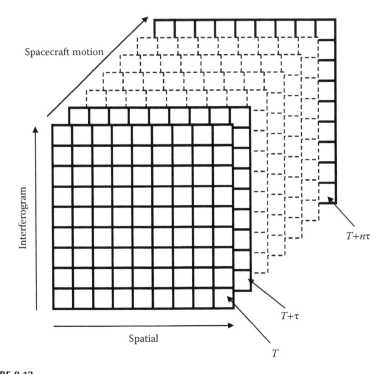

FIGURE 8.12
Organization of interferogram of a Sagnac hyperspectral camera operating in pushbroom mode.

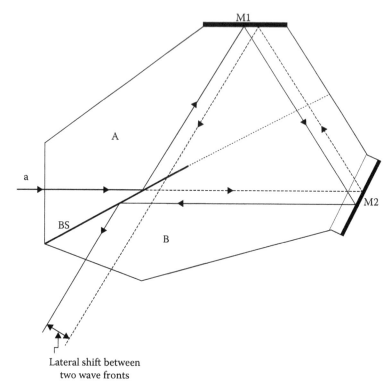

FIGURE 8.13
Schematics showing the interferometer portion of a monolithic Sagnac interferometer. The incident beam "a" is split by the beam splitter BS and reflected by mirrors M1 and M2. The beam-splitter interface extends only partially to the far edge of the interferometer assembly to facilitate the beam reflected from M2 can reach M1 unattenuated. (From J. Rafert, R. G. Sellar, and J. H. Blatt, Monolithic Fourier-transform imaging spectrometer, Appl. Opt.34, 7228–7230 (1995). With permission from OSA.)

interferometer, where the rays are split, sheared, and recombined to create an interference pattern. In a practical camera to remove the dependence on the shape and aperture of the input slit on the interference pattern, a lens is placed such that the distance between the detector plane and the lens exactly matches the focal length of the lens, referred to as the Fourier lens (Lucey et al. 1993). A cylindrical lens of appropriate focal length is kept near the Fourier lens, producing spatial information (image of the slit) in one direction and spectral information (interferogram) of each pixel in the orthogonal direction (Figure 8.13). Thus, in one integration time the data recorded in the FPA gives the interferogram of all the pixels belonging to the acquired target line. This is now similar to a dispersive spectrograph using a slit, except in the present case the spectrum is generated after FT. The spacecraft motion produces successive data frames. After carrying out Fast Fourier Transform (FFT), the spectra for each spatial pixel can be generated, thereby producing the data cube. The spectral resolution and range of a Sagnac interferometer depend on the array length and the pixel size of the detector array in the direction recording the interferogram.

Figure 8.14 shows a practical Sagnac interferometer configuration (Eismann 2012). Here, the Sagnac resides in noncollimated space. It can also reside in collimated space. In either case, it produces proper lateral shear to create the interference. One disadvantage of putting the Sagnac in noncollimated space is that any aberrations resulting from the diverging wave front propagating through the beam splitter may reduce the modulation depth of the interferogram as the interfering wave fronts will propagate through the thickness of the beam splitter differently (Michael T. Eismann, pers. comm.).

Sagnac interferometers have also been used without any input slit, referred to as the "leapfrog" configuration or as the windowing operating mode (Barducci et al. 2012).

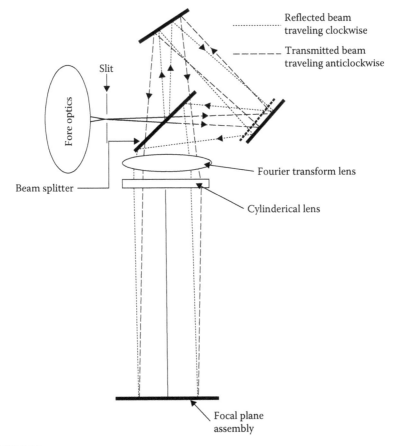

FIGURE 8.14
Schematics showing Fourier transform spectrometer using Sagnac interferometer design. (Reproduced from Eismann, M. T., *Hyperspectral Remote Sensing*, SPIE eBook series, EISBN: 9780819487889. With permission from SPIE.)

8.3.3 Filter-Based Systems

We saw in Chapters 6 and 7 that most of the multispectral scanners use interference filters for spectral selection. These are broadband filters with central wavelength at specific locations of interest. It is not practical to realize hyperspectral imagers of the required spectral resolution from a space platform by incorporating discrete band-pass filters, as a large number of filters have to be introduced in the optical path. A more convenient way is to use variable filters. A variable filter system could be realized using

1. Tunable filter systems
2. Spatially variable filter systems

The commonly used tunable filter system for space-based interferometers includes

- Acousto-optical tunable filters
- Liquid crystal tunable filters
- Fabry-Perot filters

All these select one wavelength at a time and thus basically act like a filter. Now, consider a pushbroom imager with an integration time τ; if n bands have to be selected, then each band has only $\dfrac{\tau}{n}$ time available for collection of data with corresponding degradation in signal to noise ratio. However, tunable filters could be used for imaging from geostationary platforms where integration time can be chosen as per signal collection requirements.

A practical spatially variable filter for space-based hyperspectral imaging is a wedge filter. A camera incorporating a wedge filter in the pushbroom scanning mode can provide a compact instrument. The concept of a wedge filter spectrometer was first demonstrated from an aircraft by Woody and Dermo (1994). A wedged filter is a linear variable filter (LVF) whose spectral properties vary linearly along the wedge. The wedge filter was originally designed at the Santa Barbara Research Center (U.S. Patent 4,957,371) in 1990. The filter is realized by overlaying alternate layers of dielectric films of high (H) and low (L) refractive indexes in a specific sequence. The H and L layers have a linear tapered thickness with a substantially constant slope. As the passband center wavelength for a thin-film stack depends essentially on the stack layer thicknesses, because of the tapered dielectric coatings, the center wavelength of the radiation passing through the wedge filter varies linearly along the tapered edge of the filter, with the longest wavelength at the thicker edge of the wedge (Figure 8.15a). The relative thickness of the dielectric layers and their compositions decide the spectral bandwidth and are a constant percentage of the center wavelength. In the direction parallel to the

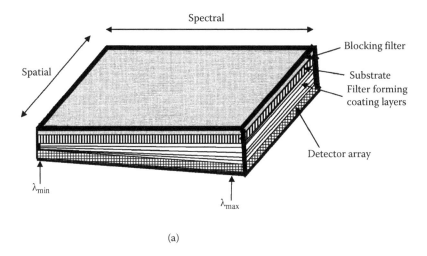

(a)

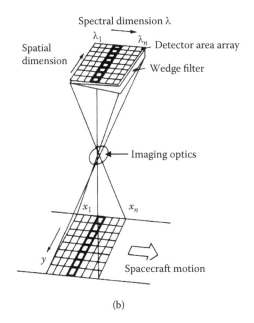

(b)

FIGURE 8.15
(a) Schematics showing wedge filter focal plane arrangement. (b) Schematics of optical layout of wedge filter spectrometer. (Reproduced from Joseph, G., *Fundamentals of Remote Sensing*, Universities Press (India) Pvt Ltd., Hyderabad, India, 2005. With permission.)

untapered edge, the center wavelength and bandwidth remains constant. To control out-of-band radiation in the wedge filter, a blocking filter whose passband covers the spectral region of interest is mounted on the opposite side of the substrate. Some of the aspects related to the filter fabrication process may be found in the work by Rosenberg et al. (1994).

To realize an imaging system, the wedge filter is coupled very close to a 2D detector array (like an area CCD array) such that the coated surface faces the array. The detector filter assembly is kept at the focal plane of the imaging optics (Figure 8.15b). Such a system when mounted on a spacecraft with the direction of the wedge along the velocity vector can generate at any instant an image of each across-track strip from x_1 to x_n (Figure 8.15b) such that each strip corresponds to a different wavelength region depending on its location at the wedge filter. Thus, we have n strips covering different regions of the scene and each corresponding to a different spectral region. As the spacecraft moves, each strip of the ground is imaged onto different positions on the wedge, thereby generating multispectral data for each strip. Thus, a frame recorded during one integration time contains all spectral bands, whereas for each ground strip only one spectral information is recorded. This is schematically shown in Figure 8.16 for six spectral channels. For clarity of drawing, let us consider only one pixel shown in the gray square. At time T, the pixel is viewed by channel λ_6. After one integration time, the same pixel is viewed by channel λ_5, and so on. For clarity, the frames are displaced horizontally. Actually, they follow the spacecraft motion. Unlike in the case of a dispersive spectrometer like a prism, where one gets the complete spectra of a pixel during one integration time, here the data have to be reorganized to produce a hyperspectral cube. As can be seen from Figure 8.16, if there are n spectral bands the first and the last band data are collected with a time difference of $(n-1) \times \tau$. As different spectral bands are not inherently registered, the attitude stability/drift of the spacecraft will influence the band-to-band registration accuracy. The main advantage of a wedge filter camera system is that one can realize an imaging spectrometer in a very compact package as there is no bulky and complex aft optics as in other hyperspectral cameras.

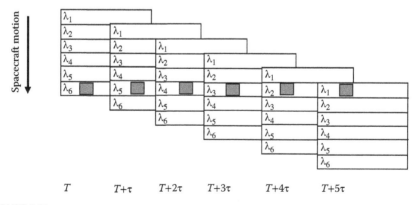

FIGURE 8.16
Schematics showing wedge filter imaging sequence. The shaded square represents a target. At time T, band 6 covers the target. (Bands 1–5 cover different strips on the ground.) During the next integration, that is, after one integration τ band 5 covers the target and the process continues. The frames are displaced horizontally for clarity.

A few other aspects are to be considered while designing a wedge filter hyperspectral camera. The bandwidth in the case of the wedge filter is a function of spot size and the variation of λ along the wedge for a given filter. Therefore, in a given configuration (i.e., for a given filter placed at a fixed distance from the focal plane) the bandwidth will depend on the F# of the optics. As the spot size increases, the bandwidth will increase; hence, with the same LVF the F/#1 system will have a larger bandwidth than the F/#4 system. Because the wedge filter is essentially an interference filter, its passband characteristics depend on the angle of incidence. Therefore, for a camera having a large FOV the imaging optics has to be telecentric. Another consideration is that the coated surface should be close to the detector array to avoid bandwidth broadening due to the finite cone angle of the incident beam.

8.4 Distortions: "Smile" and "Keystone" Effects

Two distortions associated with hyperspectral data, typically found in push-broom sensors due to optical system aberrations and misalignments, are spectral line bending (smile) in the spectral direction and chromatic distortion (keystone) in the spatial direction. These effects distort the spectral signatures of objects under study and thus could reduce classification accuracy.

As discussed in Section 3.2.2, in an imaging spectrometer employing a 2D detector array, the image of the entrance slit is aligned along the row providing the spatial information and the spectral information is dispersed along the column. In an ideal case, every pixel in a row should have the same central wavelength, λ_c, and bandwidth, $\Delta\lambda$. However, in an actual system λ_c and $\Delta\lambda$ vary with the FOV, that is, swath (Figure 8.17a). This shift in wavelength in the spectral domain, which is a function of the cross-track pixel number, is known as the smile or "frown" curve. (The name smile is given because of the general appearance of a plot of wavelength versus spatial pixels, which approximates the curve of a shallow smile with upturned ends.) Spectral smile is a measure of the deviation of λ_c from a straight line for a single wavelength band across track. The amount of spectral smile also varies with wavelength, making quantitative evaluation of this effect challenging.

The magnification change of a hyperspectral imager with wavelength is called spectral keystone. In the ideal case, the dispersed spectra for each IFOV are aligned along the column of pixels and, therefore, all spectra for all IFOVs are parallel to one another. When there is keystone distortion, the dispersed spectrum deviates from a straight line and varies with the FOV (Figure 8.17b). Spectral keystone is a measure of the spatial shift between pixels at different wavelengths. Keystone distortion results in the contamination of spectra from neighboring IFOVs.

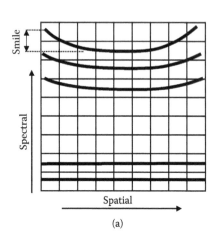

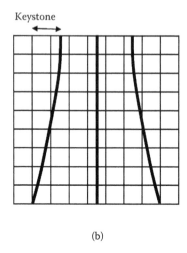

(a)   (b)

FIGURE 8.17
Schematics showing (a) smile and (b) keystone distortions. The grid represents the focal plane array. The thick lines show the pixel locations in the focal plane. The smile distortion results in a center wavelength shift, and the keystone distortion produces a band-to-band misregistration.

8.5 Hyperspectral Imaging Instruments

The precursor to hyperspectral imaging spectrometer development was the migration of the spectrometer from laboratory to field. The MSS image interpretation efforts led to the development of the first truly portable field reflectance spectrometer that could cover the spectral range of solar reflected radiance, 0.4–2.5 μm (Goetz 2009). Development of the first imaging spectrometer for Earth observation began at the Jet Propulsion Laboratory in 1980, which led to the realization of the airborne imaging spectrometer (Vane and Goetz 1988) and the airborne visible/infrared imaging spectrometer (Vane et al. 1993), which paved the way for the present-day systems for hyperspectral imaging. Encouraged by the application potential of these instruments and advancement in the FPA, a number of hyperspectral imaging systems evolved and are commercially available (Birk and McCord 1994; Vagni 2007). Spaceborne imaging spectrometers were first developed for planetary studies. The Near-Infrared Mapping Spectrometer of the Galileo mission, to explore Jupiter and its satellites, combined spectroscopic and imaging capabilities and thus introduced a new concept in remote sensing planetary experiments (Aptaker 1987). Though similar instruments were developed subsequently by France, Italy, and the Soviet Union for planetary exploration, hyperspectral imagers for Earth observation became available much later. The first spaceborne hyperspectral imager specifically designed for Earth observation was the Fourier transform hyperspectral imager (FTHSI) on board MightySat II, which was launched in 2000.

Later, in the same year the National Aeronautic and Space Administration launched the EO-1 satellite carrying two hyperspectral cameras: the Hyperion sensor and the Atmospheric Corrector. The low-cost microsatellite imaging mission of the Indian Space Research Organization IMS-1, previously referred to as TWSat, launched in April 2008 and carried a hyperspectral scanner, Hyperspectral Imager (HySI), which is a derivative of the imaging spectrometer on board India's first lunar mission Chandrayana-1. Though there are a number of operational multispectral cameras providing Earth imagery with varying spatial and spectral resolutions, hyperspectral data are not available on an assured basis. We shall briefly describe some of the spaceborne hyperspectral imagers in the following sections.

8.5.1 MightySat II: Fourier Transform Hyperspectral Imager

The FTHSI launched on board U.S. Air Force Research Laboratory's satellite MightySat II.1 is the first true hyperspectral Earth viewing imager to be successfully operated in space. The FTHSI is a monolithic Sagnac interferometer that operates in pushbroom mode. The fore optics is a Ritchey–Chretien telescope with 165-mm clear aperture that focuses the scene onto a field stop, which defines the swath and IFOV (Yarbrough et al. 2002). The one-dimensional image is then passed through the interferometer where the rays are split, sheared, and recombined to form an interferogram. From the interferometer, a Fourier lens collimates the light and a cylindrical lens images the energy onto a 1024 × 1024 2D silicon CCD array. The spatial information is collected across track, and the spectral information is collected simultaneously across the entire strip. The imager covers a spectral range from 500 to 1050 nm. A cutoff filter and CCD sensitivity prevent interference from wavelengths outside the spectral range of interest. The spectral and spatial resolutions of the spectrometer could be varied electronically by binning the CCD pixels and generating four operating modes, that is, 1024 × 1024, 1024 × 512, 512 × 1024, and 512 × 512. The instrument is designed to have 3.0° FOV and a selectable IFOV of 5.8 or 2.9 millidegrees. The instrument weighs 20.45 kg and consumes a peak power of 66 W (Otten et al. 1997; Otten et al. 1998).

8.5.2 National Aeronautic and Space Administration Earth Observation 1: Hyperion

One of the instruments carried by NASA's EO-1 satellite is the hyperspectral imager Hyperion. The Hyperion operates in pushbroom mode, covering 400–2500 nm in 220 contiguous spectral bands with a spectral resolution of 10 nm. The total spectral range is covered by two spectrometers, with overlapping spectral bands, one covering the very near infrared (VNIR) band (400–1000 nm) and the other the SWIR (900–2500 nm) band. The overlapping of the spectra helps cross-calibration of the two spectrometers. Both the spectrometers share a single set of fore optics (imaging telescope) and a common slit that defines the swath and IFOV, providing a 7.5-km swath and 30-m spatial resolution from the orbit altitude of 705 km. The imaging telescope is an off-axis TMA with an aperture of 12.5 cm. A dichroic beam splitter behind the slit reflects 400–1000 nm to the VNIR spectrometer and transmits 900–2500 nm to the SWIR spectrometer. Both spectrometers use a JPL convex grating design in a three-reflector Offner configuration and provide a spectral resolution of 10 nm. The two grating imaging spectrometers relay the slit image to two focal planes at a magnification of 1.38:1 (Lee et al. 2001). The focal plane dimension parallel to the slit axis provides the cross-track spatial image, whereas the axis perpendicular to the slit provides the spectral information for each cross-track pixel.

The VNIR channel FPA uses a custom-built 2D frame transfer CCD with 384 × 768 pixels of 20-μm pitch. For data collection, a 60-μm-pixel pitch is generated by binning 3 × 3 pixels of the 20-μm pixels of the CCD array. Of the total array, the VNIR spectrometer uses only 60 (spectral) × 250 (spatial) pixels to provide a 10-nm spectral bandwidth over a range of 400–1000 nm and swath of 7.5 km. The VNIR FPA is cooled via thermal strap by a radiator and operates at 10°C. The SWIR focal plane is 256 × 256 pixels with 60-μm pitch HgCdTe with a readout integrated circuit. Of these, only 160 pixels (spectral) × 250 pixels (spatial) are used for transmission. The SWIR detectors are cooled by an advanced cryocooler supported by a radiator and are maintained at 110 K during data collection (Pearlman et al. 2001). The data are digitized to 12 bits.

All of the mirrors in the system are made of aluminum. For thermal stability, the structure holding the optical elements is also constructed from aluminum. The electro-optical assembly, that is, the telescope, two grating spectrometers, and supporting focal plane electronics, is maintained at 20°C ± 2°C for proper operation.

8.5.3   NASA EO-1 Linear Etalon Imaging Spectral Array Atmospheric Corrector

The EO-1 spacecraft also carried another hyperspectral imager: the Linear Etalon Imaging Spectral Array Atmospheric Corrector (LAC) using wedge filter technology (Reuter et al. 2001). The LAC sensor covers a spectral range of 0.9–1.6 μm in 256 channels. The IFOV is 360 μrad. The FOV is approximately 15° to match the 185-km swath of other sensors. However, this is achieved by using three separate lenses each covering 5° with two of the lenses placed off-axis to cover the full 15°. Each subassembly consists of a focal plane formed of a wedged filter mounted to a 256 × 256 pixel InGaAs detector array placed behind the lens. Thus, a single frame consists of an effective focal plane with 768 pixels in the cross-track direction and 256 pixels in the along-track direction. One of the noteworthy designs is that the filter is placed within 200 μm of the detector array, so that bandwidth broadening, as discussed earlier, does not take place.

The LAC has a total mass of 10.5 kg and a maximum power requirement of 45 W.

8.5.4 Indian Space Research Organization Hyperspectral Imager

India's first planetary mission to the Moon, Chandrayaan-1, carried a HySI capable of generating 64 contiguous bands in the spectral range of 421–964 nm with a spectral bandwidth better than 20 nm (Kumar et al. 2009). The instrument operates in pushbroom mode with a FOV of approximately 11.4° and an IFOV of approximately 0.8 millirad providing a swath of 20 km and an IFOV of 80 m from an orbit of 100-km altitude. The camera consists of a focusing optics, wedge filter, CMOS image sensor, camera electronics, and the housing to hold it all together.

The focusing optics is an *f*/4 multielement telecentric lens assembly with an effective focal length of 62.5 mm covering a circular FOV of ±13°. The first optical element is a plane parallel glass window, which helps to reduce heat load to the powered elements. This element can also be given a neutral density coating, if required. The lens assembly has six powered elements such that a group of three lens elements are located on either side of the aperture stop (Figure 8.18). All the lens elements have a spherical surface profile.

The wedge filter carries out the spectral separation and is placed in close proximity to the focal plane area array. The focal plane area array is a 256 × 512 active pixel sensor array with 50 × 50 μm pixel size and combines analog circuitry, digital circuitry, and optical sensor circuitry into a single chip (Shengmin et al. 2009). A novel design of the detector array is the incorporation of a programmable gain amplifier to compensate for the nonuniform spectral response of the detector. The device needs only three clock signals from which it internally generates all relevant timing signals. It also has an built-in 12-bit A to D converter.

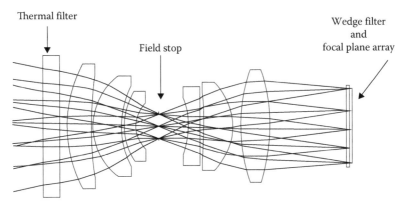

FIGURE 8.18
Schematics of the optical layout of Chandrayaan-1 Hyperspectral Imager.

The camera is mounted on the spacecraft such that the 256 elements are aligned across track and the 512 elements are aligned along track, thereby providing spatial and spectral information, respectively. To reduce the data volume, eight adjacent spectral bands corresponding to a target pixel are averaged on board to generate 64 contiguous bands. The instrument structure is of a sandwich construction with face skins of high modulus carbon fiber material and a core of aluminum alloy honeycomb and is designed to optimize the weight without compromising the functional requirements under various mechanical and temperature stresses over the mission life. The camera weighs 2.5 kg and draws 2.6 W of power.

References

Aptaker, I. M. 1987. Near-Infrared Mapping Spectrometer (NIMS) for investigation of Jupiter and its satellites. *Imaging Spectroscopy II*. ed. Vane G., Proceedings of SPIE, Vol. 834, Bellingham, WA, pp. 196–212.

Barducci, A., C. Francesco, C. Guido et al. 2012. Developing a new hyperspectral imaging interferometer for Earth observation. *Optical Engineering* 51(11): 111706.1–111706.13.

Birk, R. J. and T. B. McCord. 1994. Airborne hyperspectral sensor systems. *IEEE AES Systems Magazine* 9: 26–33.

Christensen, R., D. L. Anderson, S. C. Chase et al. 1992. Thermal Emission Spectrometer Experiment: The Mars Observer Mission. *Journal of Geophysical Research* 97: 7719–7734.

Dierking, M. P. and M. A. Karim. 1996. Solid-block stationary Fourier-transform spectrometer. *Applied Optics* 35(1): 84–89.

Eismann, M. T. 2012. *Hyperspectral Remote Sensing*. SPIE eBook Series, Bellingham, WA, EISBN: 9780819487889.

Folkman, M., J. Peariman, L. Liao, and P. Jarecke. 2001. EO-1/Hyperion hyperspectral imager design, development, characterization and calibration. *Proceedings of SPIE* 4151: 40–51.

Goetz, A. F. 2009. Three decades of hyperspectral remote sensing of the Earth: A personal view. *Remote Sensing of Environment* 113: S5–S16.

Griffiths, P., J. A. De Haseth. 2007. *Fourier Transform Infrared Spectrometry*, 2nd Edition, John Wiley & Sons, NY.

Joseph, G. 2005. *Fundamentals of Remote Sensing*. 2nd Edition. Universities Press (India) Pvt Ltd., Hyderabad, India.

Kumar, K., A. S., Arup Roy Chowdhury, A. Banerjee et al. 2009. Hyper Spectral Imager for lunar mineral mapping in visible and near infrared band. *Current Science* 96(4): 496–499.

Lee, P., S. Carman, C. K. Chan, M. Flannery, M. Folkman, and K. Iverson et al. 2001. Hyperion: A 0.4 µm–2.5 µm hyperspectral imager for the NASA Earth Observing-1 Mission, http://www.dtic.mil/dtic/tr/fulltext/u2/a392967.pdf (accessed on May 15, 2014).

Loewen, E. G. and P. Evgeny. 1997. *Diffraction Gratings and Applications*, Marcel and Dekker Inc, New York, NY.

Lucey, P. G., K. A. Horton, T. J. Williams et al. 1993. SMIFTS: A cryogenically cooled, spatially modulated imaging infrared interferometer spectrometer. *Proceedings of SPIE* 1937:130–141.

Meigs, A. D., L. J. Otten, and T. Y. Cherezova. 2008. Ultraspectral imaging: A new contribution to global virtual presence. *Aerospace and Electronic Systems Magazine, IEEE* 23(10): 11–17.

Pantazis, M. and M. M. McKerns. 2000. Pushbroom imaging spectrometer with high spectroscopic data fidelity: Experimental demonstration. *Optical Engineering* 39(3): 808–816.

NASA. 1987. *Earth observation system—Instrument panel report-HIRIS*, Vol. 2c.

Prieto-Blanco, X., C. Montero-Orille, B. Couce, and R. de la Fuente. 2006. Analytical design of an Offner imaging spectrometer. *Optics Express* 14(20): 9156–9168.

Otten, L. J., A. D. Meigs, B. A. Jones, P. Prinzing, and D. S. Fronterhouse. 1998. Payload qualification and optical performance test results for the MightySat II.1 Hyperspectral Imager. *SPIE* 3498: 231–238.

Otten L. J., A. D. Meigs, B. A. Jones et al. 1997. Engineering model for the MightySat II.1 hyperspectral imager. *Proceedings of SPIE* 3221:412–420.

Pearlman, J., S. Carman, C. Segal, P. Jarecke, P. Clancy, and W. Browne. 2001. Overview of the Hyperion Imaging Spectrometer for the NASA EO-1 mission. *Geoscience and Remote Sensing Symposium. IGARSS '01* 7: 3036–3038.

Rafert J. B., R. G. Sellar, and J. H. Blatt. 1995. Monolithic Fourier transform imaging spectrometer. *Applied Optics* 34(31):7228–7230.

Reuter, D. C., G. H. McCabe, R. Dimitrov et al. 2001. The LEISA/Atmospheric Corrector (LAC) on EO-1. *Proceedings of the International Geoscience and Remote Sensing Symposium. IEEE* 1: 46–48.

Rosenberg, K. P., K. D. Hendrix, D. E. Jennings, D. C. Reuter, M. D. Jhabvala, and A. T. La. 1994. Logarithmically variable infrared etalon filters. *Proceedings of SPIE* 2262: 223–232.

Sellar, R. G. and G. D. Boreman. 2003. Limiting aspect ratios of Sagnac interferometers. *Optical Engineering* 42(11): 3320–3325.

Sellar, R. G., R. Branly, A. I. Ayala et al. High-Efficiency HyperSpectral Imager for the Terrestrial and Atmospheric MultiSpectral Explorer, http://commons.erau .edu/cgi/viewcontent.cgi?article = 1170&context = space-congress-proceedings (accessed on March 29, 2014).

Shengmin, L., Chi-Pin Lin, Weng-Lyang Wang et al. 2009. A novel digital image sensor with row wise gain compensation for Hyper Spectral Imager (HySI) application. *Proceedings of SPIE* 7458: 745805.1–745805.8.

Shippert, P. 2004. Why use hyperspectral imagery? *Photographic Engineering and Remote Sensing* 70: 377–380.

Silny, J., S. Schiller, M. David et al. 2010. Responsive Space Design Decisions on ARTEMIS. 8th Responsive Space Conference. AIAA-RS8-2010-3001, http:// www.responsivespace.com/Papers/RS8/SESSIONS/Session%20III/3001_ Silny/3001P.pdf (accessed on May 15, 2014).

Silverglate, P. R. and D. E. Fort. 2004. System design of the CRISM (Compact Reconnaissance Imaging Spectrometer for Mars) hyperspectral imager. *Proceedings of SPIE* 5159: 283–290.

Silverman, S., D. Bates, C. Schueler et al. 1999. Miniature Thermal Emission Spectrometer for the Mars 2001 Lander. *Proceedings of SPIE* 3756: 79–91.

Vagni, F. 2007. Survey of Hyperspectral and Multispectral Imaging Technologies, North Atlantic Treaty Organisation, Research and Technology Organisation. TR-SET-065-P3, http://ftp.rta.nato.int/public/PubFullText/RTO/TR/RTO-TR-SET-065-P3///$$TR-SET-065-P3-ALL.pdf (accessed on May 15, 2014).

Vane, G. and A. F. H. Goetz. 1988. Terrestrial imaging spectroscopy. *Remote Sensing of Environment* 24: 1–29.

Vane, G., R. O. Green, T. G. Chrien, H. T. Enmark, E. G. Hansen, and W. M. Porter. 1993. The airborne visible/infrared imaging spectrometer (AVIRIS). *Remote Sensing of Environment* 44:127–143.

Wolfe, W. L. 1997. *Introduction to Imaging Spectrometers*. SPIE Press, Bellingham, WA [doi:10.1117/3.263530].

Woody, L. M. and J. C. Demro. 1994. Wedge Imaging Spectrometer (WIS) hyperspectral data collections demonstrate sensor utility. *Proceedings of the International Symposium on Spectral Sensing Research (ISSR)* 1:180–190.

Yarbrough, S., T. Caudill, E. Kouba et al. 2002. MightySat II.1 Hyperspectral imager: Summary of on-orbit performance. *Imaging Spectrometry VII*. ed. Michael R. D. and S. S. Shen. *Proceedings of SPIE* 4480:186–197.

9

Adding the Third Dimension: Stereo Imaging

9.1 Introduction

The cameras we discussed so far are primarily meant to produce two-dimensional images. However, mapping and monitoring of the Earth's surface is essentially a three-dimensional problem. Therefore, it is necessary to supplement traditional two-dimensional thematic mapping with information about the third dimension, which is height information. Height information is important for many applications such as geology, soil erosion studies, disaster assessment, and so on. Terrain elevation information enables generation of a digital elevation model (DEM), which is essential for mapmaking. The well-established method to measure the third dimension is stereoscopy. Stereoscopy is based on binocular vision, the same principle by which we perceive depth. In binocular vision, each of the eyes (which are about 65 mm apart) produces image from different angles (viewpoints), thereby generating slightly different spatial relationships between near and distant objects. The brain fuses these different views into a three-dimensional impression. The principle of binocular vision can be used to produce three-dimensional data from images. This requires photographs, which are taken from two locations. Conventionally, this is achieved by vertical photographs with an overlap (stereo pairs). The overlapped area is essentially viewed by the same camera but displaced along the flight path. Due to this different view angle between the two photographs, there is a displacement of the object in the second photograph compared to the first (referred to with respect to the principal point). This apparent shift due to the changed view angle is called parallax. From the stereo pairs, one can estimate differential parallax, dp, that is, the difference in absolute parallax between the top and bottom of the object (Figure 9.1). Then the height, h, is given as per the following equation:

$$h = \frac{dp}{B/H}$$

(9.1)

where
dp—differential parallax
B—length of airbase
H—flight altitude

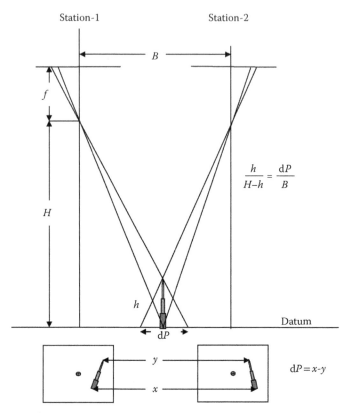

FIGURE 9.1
Geometry of a stereoscopic pair of photographs. *B*, the base separation; *H*, the flight altitude; *f*, focal length of the lens; *h*, height of the object; d*P*, the difference in parallax from the top and bottom of the object.

The accuracy of the height determination is dependent on the base to height ratio, B/H, and the accuracy with which the parallax can be measured from the stereo pair. The National Map Accuracy Standards (NMAS) of the United States has defined the maximum allowable root mean square error in height [RMSE (z)] and plan [RMSE (x,y)] measurements, at a given map scale (Table 9.1). As seen from Table 9.1, RMSE (z), which is the total error in the height measurement of points in image data, dictates the contour interval applicable to a particular map scale. The planimetric error and the height error derived from stereo pairs are dependent on the pixel size. From Table 9.1, one can infer that if one can achieve a planimetric accuracy of ±1.5 pixel then for the 1:25,000 scale map, we require a spatial resolution of 5 m. Therefore, one can appreciate the need for high-resolution images for preparing large-scale topographic maps. Figure 9.2 gives, for a spherical Earth, the error in stereoscopic height δh determination, as a function of B/H for different values of parallax measurement error, δP (Wells 1990). As can be seen from the graph, if the differential parallax can be measured with accuracy better than one pixel the improvement in accuracy of height determination is marginal as the base to height ratio is increased beyond about 0.6.

The task of obtaining three-dimensional space information from two-dimensional airborne/spaceborne imagery by digital photogrammetry is really in the realm of processing the data to generate the differential parallax. Once the stereo pairs are acquired, the important steps to generate DEMs are camera modeling (which enables transformation between image and ground), feature extraction, image matching, elevation determination, and interpolation. The details

TABLE 9.1

Maximum Allowable RMSEs for Topographic Maps Meeting the Requirements of the U.S. NMAS

Map Scale	Planimetric RMSE (x, y) (m)	Spot Height RMSE (z) (m)	CI = 3.3 × RMSE (z) (m)
1:500,000	±150	±30	100
1:250,000	±75	±15–30	50–100
1:100,000	±30	±6–15	20–50
1:50,000	±15	±6	20
1:25,000	±7.5	±3	10

Source: Wells, N. S., *Technical Report 90032*, Royal Aerospace Establishment, Farnborough, United Kingdom, PDF Url: ADA228810, 1990.

CI, contour interval.

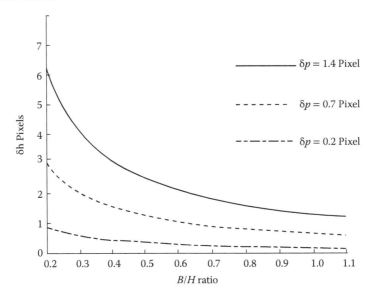

FIGURE 9.2
Height error as a function of B/H (for a spherical Earth), for various values of parallax (δp) measurement error. (From Wells, N. S., Technical Report 90032, Royal Aerospace Establishment, Farnborough, United Kingdom, PDF Url: ADA228810, 1990.)

of the procedure involved to generate DEM are beyond the scope of this book. However, to appreciate the task involved we shall briefly describe the various steps leading to height determination from a stereo pair. Once the two overlapping images are received, the first task is to find the exact overlap area between the two imageries and the corner coordinates of the overlap area. The overlapped images are then registered such that the left and right images have common orientation and the matching features appear along a common axis, referred to as epipolar registration. The accuracy of parallax determination is dependent primarily on the accuracy of identification of common points (called conjugate points or analogous points) in the stereo pairs. In the analytical mode of height determination (in the bygone era), a human operator performs identification of conjugate points to determine the differential parallax. In the digital mode (soft copy photogrammetry), it is carried out through computers in a semiautomatic or fully automatic way, a process called *image matching*. Due to various reasons, such as temporal changes of the stereo data sets, different view or look angles, various imaging conditions, sensor-specific effects in images, scale changes between images, and so on, image matching could result in wrong correlation of points. Wrong correlation leads to feature or point mismatch, and hence these wild points (often referred to as *blunder*) are to be removed before elevation extraction.

The base to height ratio can be computed from the view angle of the camera generating the stereo pairs from the following equation:

$$\frac{B}{H} = \tan\theta_1 + \tan\theta_2 \tag{9.2}$$

where θ_1 and θ_2 are the two view angles.

Two-image stereo reconstruction is the normal mode of DEM generation utilized by most users. From the data reduction point of view, a triplet can give, in general, better accuracy compared to using two overlapping images. Use of image pairs from three cameras can improve the reliability of image matching utilizing the redundant multiview image information and also reduces the shadow area if the third view is near vertical, thereby reducing the occlusion problem in matching.

From the sensor designer's perspective, the camera for stereo imaging is not different from what we have discussed earlier. In this chapter, we discuss different possibilities for generating stereo pairs from satellites and describe some of the cameras specifically designed for stereoscopy.

9.2   Stereo Pair Generation Geometries

From a space-based imaging system, stereo viewing (i.e., generating two images of the same point with different view angles) can be achieved in different ways. Broadly, this could be achieved by

1. Generating stereo pairs by imaging from two different orbits— *across-track stereoscopy*

2. Viewing at different view angles along the same orbit—*along-track stereoscopy*

9.2.1 Across-Track Stereoscopy

Across-track stereo images are those taken by the same sensor from multiple orbits. Across-track stereoscopy is possible in two modes. In the case of the Landsat MSS/TM, IRS LISS 1/2/3 cameras, and other similar systems, images from two adjacent orbits have a certain common area—sidelap (Figure 9.3). As this area is viewed from two different imaging stations, the view angles are different and hence the overlapped area from the two images forms a stereo pair. However, the B/H for the cameras mentioned earlier is only approximately 0.2 at the equator and progressively decreases as one goes to higher latitudes, while the overlap increases (Table 9.2). In addition, the sidelap covers only a very small area and hence cannot be used for full coverage of Earth, unless the swath is wide as in the case of the IRS Resourcesat AWiFS camera.

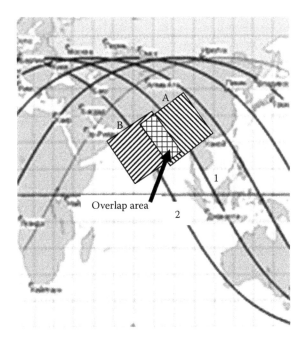

FIGURE 9.3
Schematics showing across-track stereoscopy from adjacent orbital tracks. From the track 1 and 2 image frames, A and B are generated. As the overlap region is viewed from different angles, it forms a stereo pair. The overlap area depends on the separation between adjacent tracks, swath, and the latitude.

TABLE 9.2

B/H Values for Generating Across-Track Stereo Pair from the Adjacent Paths for
Resources at (817 km) LISS-3 and AWiFS Cameras for Various Values of Latitude
and the Corresponding Overlap

		Sidelap (km)	
Latitude	B/H	LISS-3	AWiFS
0	0.144	23.6	618.8
20	0.135	30.7	625.9
40	0.11	51.8	650.1
60	0.072	89	686.4
80	0.025	123.1	728.4

Source: Gopalkrishna B and Roy Sampa, SAC/ISRO.

For the aforementioned satellite systems, the B/H and overlap area are
fixed unless one maneuvers the spacecraft attitude. SPOT-1 was the first
civilian mission specifically designed to point the camera optical axis across
the orbital track. Such a design helps to improve the revisit capability of a
particular target and also facilitates generation of across-track stereo pairs.
The IRS-1C/1D PAN camera also has a similar capability. The IRS-1C/1D
PAN camera can be steered across track to approximately ±26°, thereby generating imagery of the same area from two different orbits to produce stereo pairs (Figure 9.4). Thus, depending on the cross-track steering angle it is
possible to generate stereo pairs of B/H up to approximately 0.9 (Figure 9.5).
However, such a scheme has the disadvantage that the stereo pairs are produced at different dates, which could affect the parallax extraction due to
varying atmospheric conditions including clouds, and radiometric changes
due to temporal variations between the observations. In addition, oblique
viewing at different angles produces pairs with different spatial resolutions.
Notwithstanding these limitations, various software has been developed for
the restitution of SPOT stereo pairs (Gugan and Dowman 1988). SPOT in theory can produce maps to meet the accuracy standards of 1:50,000 scale mapping, that is, RMSE accuracies of ±15 m in plan and 6 m in height, for a 20-m
contour interval. Jacobsen (1999) carried out investigations with three IRS-1C
images (two off-nadir $B/H \sim 0.8$ and one nadir) over Hanover, Germany, and
obtained an accuracy of ±1.1 pixels (6.5 m) in planimetry and elevation.

When considering regions prone to frequent cloud cover or rapidly variable weather systems, suitable stereo pairs could be separated in time by
weeks. Coregistration of such images is often made difficult because the
scene illumination, atmospheric conditions, and so on could have altered for
the second image compared to the first image. This is the major drawback
of relying on different orbits to generate stereo pairs. To take care of these
issues, a more convenient and practical method is to generate stereo images
in the same orbit, that is, along-track stereoscopy.

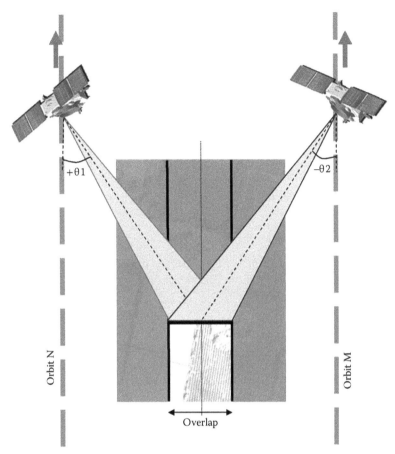

FIGURE 9.4
Viewing geometry for generating across-track stereo pairs by SPOT/IRS-1C PAN.

9.2.2 Along-Track Stereoscopy

Along-track stereoscopy can be carried out in different modes. There could be two cameras mounted such that the optical axis of one points forward (fore camera) and the other backward (aft camera) with respect to the nadir (Figure 9.6). Thus, the fore camera images ahead of nadir, whereas the aft camera images behind nadir. For the time being, let us ignore the Earth's rotation. Here, a strip imaged by the fore camera at an angle $+\theta$ with respect to nadir will be viewed after a time t by the aft camera at an angle $-\theta$, thereby generating two images with different look angles. The time t depends on the angle between the optical axes and the subsatellite velocity. The angle between the cameras also decides the B/H.

As the fore and aft cameras see a strip within a short period of time, the radiometric variations between the images are minimal and thus increase

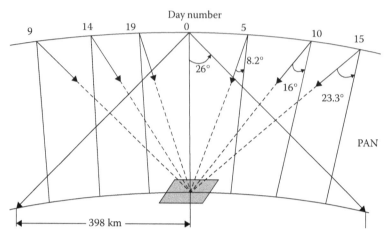

FIGURE 9.5
Stereo pair generation possibilities for IRS-1C/D PAN and LISS-4 due to off-nadir pointing. If a target is seen at day 0, the same target can be seen with a maximum *B/H* of 0.86 after 9 and 15 days, thereby producing a triplet for height extraction. (From NRSA, National Remote Sensing Agency, Hyderabad, India, *IRS1 C Data User's Handbook*, NDC/CPG/NRSA/IRS1-C/HB/Ver 1.0/Sept.1995, 1995.)

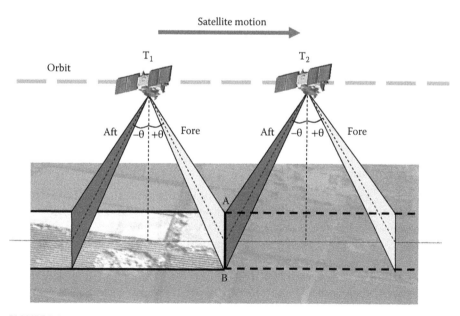

FIGURE 9.6
Schematics showing along-track stereoscopy. At time T_1, the strip "AB" is viewed by the fore camera. As the satellite moves, after time *t* (at time T_2) the same strip is seen by the aft camera. Thus, the same strip is seen at two different angles to produce a stereo pair.

the correlation success rate in image matching is increased. This is one of the advantages of along-track stereoscopy compared to across-track stereoscopy. Although each camera has the same swath, do they cover the same area to produce a stereoscopic pair? A strip on the Earth is first imaged by the fore camera, and the aft camera sees the same latitude after a certain time delay. The time difference is a function of the difference in the look angles between the fore and aft cameras. In the absence of Earth's rotation, the strip covered by the aft camera coincides with that of the fore camera. But due to the Earth's rotation, there will be a shift in the traces covered by the two cameras. The shift is a function of latitude, and at the equator it will be maximum. Therefore, only a fraction of the targets recorded in the fore-looking view will be stereo-scopically recorded by the aft telescope. This problem is circumvented by giving appropriate yaw rotation to the spacecraft, which is a function of latitude. This involves dynamic control of the yaw axis of the satellite such that, ideally, a target observed by the nth pixel in the image line of the fore-looking telescope is also observed by the nth pixel in the aft-looking telescope. The required yaw rotation to compensate for the Earth's rotation is independent of the image strip swath (Nagarajan and Jayashree 1995). If a stereo pair is not required, the spacecraft can be maneuvered such that the image strips will fall side by side so that wider swath images are obtained by the two cameras.

A dedicated satellite for along-track stereo imaging—Stereosat—using three linear arrays operating in the fore, aft, and nadir was proposed as early as 1981 (Welch 1981). However, the first Earth observation satellite to utilize the principle of along-track stereo is the Japan Earth Resources Satellite (JERS-1) (aka Fuyo 1), launched in 1992. The optical sensor (OPS) on board JERS-1 consists of two independent radiometers: the very near infrared (VNIR) band instrument and the shortwave infrared instrument. One of the bands (0.76–0.86 μm) of the VNIR instrument has two linear arrays, one looking at nadir and the other viewing 15.3 forward in flight direction, thereby making a stereo pair give two-line stereoscopic along-track imaging capability with a B/H ratio of 0.3. With a pixel size of 18 m across track and 24 m along track, the instrument could produce height information with an error better than 20 m root mean square (Westin 1966). The first high-resolution satellite specifically designed for along-track stereoscopy is the Indian satellite CARTOSAT1, launched in 2005. We shall now discuss the design of CARTOSAT1.

9.3 Along-Track Stereo Using Multiple Telescopes

9.3.1 IRS CARTOSAT 1

The CARTOSAT1 is a three-axis body–stabilized remote sensing satellite providing along-track stereo imaging capability for cartographic applications. To generate stereo pairs, the satellite carries two panchromatic cameras mounted with a tilt of +26° (fore) and –5° (aft) from the yaw axis (local vertical) in the yaw-roll plane. Because of the tilt, the ground resolution and swath are slightly different for both the cameras. The fore camera provides an across-track resolution of 2.452 m and covers a swath of 29.42 km, whereas the aft camera provides an across-track resolution of 2.187 m with a swath of 26.24 km. From 618-km orbit, the time difference between the acquisitions of the same scene by the two cameras is about 52 seconds. The platform is continuously steerable about the spacecraft yaw axis to compensate for the Earth's rotation, thus allowing both fore and aft cameras to look at the same ground strip when operated in stereo mode. The stereo pairs have a swath of about 26 km and a fixed B/H ratio of 0.62. Apart from stereo mode, the satellite can also be operated in wide swath mode. When operated in this mode, the satellite can be maneuvered such that image strips will fall side by side so as to produce a combined swath of about 55 km. By giving roll bias, the satellite can be tilted up to ±23° in the across-track direction, thereby providing a capability to revisit a target more frequently compared to the temporal resolution. By exercising roll bias, the maximum wait period to view an area again at the equator is 11 days. As the satellite moves toward the poles, the paths become closer to each other and, hence, more paths can be used to view repeatedly a given region at high latitudes (NRSC 2006).

The optical system consists of an off-axis three-mirror anastigmatic (TMA) telescope, similar in design to the IRS1C/D PAN camera (Chapter 6), that is, an off-axis concave hyperboloid primary mirror, a convex spherical secondary mirror, and an off-axis concave oblate ellipsoidal tertiary mirror. The telescope design is optimized to have a rectangular field of view of ±1.3° (cross track) and ±0.2° (along track) with 1945-mm effective focal length (EFL), and an aperture ratio of F/4.

The mirrors are made of Zerodur and are lightweighted by scooping out material from the rear. The weight reduction is optimized based on computer simulation to have adequate stiffness so that the natural frequency is above 200 Hz. Three flat regions on the periphery of the mirrors are bonded to bipod flexure elements (mirror fixation devices [MFDs]) using a room temperature curing adhesive. The MFDs are made of Invar and are designed to hold the mirrors such that the reflective surface optical quality is not degraded, under the influence of thermal, vibrational, and assembly stresses (Subrahmanyam et al. 2006). The MFDs are fixed to suitably designed interface rings, which are mounted onto appropriate locations onto the main telescope structure. Figure 9.7 shows the schematics of the CARTO1 EO module main structure. The main structure—EO module—holds the primary, secondary, and tertiary mirror subassemblies and detector head assembly along with stray light suppression baffles at specified locations and stray light cover. To have better structural integrity, the main structure is fabricated from a monolithic Invar piece. Based on detailed structural analysis, the weight of the structure is optimized to have adequate stiffness so that the natural frequency

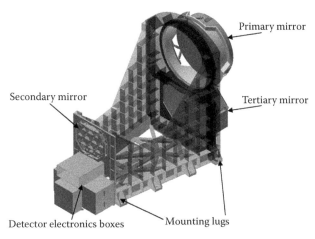

FIGURE 9.7
Schematics of CARTO1 electro-optics module main structure supporting the mirrors, focal plane assembly, and associated electronics.

is > 80 Hz, and the structure maintains alignment integrity and the mirror surface contour under various stress loads.

The detector used is a 12 K silicon linear array charge-coupled device (CCD) with a pixel size of 7 × 7 μm, which is similar to the type used in LISS-4. The device has eight output ports, and the odd and even pixel rows are physically separated by five lines, that is, 35 μm. The spectral selection is carried out by an interference filter placed in front of the CCD, with a bandwidth of 0.5–0.85 μm. The CCD temperature is controlled within 20°C ± 1°C. The temperature-controlling arrangement is similar to that described for the LISS-4 camera. As in the earlier IRS cameras, the onboard calibration of the detectors is carried out using light-emitting diodes. The video output is digitized to 10 bits, thereby allowing full dynamic range up to 100% reflectance at all latitudes and seasons without changing the gain. Both the fore and aft cameras are identical in design in terms of optics, focal plane assembly, electronics, and so on. The digital video data for each camera are about 336 Mbps. To be compatible with the data transmission system, the digital video data are compressed and encrypted and transmitted to the ground at a data rate of 105 Mbps for each camera through two X-band QPSK carriers. The compressed and formatted data can also be stored in a 120-Gb solid-state recorder. The data compression ratio is 3.2:1, and the compression type is JPEG (NRSC 2006). The broad specifications of the CARTO1 camera are given in Table 9.3.

The CARTO1 satellite, weighing 1560 kg at liftoff, was launched into a 618-km polar Sun-synchronous orbit by the PSLV on May 5, 2005 from Indian Space Research Organization's (ISRO's) launch complex at Sriharikota, Andhra Pradesh. The local time for equatorial crossing is 10:30 am. The distance between adjacent ground traces is 21.46 km, and the adjacent path scenes at the equator have a sidelap of 5.3 km and 8.5 km for the aft and fore

TABLE 9.3

Major Specifications of CARTO1 Cameras

Parameter		Value
Optics	Design	Off-axis TMA
	EFL (mm)	1945
	F/no	4
	Field of view (degree)	
	Across track	±1.3
	Along track	±0.2
Detector	Type	Linear 12,000-element CCD
	Pixel size (μm × μm)	7 × 7
	Pixel layout	Odd and even separated by 5 pixels
	Number of ports	8
Spectral range (μm)		0.5–0.85
Tilt angle (degree) fore/aft		+26/–5
Geometric characteristics from 618 km		
Spatial resolution (m) fore/aft		2.5/2.2
Swath (km) fore/aft		30/27
GSD (m)		2.5
Base to height ratio (B/H)		0.62
Radiometric properties		
Mean saturation radiance (mw·cm^{-2}·μ^{-1}·sr^{-1}) fore/aft		55.4/57.7
Quantization (bits)		10
SNR at saturation radiance		345
Data rate per camera (Mbps)		338
Compression		JPEG, 3.2
System SWR @70 lp/mm		>20

SNR: Signal to noise ratio. GSD: Ground sampling distance, SWR: square wave response.

camera scenes. The satellite covers the entire globe between 81° north and 81° south latitudes in 1867 orbits with a repetitivity of 126 days.

9.3.2   SPOT High-Resolution Stereo Camera

The SPOT-5, launched in 2002, carries a dedicated payload—Haute Résolution Stéréoscopique (HRS)—for along-track stereoscopy. The two HRS cameras, HRS1 and HRS2, operating in the spectral band 0.49–0.69 μm are tilted by ±20° about the nadir. Therefore, during one satellite pass the forward-looking telescope acquires images of the ground at a viewing angle of 20° ahead of the vertical and about 90 seconds later the aft-looking telescope images the same strip at an angle of 20° behind the vertical, thereby generating near simultaneous stereo pairs with a B/H ratio of about 0.8 (Bouillon et al. 2006). The HRS has a swath of 120 km, and a continuous strip of 600-km length can be covered stereoscopically. The instantaneous geometric field of view is 10 m, but the sampling rate is 5 m along the track. The swath of 120 km allows for observation of the same area every 26 days. SPOT-5 also carries two high-resolution geometrical (HRG) instruments with 5/2.5 m spatial resolution in panchromatic mode. Along with the vertical viewing HRG camera of 5-m resolution, it is possible to generate a stereoscopic triplet (Kornus et al. 2006).

From the instrument design point of view, the HRS is the first Earth observation camera using a refractive collecting optics for 10 m instantaneous geometric field of view (Figure 9.8). Each camera has a 11-element lens assembly with a focal length of 580 mm and useful aperture of 150 mm. The detector is a Thomson TH7834 CCD, with 12,000 elements and a pixel dimension of 6.5 μm × 6.5 μm. The CCD is controlled to 24°C. The optical assembly is controlled to ±0.5°C, thereby avoiding the need for refocusing in orbit. The main structure is made of Al honeycomb/carbon fiber reinforced plastic (CFRP), with a low coefficient of thermal expansion and coefficient

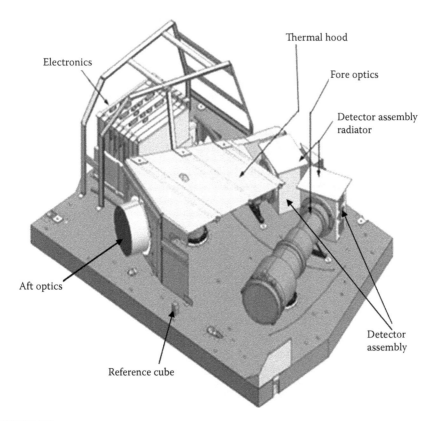

FIGURE 9.8
Schematic illustration of the Haute Résolution Stéréoscopique instrument. (Adapted from SPOT-5 eo portal, https://directory.eoportal.org/web/eoportal/satellite-missions/s/spot-5, accessed on June 2, 2014.)

of moisture expansion. The temperature stability is achieved by having a thermal enclosure, which is radiatively coupled to the optical assembly. The thermal enclosure temperature is controlled by a combination of active and passive thermal control systems (Planche et al. 2004). The satellite is placed in a Sun-synchronous orbit of 832 km with equatorial crossing at 10:30 am. The nominal pointing used by SPOT-5 is yaw steering mode, that is, controlling the satellite's yaw axis to counter the effects of Earth's rotation to ensure optimal overlap between two HRS images acquired 90 seconds apart.

9.3.3 Advanced Land Observation Satellite Stereo Mapping Camera: PRISM

The Advanced Land Observation Satellite (ALOS)—DAICHI—of Japan, launched in 2006, carries a dedicated instrument for stereoscopic observation: Panchromatic Remote Sensing Instrument for Stereo Mapping (PRISM). PRISM has three independent optical cameras operating in the 0.52–0.77 μm spectral band for viewing the Earth, each in different directions, nadir, forward, and backward. The fore and aft telescopes are inclined by about +24° and –24° from nadir. Such a scheme allows for the acquisition of three quasi-simultaneous images per scene, forward, nadir, and backward views, in the along-track direction. The nadir-viewing telescope covers a width of 70 km, whereas the fore and aft telescopes have a swath of 35 km (Figure 9.9).

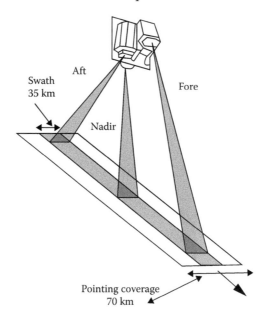

FIGURE 9.9
The triplet imaging geometry of the Panchromatic Remote Sensing Instrument for Stereo Mapping. (Adapted from Jaxa portal, http://www.eorc.jaxa.jp/ALOS/en/about/prism.htm, accessed on May 15, 2014.)

PRISM has a spatial resolution of 2.5 m at nadir. Each scene has three stereoscopic viewing capabilities, that is, the forward-nadir, forward-backward, and nadir-backward images. The fore and aft images form a stereo pair with a base to height ratio of 1.0.

The camera telescope has a three-mirror off-axis anastigmatic design to cover a 7.6° field of view in the cross-track direction. The focal length of each telescope is about 2000 mm. To maintain precise alignment among the three mirrors and to ensure stability of the focal plane assembly location, PRISM uses a specially fabricated CFRP truss structure whose temperature is controlled to ±3°C. The structure consists of an inner truss and an outer truss. The inner truss supports mirrors and the focal plane assembly. The outer truss supports the inner truss and is attached to the spacecraft (Figure 9.10). The three telescopes are mounted on both sides of the optical bench with the required orientation of the optical axis. To reduce the attitude error, high-precision star trackers and inertial reference units are also mounted on the bench (Osawa and Hamazaki 2000).

The focal plane assembly consists of linear CCD arrays with a pixel size of 7 μm for the cross-track direction and 5.5 μm for the along-track direction. As a single CCD of a requisite number of CCD elements is not available, each focal plane array is made by butting smaller length CCDs, each having 4960 elements with overlap (Kocaman and Gruen 2008). The data are digitized to 8 bits per pixel and have a data rate of about 960 Mbps, in the triplet mode. To reduce the downlink data rate from ALOS to ground stations, a lossy JPEG compression based on DCT quantization and the Huffman coding technique is performed. Compression can be selected as either 1/4.5 or 1/9, which gives a downlink data rate of 240 and 120 Mbps, respectively (Tadono et al. 2009).

To compensate for the Earth's rotation, each radiometer will use an electrical pointing function, within +/−1.5° (i.e., selecting the pixels along the array) and thus can provide fully overlapped three-stereo (triplet) images of 35 km width without yaw steering of the satellite.

9.4 Stereo Pairs with Single Optics

The three systems that we described in Section 9.3 require separate optics to generate nearly simultaneous stereo pairs. It is also possible to generate stereo pairs by using a single optics and multiple CCD arrays. We shall discuss some of the spaceborne cameras capable of generating stereo pairs using single optics.

9.4.1 Monocular Electro-Optical Stereo Scanner

The ISRO's SROSS satellite, launched in 1988, carried an Earth observation camera—Monocular Electro-Optical Stereo Scanner (MEOSS)—designed by DFVLR, Federal Republic of Germany. The MEOSS generates

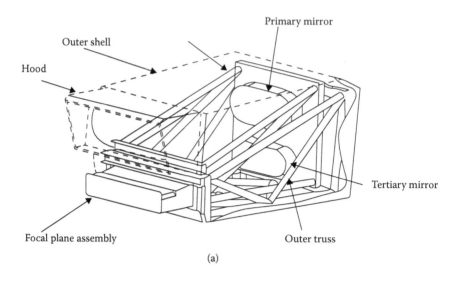

(a)

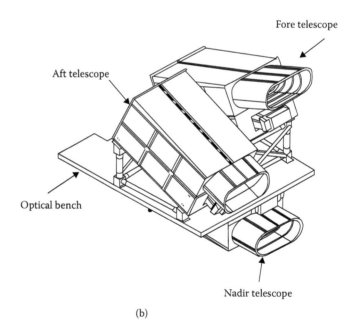

(b)

FIGURE 9.10
Schematics of the mechanical configuration of the Panchromatic Remote Sensing Instrument for Stereo Mapping EO module. (a) Single telescope and (b) three telescope assembly on the optical bench. (Adapted from Osawa, Y., PRISM, AVNIR-2, PALSAR—ALOS's major mission instruments at ALOS Symposium, http://www.eorc.jaxa.jp/ALOS/conf/symp/2001/5.pdf, accessed on May 15, 2014, 2001.)

along-track, threefold stereo imaging using a CCD line scanner. The camera consists of a single lens at whose focal plane three CCDs each having 3456 pixels are mounted perpendicular to the flight direction on a common plate (Lanzl 1986). The three CCD arrays view +23° and –23° (forward and backward) as well as nadir, leading to threefold stereoscopic images. This allows for a nearly simultaneous generation of all three images of a stereo triplet (Figure 9.11).

Such an arrangement apparently looks very attractive as only one optics needs to be used, thereby allowing for a compact instrument with less weight compared to a design having three separate optical heads. However, the lens has to be designed for a field of view of ±23°. The Terrain Mapping Camera (TMC), realized for India's first lunar mission Chandrayaan1, has adopted an innovative design to reduce the field of view of the lens but still can realize a B/H of 1.

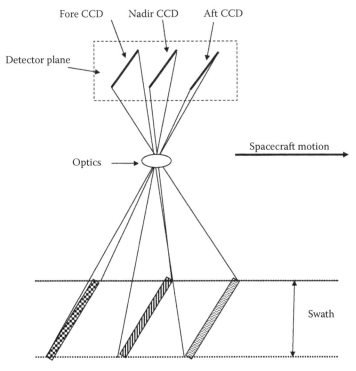

FIGURE 9.11
Schematic of Monocular Electro-Optical Stereo Scanner focal plane configuration. Three charge-coupled devices (CCDs) are placed parallel in the focal plane. Each of the CCDs images with different view angles, thereby generating stereo pairs.

9.4.2 Chandrayaan1 Terrain Mapping Camera

TMC images the lunar surface in pushbroom mode in the panchromatic spectral band of 0.5–0.75 μm. It is a triplet-forming stereo camera realized with a single optics, giving a *B/H* of 1 between fore and aft imagery. The required along-track field angle of ±25° is achieved with the help of a pair of plane mirrors placed strategically on either side of the 0° field (Kiran Kumar et al. 2009). The mirrors fold the incoming beams entering at ±25° such that the lens assembly receives the same at a reduced angle of ±10° (Figure 9.12). Therefore, the focusing optics needs to be corrected only for ±10° instead of ±25°, and the lens can be realized with spherical elements and good image quality. The TMC focusing optics has an eight-element refractive F/4 lens with field of views of ±5.7° and ±10° in the across-track and along-track directions, respectively. With this configuration, good image quality is achieved with minimum size and weight.

To realize the fore, nadir, and aft views, three parallel linear detectors are kept in the focal plane such that they are at right angles to the velocity vector of the satellite. The TMC detector head is a monolithic complementary metal–oxide–semiconductor linear array active pixel sensor (APS). The APS has integrated sensing elements, timing circuitry, video processing, and 12-bit analog to digital conversion. The camera has 5-m spatial resolution from 100-km altitude. Four programmable gains are provided for imaging low illumination regions. The gain can be set through ground commanding. The digital data are losslessly compressed for transmission; a compression bypass mode is also available. The operating temperature limit of the instrument is 10°C–30°C and

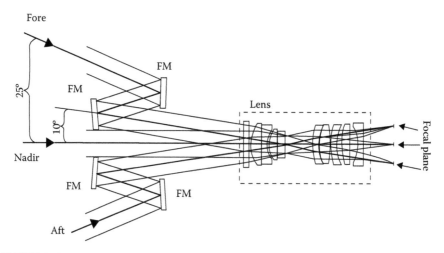

FIGURE 9.12
Optical schematics of Terrain Mapping Camera. FM, fold mirrors. Judicial placement of the fold mirrors reduces the lens field of view requirement compared to the actual view angles of the fore and aft charge-coupled devices.

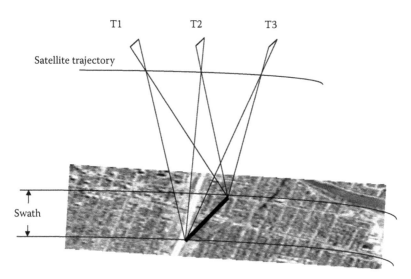

FIGURE 9.13
Schematics showing along-track stereo generation by tilting satellite to produce fore (T1), nadir (T2), and aft (T3) images of the same strip.

is maintained with a passive thermal control technique augmented with heaters. The camera requires 1.8 W regulated supply and weighs 6.3 kg.

9.4.3 Along-Track Stereo by Satellite Tilt

Another method of having along-track stereoscopy using a single optics is to have the satellite itself tilted in the along-track direction so that a scene can be viewed from different angles.

With the launch of the new generation of high-resolution (spatial resolution ≤ 1 m) Earth imaging satellites, starting with IKONOS in 1999, a new era of generating along-track stereo imagery began. These satellites are very agile and capable of pointing the optical axis along and across track in a short time and stabilizing to take images. Therefore, it is possible to take images of a strip with different look angles during one orbital pass. In principle, it is possible to produce a triplet (Figure 9.13). Thus, three stereo pairs can be generated: fore nadir, aft nadir, and fore aft.

Some of the other agile satellites with off-nadir viewing capability include QuickBird, launched in 2001 with 0.6-m resolution, GeoEye, launched in 2008 with a resolution of 0.41 m, Worldview2, launched in 2009 with 0.46-m resolution, and CARTOSAT2B, launched in 2010 with a spatial resolution of 0.8 m. The list is only representative of a number of submeter systems on agile platforms, and the resolution mentioned is for the panchromatic band. As such systems have already been described in Chapter 7, the details of the systems are not repeated here.

References

Bouillon, A., M. Bernard, P. Gigord, A. Orsoni, V. Rudowski, and A. Baudoin. 2006. SPOT 5 HRS geometric performances: Using block adjustment as a key issue to improve quality of DEM generation. *ISPRS Journal of Photogrammetry and Remote Sensing* 60(3): 134–146.

Gugan, D. J. and I. J. Dowman. 1988. Topographic mapping from SPOT imagery. *Photogrammetric Engineering and Remote Sensing* 54(10): 1409–1414.

Jacobsen, K. 1999. Geometric and information potential of IRS-1C PAN-images. Geoscience and Remote Sensing Symposium. IGARSS'99. *Proceedings of IEEE International* 1: 428–430.

Jaxa portal. http://www.eorc.jaxa.jp/ALOS/en/about/prism.htm (accessed on May 15, 2014).

Kiran Kumar, A. S., A. Roy Chowdhury, A. Banerjee et.al. 2009. Terrain Mapping Camera: A stereoscopic high-resolution instrument on Chandrayaan-1. *Current Science* 96(4): 492–495.

Kocaman, S. and A. Gruen. 2008. Orientation and self-calibration of ALOS PRISM imagery. *The Photogrammetric Record* 23: 323–340.

Kornus, W., R. Alamús, A. Ruiz, and J. Talaya. 2006. DEM generation from SPOT-5 3-fold along track stereoscopic imagery using autocalibration. *ISPRS Journal of Photogrammetry and Remote Sensing* 60:147–159.

Lanzl, F. 1986. The Monocular Electro-Optical Stereo Scanner (MEOSS) satellite experiment. *International Archives of Photogrammetry and Remote Sensing* 26–1: 617–620.

Nagarajan, N. and M. S. Jayashree. 1995. Computation of yaw program to compensate the effect of Earth rotation. *Journal of Spacecraft Technology* 5(3): 42–47.

NRSA. 1995. IRS1 C Data User's Handbook, National Remote Sensing Agency, Hyderabad, India. NDC/CPG/NRSA/IRS1-C/HB/Ver 1.0/Sept.1995.

NRSC. 2006. *CartoSat-1 Data User's Handbook*. National Remote Sensing Agency, Hyderabad, India. http://www.nrsc.gov.in/pdf/hcartosat1.pdf (accessed on June 21, 2014).

Osawa, Y., and T. Hamazaki. 2000. Japanese spaceborne three-line-sensor and its mapping capability. *International Archives of Photogrammetry and Remote Sensing* III (B4): 779–782, Amsterdam, Netherlands. http://www.isprs.org/proceedings/III/congress/part4/779_ III-part4.pdf (accessed on May 15, 2014).

Osawa, Y. 2001. PRISM, AVNIR-2, PALSAR—ALOS's major mission instruments at ALOS Symposium. http://www.eorc.jaxa.jp/ALOS/conf/symp/2001/5.pdf (accessed on May 15, 2014).

Planche, G., C. Massol, and L. Maggiori. 2004. HRS camera: A development and in-orbit success. *Proceedings of the 5th International Conference on Space Optics* (ICSO 2004), Toulouse, France. ed. Warmbein B. ESA SP-554, ESA Publications Division, Noordwijk, Netherlands, ISBN 92-9092-865-4: 157–164.

SPOT 5 eo portal. https://directory.eoportal.org/web/eoportal/satellite-missions/s/spot-5 (Accessed on June 2, 2014).

Subrahmanyam, D., S. A. Kuriakose, P. Kumar, J. Desai, B. Gupta, and B. N. Sharma. 2006. Design and development of the Cartosat payload for IRS P5 mission. *Proceedings of SPIE* 6405: 640517.1–640517.10.

Tadono, T., M. Shimada, H. Murakami, and J. Takaku. 2009. Calibration of PRISM and AVNIR-2 Onboard ALOS "Daichi." *IEEE Transactions on Geoscience and Remote Sensing*. 47(12): 4042–4050.

Wells, N. S. 1990. The stereoscopic geometry of the remote sensing 'optical mapping instrument'. *Royal Aerospace Establishment Technical Report 90032*. PDF Url: ADA228810.

Westin Torbjörn. 1996. Photogrammetric potential of JERS1 OPS, ISPRS Commission IV, WG2. *International Archives of Photogrammetry and Remote Sensing* I (B4), http://www.spacemetric.com/files/references/jers.pdf (accessed on May 15, 2014).

Welch, R. 1981. Cartographic potential of a spacecraft line-array camera system: Stereosat. *Photogrammetric Engineering and Remote Sensing* 47(8):1173–1185.

10

Journey from Ground to Space

10.1 Introduction

In this chapter, we shall discuss various aspects involved in qualifying a payload to make it fit for use in space. This is not a "cookbook" for space qualification, but the intention is to familiarize the reader with various aspects involved in realizing "flight-worthy" space hardware.

Building space hardware is a challenging task for many reasons. The satellite/payload has to survive severe mechanical stress during launch and operate in the hostile environment of space dealing with elements such as hard vacuum, radiation, extreme temperatures, and temperature fluctuation. The system is expected to function for a long duration with very little help if any malfunctioning happens. To cap it all, the space systems are expected to *do more with less*—less mass, less volume, less power! Before we discuss the aspects regarding the realization of space hardware, it is useful to briefly review the environment in which the satellite has to survive during launch and operation.

10.2 Launch Environment

A spacecraft is subjected to very large dynamic forces of various nature from its launch vehicle starting from liftoff through final engine shutdown at orbit insertion. High levels of acoustic noise are generated during the initial phase of launch and the subsequent ascent phase due to turbulence-induced pressure fluctuations of the surrounding air. The fluctuating external pressure field causes an oscillatory response of the rocket structure, which is transmitted through the spacecraft fairing (attachment ring) in the form of random vibration (Petkov 2003). Random vibration input occurs over a broad frequency range, from about 10 to 2000 Hz. The exact magnitude of these is dependent on the type of launch vehicle used. These vibrations can cause stress concentrations on the structure, which if not taken care of in the design, can lead to eventual failure. Therefore, it is necessary to model the acoustic environment and its transmissibility to satellite and other subsystems by coupled load analysis.

The spacecraft and associated subsystems are also subjected to shock. The ignition of pyrotechnic devices used to separate structural systems from a launch vehicle (such as strap-on rockets and heat shields) induce transients characterized by high peak acceleration, high-frequency content, and short duration (Arenas and Margasahayam 2006). The spacecraft and payload structure should be designed so as to withstand the anticipated shock environment.

10.3   Space Environment

A satellite when placed in orbit is subjected to an environment entirely different from what it has experienced on Earth. The space environment includes extreme temperature, hard vacuum, energetic particle and electromagnetic radiation, plasmas, micrometeoroids, and space debris. Most of these can cause serious problems for space systems and the payload designer needs to take them into account during the development of the hardware. In this section, we shall briefly describe how these environments affect space hardware.

10.3.1   Thermal Environment

External radiation that contributes to heating up the satellite includes

- Direct solar flux, which outside of the atmosphere is approximately 1375 watts/m²
- Solar flux reflected by the Earth and clouds (albedo)
- Infrared (IR) energy emitted from Earth

In addition to the external sources, the heat dissipated by electrical and electronic equipment also contributes to the thermal environment of the satellite. Because of the hard vacuum at the orbital altitude, the satellite's temperature is controlled by the abovementioned input heat load and the radiative exchanges with deep space that can be considered to be about 4°K. The heat input itself can vary over an orbit and thus the satellite temperature can fluctuate to extreme hot and cold temperatures if proper thermal control is not adopted. A spacecraft contains many components that will function properly only if they are maintained within specified temperature ranges. The goal of spacecraft thermal control (STC) design is to maintain the temperature of all subsystems within the desired range of temperature at all phases of the spacecraft mission. Passive thermal control techniques, such as thermal coating, thermal insulation (multilayer insulation [MLI]), heat sinks, and phase change materials, and active thermal control techniques such as electrical heaters and cryocoolers, are used to maintain the temperature of different subsystems within the desired temperature limits. Apart from this, surface treatments such as gold/silver plating and anodizing are also carried out on subsystems to improve the surface emission/absorptance properties. Proper thermal modeling is the starting point for adopting thermal control. This is carried out by suitable mathematical models that take into account the surface properties of the materials and the radiation exchange taking place between different subsystems. More details can be found in Clawson et al. (2002).

10.3.2 Vacuum

The satellite orbit is characterized by hard vacuum, though it still contains some residual particles. The typical pressure in the Low Earth Orbit (LEO) environment is below 10^{-8} torr, and lower than this pressure for higher orbits. Such a hard vacuum environment produces three potential problems that can affect the functioning of the space system. These are outgassing, cold welding, and restricted mode of heat transfer.

Release of gases from material is called outgassing. Most materials outgas to some extent in a vacuum environment. The outgassing can have two effects on space systems. Composites such as graphite fiber-reinforced epoxy are often selected as structural material because of their high stiffness to weight ratio. Unless specifically taken care of, these materials are hygroscopic and absorb water vapor from air. The water vapor will be released in orbit under vacuum. With time, the release of water vapor can produce dimensional changes to the composite material affecting the performance of the subsystem where the composites are used. The second issue is that the vapor produced by outgassing can get deposited on optical surfaces, thereby affecting its performance. The outgassing problem is dealt with by selecting only those materials having low outgassing properties specified in terms of total mass loss (TML < 1%) and collected volatile condensable material (CVCM < 0.1%). Ultralow outgassing properties are preferred in the vicinity of optical components where CVCM should be as low as 0.01%. The camera designer should consider this aspect while selecting materials and also present this as one of the requirements to the satellite builders.

Another problem created by hard vacuum is cold welding. If two clean metal surfaces are brought together in a hard vacuum (to keep them clean), molecules of metals diffuse into each other, without being heated, resulting in one solid piece of metal; this phenomenon is called cold welding. On Earth, even those so-called "clean surfaces" are generally covered by physically or chemically absorbed layers or other contaminants such as oxide and nitride layers, which act as natural protection layers against cold welding. In a hard vacuum space environment, once these layers are removed by wear or outgassing, they are not rebuilt and the exposed clean metal surfaces show a higher susceptibility to cold welding. Wear on the surface of a mechanism (e.g., relay contacts) can happen due to impacts during repeated closing, which can eventually remove the mechanism's coated surface layers and produce a pure metal surface, thereby increasing the tendency of these contacting surfaces to "cold weld" to each other. Vibrations occurring during launch or due to other disturbances such as the movement of antennas in space can lead to small oscillating movements in the contact resulting in a special wear process that is referred to as "fretting." This lateral motion can also cause surface destruction and may lead to cold welding. A typical opening/closing mechanism can fail if the adhesion force exceeds the force available to open this mechanism. Gears, hinges, and a variety of mechanisms can be subjected to cold welding due to impact or fretting (Merstallinger et al. 2009). Cold welding is believed to be the cause of some of the observed failures in certain satellite systems. These include the Galileo spacecraft's partially unfurled antenna and failure of a calibration mechanism in one of the Meteosat satellites (Merstallinger et al. 1995). The aspect of cold welding relevant to opening and closing of mechanisms placed on spacecraft needs special attention. The possibility of cold welding can be avoided only by ensuring that atomically clean metal surfaces do not come in contact. Metal-to-metal contact may be avoided by using nonmetal separators such as solid film lubrication between metal components. The surfaces may be anodized to avoid any contact between pure metal surfaces.

The third issue associated with operating in vacuum is that there is no convection and heat transfer can take place only through conduction and radiation. This constraint should be taken into account in the design of thermal management of the spacecraft.

10.3.3   Radiation Environment

The primary radiation fields in a satellite orbit are the galactic cosmic rays (GCR) and the Sun. The Sun is an active source of radiation emitting electrons, protons, heavy ions, and photons. Apart from a steady flow of radiation from the Sun, solar flares occurring during times of high solar activity also contribute varying quantities of electrons, protons, and lower energy heavy ions. Galactic cosmic rays are high-energy charged particles considered to originate outside our solar system, primarily consisting of protons (~87%), alpha particles (~12%), and heavy ions.

Apart from the galactic and solar cosmic ray particles, the other charged particles reaching the Earth are those trapped in the Earth's magnetic field in two zones encircling the Earth, called the Van Allen radiation belts. The Van Allen radiation belts are centered along the Earth's magnetic equator in two bands. The inner belt extends from roughly 1000 to 6000 km above the Earth, consisting mostly of protons with energies greater than 10 Mev and much lesser number of electrons with an average energy of 1–5 Mev. There is an outer belt, from roughly 15,000 to 25,000 km above the Earth primarily consisting of electrons having an average energy between 10 and 100 Mev. Most of the particles that form the belts come from the solar wind, a continuous flow of particles emitted by the Sun in all directions. Because the Earth's magnetic axis is tilted with respect to the Earth's rotational axis by an angle of ~11 degrees, the inner Van Allen belt is closest to the Earth's surface over the South Atlantic Ocean. This is usually referred to as the South Atlantic Anomaly and it roughly lies in about the 0- to 50-degree latitude belt over the South American region. This means that if a satellite passes over the South Atlantic, it will be closer to the radiation belts and receive a larger-than-average dose of radiation.

When energetic charged particles interact with spacecraft materials and are decelerated, secondary charged particles and x-rays are emitted, which also adds to the total radiation field to which the satellite is subjected.

The radiation energy density absorbed by a material is called *dose*. When a mass "dm" is irradiated and if "dE" is the mean energy absorbed, then the dose is expressed as

$$D = \frac{dE}{dm} \tag{10.1}$$

The unit of absorbed dose in a radiation environment is expressed in rad (radiation absorbed dose). One rad is equal to an absorbed dose of 100 ergs per gram or 0.01 joule per kilogram (J/kg). The material is always specified in parentheses, for example, rad (Si). The international system of units (SI) for absorbed dose is referred as "gray" (GY); 1 GY = 100 rad. The term used to quantify radiation field is *fluence*. The fluence (Φ) is expressed in terms of the number of radiation particles (N) incident per unit area around a point.

The radiation environment the spacecraft experiences depends on orbital altitude and inclination, whereas total radiation dose depends on the radiation environment as well as the mission life, and assumptions made about solar flares expected during the mission life. In the LEO of 100- to 1000-km-altitude range, the typical dose rate is about 0.1 krad (Si) per year. Satellites that pass near the poles, where geomagnetic shielding is not effective, are subjected to larger flux of cosmic ray and solar flare particles. While calculating the total dose, the higher radiation dose accumulated during the passage through these regions, though for a short period compared to the total orbit period, also needs to be considered. Satellites in Geosynchronous Earth Orbit (GEO) are exposed to the outer radiation belts, solar flares, and cosmic rays. The dose rate is of the order of 10 krad (Si) per year (Petkov 2003).

In addition, satellites are affected by the presence of plasma and atomic oxygen. Apart from the solar radiation, the fraction of the incident sunlight reflected off the Earth, and the Earth's emitted infrared radiation also contributes to the radiation field of the satellite.

The performance of an electronic circuit can be seriously affected in the radiation environment, which can lead to performance degradation or catastrophic failure of the hardware. The major impact of radiation on electronic circuits is to the semiconductor devices. The basic radiation damage mechanisms affecting semiconductor devices are due to ionization and atomic displacement. In ionization, charge carriers (electron-hole pairs) are produced, which diffuse or drift to other locations where they may get trapped, leading to additional concentrations of charge and parasitic fields, thereby affecting the device performance. In displacement damage, incident radiation dislodges atoms from their lattice site thereby altering the electronic properties of the crystal. The impact of radiation on semiconductor devices can be classified into two main categories. They are *total ionizing dose* (TID) and *single-event effect* (SEE).

The impact due to TID is the cumulative long-term degradation of the device when exposed to ionizing radiation. The total amount of radiation energy absorbed depends on the material. In the case of semiconductors, the materials to be considered are primarily silicon and its dioxide and increasingly, compound semiconductors such as gallium arsenide (GaAs). The cumulated dose can induce a degradation/drift of the device electrical and functional parameters, such as leakage current, threshold voltages, and gain. As the dose accumulates, these changes push the component parameters outside of the design range for the circuits in which they are used and can lead to the malfunction of the circuit. Thus, TID is cumulative and its effect is felt after the device has been exposed to radiation for some time. The effect of TID is also dependent on the dose rate (Dasseau and Gasiot 2004). In semiconductor detectors, TID causes an increase in dark current and consequently affects NEΔR.

The SEE is a localized phenomenon, which is produced by a single incident ionizing particle that deposits enough energy onto a sensitive region of a device, resulting in an individual malfunction. In contrast to TID, which is related to long-term response of the devices, the SEE is a short-time response produced by a single ionizing particle. Although the TID uniformly affects the whole device as it is the outcome of several particles randomly hitting the device, the effect due to SEE affects only the localized part of the device corresponding to the position where the particle strikes. The SEE is observable in different ways such as the following:

Single event upset (SEU): This is a change of logical state caused by passage of radiation, for example, bit flips in memory circuits. SEU is usually reversible and does not cause permanent damage to the semiconductor function. Such effects, which are not totally destructive, are referred to as soft errors as against hard errors, which are not recoverable and cause permanent damage (Joint Electron Device Engineering Council)

Single-event latch-up (SEL): An abnormal high-current state in a device caused by the passage of a single energetic particle through sensitive regions of the device structure, resulting in the loss of device functionality. SEL may cause permanent damage to the device (if the current is not externally limited). If the device is not permanently damaged, power cycling of the device ("off" and back "on") may restore normal operation.

Single-event burnout (SEB): Here, the device is permanently damaged due to the passage of high current. This is a hard error that is not recoverable.

A reader interested in more details on radiation damage on semiconductor devices may refer to Janesick (2001).

Some of the semiconductor devices are specially designed and manufactured to reduce the impact of radiation damage—these devices are referred to as radiation-hardened devices. External protection from radiation for TID using a molybdenum, tantalum, or tungsten metal sheet pasted on the semiconductor device is also used. The overall circuit design should be taken care of such that a single error does not jeopardize the function of the device. To mitigate the radiation effects, especially SEU, software logic such as majority voting logic and triple modular redundancy are incorporated in recent designs.

We briefly covered in Chapter 3 the effect of the space environment on optical components. Farmer (1992) gives a review of issues related to the reliability and quality of optics in space environment.

Apart from the radiation damage, space debris and micrometeorites can damage the surface exposed directly to space.

10.4 Space Hardware Realization Approach

Space hardware realization evolves through several phases with critical evaluation at the completion of each phase. The way a space-qualified system is developed is very much standard across most space organizations/engineering companies. We shall discuss in this section the steps followed in realizing a space-qualified Earth observation camera in the broad framework of practices followed for the Indian Remote Sensing Satellite program. The performance specifications of the camera payloads are generally generated by a national committee consisting of all stakeholders. A set of environmental test specifications is generated by the spacecraft project. The executive of the payload (say, the project manager) has to deliver the payload to the satellite project meeting the prescribed specifications under the specified environmental conditions, within the time schedule and allotted budget. A multidisciplinary team supports the project manager in the following broad areas, each led by a project engineer (PE) (the nomenclatures PM, PE, etc. could vary from project to project depending on the size of the task).

- Optics
- Focal plane and associated systems
- Electronics
- Mechanical
- System integration and characterization
- Quality (or product) assurance

An independent quality assurance team oversees the quality and reliability aspects and reports to the head of the institution.

The task of the project manager and his team is to evolve a design such that the hardware meets the specification with adequate margin through all the specified environmental tests. The first task is to study various configurations for each of the subsystems so that the total system meets the mission goal. In doing so, one takes into account past experiences with similar subsystems for space use. The study could also address whether some subsystems already flown in earlier missions can be adopted as is or with marginal modifications, which could reduce cost and development time. However, a proven technology may not be optimal. Therefore, it is also important to address the availability of potential newer technology that can meet the mission goals more efficiently than a proven conventional system. During the early 1980s, the decision to have CCD-based cameras for IRS was in a sense a calculated "risk" when optomechanical scanners were the flight-proven technology for Earth observation. The design should address the fabrication issues likely to be encountered in the realization of the hardware. For example, the design on paper may give excellent performance, but if the tolerances are so stringent that it is not possible to fabricate or fabrication is exorbitantly expensive or if proper operation of the camera requires very tight temperature control, then the design cannot be implemented in practice. These are conceptual designs using analytical tools wherever required. This is referred to as the conceptual design phase where all possible options are considered to arrive at a baseline configuration. Following this, a baseline design document is prepared highlighting the trade-off studies carried out, which will be reviewed by an independent expert team. The review team assesses the adequacy of the system configuration to meet the mission requirements.

Once the baseline configuration has been vetted by the review team, the payload team starts development of the different subsystems, that is, the breadboard model (BBM). This is a developmental model for the designers to verify the design concept of the baseline configuration and to ensure that the system as a whole meets the performance goals. However, to control time and cost, the BBM uses commercial components. The mechanical design is as per the proposed baseline configuration, but may not use the final material to reduce time and cost. For example, if carbon fiber reinforced polymer (CFRP) is planned for a metering structure of the telescope, because of its long lead time for procurement, an alternate material could be used. Similarly, the redundant circuits that will be in the flight model (FM) could be avoided in the BBM, if the functional testing is not compromised. At this stage, the Reliability & Quality Assurance (R&QA) plan, the test philosophy, and the interface to spacecraft are also worked out. Along with the BBM development, the design and realization of ground support equipment (GSE) required to align/integrate and characterize the payload is also initiated. The BBM is not subjected to mechanical environmental test. However the electronics could be subjected to qualification level temperature extremes, to ensure design margins keeping in mind the limitations of the components used. The design and test results are subjected to a formal review—Preliminary Design Review (PDR). Since based on the PDR review the configuration is expected to be frozen, the designer should take utmost care that all the design and fabrication issues are well understood and correction implemented wherever required before formal PDR is initiated. The author used to conduct internal reviews, before PDR, for all subsystems involving experts within the team directly not involved in the design. Instead of formal presentations it used to be a "roundtable review" where the reviewers are encouraged to ask probing questions and the designer has to respond. We have found that it is a very effective way to get into the bottom of design problems, if any.

After successful completion of the PDR, the recommendations are incorporated in the design to fabricate the EM. (Depending on the confidence of the design, to reduce delay due to component availability, action could be initiated for procurement of the components as soon as functional tests on the BBM are satisfactorily completed, without waiting for the PDR.) The components used for the EM realization are identical to those intended to be used for the FM in terms of electrical and environmental (except for radiation environment) specifications, but can be of a lower quality grade than flight grade components. For example if the flight requires radiation-hardened devices the EM could use components of the same electrical quality but they need not be radiation hardened. Except for the lower quality standard of components used for electrical parts, materials, etc. the EM is fully representative of the FM in form, fit, and function. The EM is subjected to full qualification level tests (Section 10.5).

The EM is used to verify all of the satellite's electrical interfaces, and to demonstrate that the payload can meet the required performance goals after integrating to the satellite. Along with the development of EM, the procedure for integration, alignment, and characterization is established, thereby ensuring the adequacy of GSE. The EM test results form a database for future reference. Once satisfactorily integrated with the satellite, the results are subjected to a formal review—critical design review (CDR). The purpose of the CDR is to ensure that the design has been carried out correctly and that the payload will fulfill the performance requirements. The CDR team assesses the action taken on the PDR action items and evaluates the integrated performance of the subsystem with the satellite to meet the set mission goals.

After the successful completion of the CDR, the designer is authorized to fabricate the FM. The FM, after satisfactory completion of performance tests, is subjected to acceptance-level environmental tests and delivered to the satellite project for integration with the satellite.

The roadmap of realizing space hardware is similar for most satellite projects, though the nomenclature may differ from space agency to space agency. Basically, the project activities are split into different phases so that the project can be gradually developed from initial concept to final deliverable hardware with periodic monitoring of the mission goal compliance. Table 10.1 outlines typical NASA project life-cycle phases.

TABLE 10.1

Typical NASA Project Life-Cycle Phases

Phase	Purpose	Typical Output
1. Pre-Phase A: Concept Studies	To produce a broad spectrum of ideas and alternatives for missions from which new programs/projects can be selected. Determine feasibility of desired system, develop mission concepts, draft system-level requirements, and identify potential technology needs	Feasible system concepts in the form of simulations, analysis, study reports, models, and mockups
2. Phase A: Concept and Technology Development	To determine the feasibility and desirability of a suggested new major system and establish an initial baseline compatibility with NASA's strategic plans. Develop final mission concept, system-level requirements, and needed system structure technology developments	System concept definition in the form of simulations, analysis, engineering models, and mockups and trade study definition
3. Phase B: Preliminary Design and Technology Completion	To define the project in enough detail to establish an initial baseline capable of meeting mission needs. Develop system structure end product (and enabling product) requirements and generate a preliminary design for each system structure end product	End products in the form of mockups, trade study results, specification and interface documents, and completion prototypes
4. Phase C: Final Design and Fabrication	To complete the detailed design of the system (and its associated subsystems, including its operations systems), fabricate hardware, and code software. Generate final designs for each system structure end product	End product detailed designs, end product component fabrication, and software development
5. Phase D: System Assembly, Integration and Test, Launch	To assemble and integrate the products to create the system, meanwhile developing confidence that it will be able to meet the system requirements. Launch and prepare for operations. Perform system end-product implementation, assembly, integration and test, and transition to use	Operations-ready system end product with supporting related enabling products
6. Phase E: Operations and Sustainment	To conduct the mission and meet the initially identified need and maintain support for that need. Implement the mission operations plan	Desired system
7. Phase F: Closeout	To implement the systems decommissioning/disposal plan developed in Phase E and perform analyses of the returned data and any returned samples	Product closeout

Source: NASA/SP-2007-6105. 2007. *NASA Systems Engineering Handbook.* http://www.acq.osd. mil/se/docs/NASA-SP-2007-6105-Rev-1-Final-31Dec2007.pdf (accessed on May 15, 2014.

10.4.1 Model Philosophy

What was presented earlier is the workflow from the conceptual level to the delivery of a flight-worthy payload to the satellite project for final integration with the satellite FM. Every satellite project has a model philosophy to ensure all the subsystems and the satellite bus are tested to develop confidence that the total system will perform as per the planned mission goals. In the early years of satellite building, a number of hardware models were developed to ensure satellite design is robust. To qualify the structural design and to correlate with the mathematical model, a *structural hardware model* is developed, which includes mechanical dummies (to simulate mass) of all the subsystems. A thermal model fully representing the thermal characteristics is realized to qualify thermal design and correlate with the mathematical model. As a cost-saving measure, it is possible to have a single-structural thermal model (STM), combining the objectives of the structural model and thermal model. The analytical capability has substantially improved over time and one could theoretically model with better confidence thereby reducing the actual hardware models.

The number of models to be developed for the payload depends on the heritage of the payload design. In general, we can consider the following levels of maturity:

1. A new design, that is, no previous flight heritage
2. Hardware substantially modified from a previous flight instrument
3. Previously qualified and flown successfully, but with marginal modifications to take care of some anomaly seen during the mission and/or small addition to improve performance
4. Repeating identical hardware

The confidence level of the product increases as one moves from (1) to (4). We shall explain this aspect with examples. The LISS 1 & 2 on board IRS-1B were identical to that flown in IRS-1A and fall in category (4). In such cases, only the FM needs to be fabricated, which is tested to acceptance level. In the case of (3), depending on the confidence of the modifications carried out through analytical evaluation, a proto-flight model (PFM) approach could be adopted. In PFM, all quality aspects required for a flight hardware is followed undergoes various tests such that the magnitude of the test levels are same as that used for qualification tests but for a duration corresponding to the acceptance test. In the case of (2), it is desirable to go through an EM that is tested to qualification level and then produce the flight hardware. Category (1) goes through all the three models—BBM, EM, and FM—as discussed earlier.

The model philosophy varies from project to project and what is given above is an indicative scenario. As the number of models increases, the cost and schedule increase. Therefore, a judicial decision has to be made whether, by adding more models, the confidence on the reliability of the hardware has increased.

10.5   Environmental Tests

Environmental tests are one of the vital steps to demonstrate beyond doubt that the hardware will survive and function as expected under the antici-pated environmental conditions. The important environmental conditions to which the payloads are tested include

- Mechanical: vibration, shock
- Thermal: temperature extremes and fluctuations
- Vacuum: hard vacuum achievable in the laboratory
- Electromagnetic: electromagnetic compatibility (EMC), electromag-netic interference (EMI), electrostatic discharge (ESD)

In a fully integrated satellite, there are conducted and radiated electromag-netic fields and, therefore, EMC tests are carried out to ensure that the EM fields do not produce EMI that affects the functioning of the payload. It is also important that the payload should not produce an electromagnetic field beyond a certain limit to hinder the operation of other subsystems of the satellite. Satellites in geostationary orbit are prone to ESD due to charging of the satellite in the plasma environment, which could produce status changes in the electronic logic circuits.

There should be a conscious effort right from the initial design to incorpo-rate practices to take care of EMC/EMI/ESD. Some of the important consid-erations to take care of these aspects include grounding schemes and harness design/layout (Hariharan et al. 2008). Guidelines for grounding architecture can be found in NASA-HDBK-4001. Some practical "tips" can be found in the lecture notes of Åhlén (2005).

10.5.1   Mechanical Tests

The objective of the mechanical tests is to verify the capability of the satellite structure and components to withstand various mechanical loads such as vibration and shock loads introduced primarily during launch. The vibration levels to which the payload will be subjected to by the launch vehicle vibro-acoustic environment are estimated using the mathematical models of the spacecraft-launcher through coupled load analysis, which forms the basis for test level. This is supplied by the satellite project. The vibration test uses two excitation methods—sinusoidal vibration and random vibration. In sinusoi-dal vibration, the amplitude of the vibration load (typically specified as dis-placement or acceleration), is characterized by an oscillating motion wherein loads at varying sinusoidal frequencies are applied to excite the structure. However, they are of discrete amplitude, frequency, and phase at any instant in time. Initially, a low-level sinusoidal excitation is given to find out frequencies at which resonances take place. At such resonance, the structural response of the hardware would exceed the design capability. To avoid overtesting, the acceleration input levels around resonant frequencies are reduced, provided the coupled load analysis does not indicate abnormally a high input level at that frequency. This is referred to as notching. The payload is vibrated as per the specified levels in three perpendicular axes sequentially.

The most realistic vibration environment for a payload that can be reproduced in laboratory is the random vibration. Unlike a sine excitation, in random vibration numerous frequencies may be excited at the same time and the motion is nondeterministic, that is, the future behavior cannot be precisely predicted. In random vibration, the instantaneous amplitudes vary in a random manner, usually assumed to be according to a "normal" (Gaussian) curve. These random excitations are usually described in terms of a power spectral density (PSD) function. The unit of a PSD is acceleration squared per unit bandwidth (g^2/Hz). Here, g is the root mean square (RMS) value of acceleration over the frequency. PSD represents how the average power is distributed as a function of frequency. The PSD is plotted as a function of frequency, which gives the RMS acceleration of the random signal per unit bandwidth at any frequency. Random vibration testing helps to demonstrate that hardware can withstand the broadband high-frequency vibration environment. The tests are conducted on a special vibration machine such as an electrodynamic shaker.

A typical PSD plot for exercising random vibration is given in Figure 10.1. Qualification test levels are increased typically 3–6 dB above flight acceptance level (which are the expected values based on calculation) to make sure the design is not marginal. The tests are done sequentially in three mutually perpendicular directions. A low-level random vibration test is carried

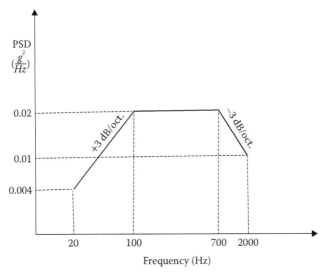

FIGURE 10.1
A typical power spectral density plot for exercising random vibration.

out before and after the full level acceptance/qualification test to verify any change in the response plots, which is an indication of damage/failure.

As discussed earlier, the satellite is subjected to mechanical shocks, which are characterized by high acceleration for very short impact duration. A shock test is carried out to demonstrate the ability of a specimen to withstand the shock loads during launch and operation. Shock tests can be carried out using a drop shock tester or simulating the shock in a vibration shaker system. The shock test is usually done as part of the qualification test only. NASA-STD-7003A gives pyro shock test criteria.

During the launch phase, the payloads are not switched "ON" and hence during the mechanical tests the packages are not powered. After each test, the package under test is checked for any mechanical damage such as loose screws and structural distortions, and the functional performance is also checked.

NASA technical handbook NASA-HDBK-7005 gives the dynamic environments that a spacecraft might be exposed to during its service life from the completion of its manufacture to the completion of its mission and the equipment and procedures used to test a spacecraft and its components. More details on mechanical testing also can be found in ECSS (2013).

10.5.2 Thermovacuum Test

The thermovacuum (TV) test demonstrates the capability of the payload through the operating phase in space wherein the satellite is subjected to vacuum and temperature variations. Because the TV test is expensive, it is preceded by thermal tests at atmospheric pressure. These include the hot and cold storage temperature test, humidity test, and operational temperature test. The storage and humidity tests demonstrate the ability of the design to withstand the environment that might be encountered in transportation and storage of the payload and are conducted as a part of qualification tests only.

The TV test is performed at a vacuum of 10^{-5} torr or better and the payload is subjected to a hot and cold cycle at least 15°C above and below the expected high and low temperatures during the qualification test and 10°C above and below during the acceptance test. A typical temperature profile followed during the TV test for the IRS payload camera is shown in Figure 10.2. This is an active test, that is, the equipment is powered and the performance is monitored periodically. Each electronics subsystem of the payload is initially tested separately to find out any performance deviation and then the integrated payload is finally tested.

During the normal flight sequence, the launch environment is followed by vacuum and potential temperature extremes. To assure that this sequence is followed, the dynamic tests are performed before thermal-vacuum tests. In addition, it has been observed that failure that might be induced by dynamic tests may not be revealed until the thermal vacuum test (NASA 1999). Therefore, it is preferable to perform the thermal-vacuum test after dynamic tests have been performed.

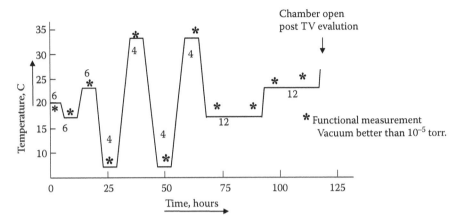

FIGURE 10.2
Typical temperature profiles during thermal vacuum test. The temperature range is project dependent. What is shown is for a typical IRS camera acceptance test. The numbers on the graph represent duration of exposure in hours.

10.5.3 Electromagnetic Interference/Compatibility Tests

Satellite subsystems have varying electrical signals of different frequencies and the test ensures that the design is such that the payload function is not degraded. It is also necessary to ensure that the payload will not interfere with operation of other subsystems through spurious emission. These tests are generally conducted as per MIL-STD-461 and MIL-STD-462 or as per any other approved specifications of the satellite project.

10.5.4 Environmental Test Levels

In general, the environmental tests are carried in two phases. One of the models goes through qualification level tests. The qualification tests are intended to subject the payload to loads more severe than expected in the laboratory, transportation, launch, and finally in-orbit operation conditions. Here, various stress levels are more than the anticipated analytical model values. This is to establish margins over the uncertainty in the calculations carried out using mathematical models. *Qualification testing, however, should not create conditions that exceed applicable design safety margins or cause unrealistic modes of failure.* As mentioned earlier, some of the tests, such as storage and humidity tests, are carried out to ensure that the payload function does not degrade in the ground environment. Successful completion of the test gives confidence that the design and fabrication of the payload is such that it will perform in orbit without degradation. Qualification units are never flown as they are exposed to harsh test levels with margins well above the maximum expected flight levels.

After successfully undergoing qualification tests, the FMs undergo an acceptance test.

Acceptance tests are intended to demonstrate the flight-worthiness of each deliverable item. The test is to locate latent material and workmanship defects, if any, in an already proven design and establish successful performance in an environment similar to that faced during the mission. The acceptance test levels are lower than the respective qualification levels, but could be higher than the predicted maximum environmental levels. The satellite project specifies the test levels each subsystem has to undergo based on expected levels from modeling. The test matrix for the environmental tests is given in Table 10.2.

In addition to these standard tests, some of the subassemblies, especially employing moving mechanisms (such as the scan mirror assembly), lamps, etc., may undergo offline accelerated life testing to ensure that they perform as per the design throughout the mission life.

10.5.5   Ground Support Equipment/Facilities

For proper evaluation of the performance and handling of the payload, a number of test/support facilities and equipment, other than what we described earlier, are required during the different phases of payload development. These can be broadly classified into mechanical ground support equipment (MGSE), electrical ground support equipment (EGSE), and clean room laboratory. MGSE is required essentially for handling the payload for its movement to different test facilities and final transportation to integration with the satellite. These include various handling fixtures and the transportation container. Transportation of payload may have to take place via road, rail, or air. The container should be designed such that the shock and vibration encountered during transportation transmitted to the payload should be much less than the acceptance test levels. It should also offer protection against contaminants such as moisture, rain, and dust. The containers need

TABLE 10.2

Test Matrix for the Environmental Tests

Parameters	Qualification Test	Acceptance Test
Sine vibration	√	-
Random vibration	√	√
Shock	√	-
Cold storage	√	-
Hot storage	√	-
Humidity	√	-
Thermovacuum	√	√
Operational temperature	√	√
EMI/EMC	√	-

to be fitted with appropriate sensors to monitor the shock and vibration level experienced by the payload during transportation.

The EGSE has many functions to perform during the final integrated performance evaluation of the payload and requires different equipment. When a payload is finally integrated with the satellite, for proper operation, it receives power and other electrical interface signals from the satellite. Therefore, during the test and evaluation of the payload, one requires equipment usually referred to as a Satellite Interface Simulator (SIS) to simulate the interfaces such as power, telecommand, and telemetry of the satellite. In addition, the final output data from the payload has to be analyzed. The data from the payload is in digital form and is accessed by the checkout computer that analyses the data to generate various performance parameters, such as the MTF and S/N. In Section 6.6.5, we already discussed the configuration of a typical checkout system.

It is imperative that all the support instrumentation directly interfacing with the payload should also undergo design review and test and evaluation to ensure any failure of GSE does not propagate and damage the flight hardware. The measurement accuracy of the test instruments should be much better than the corresponding specification of the payload and they should be calibrated before starting the T&E of payload.

10.5.6 Contamination Control

Performance of electro-optical sensors is affected by surface contamination. There are two broad categories of surface contaminants: particulates and molecular film. The instruments working in the ultraviolet region are particularly vulnerable to the latter type. The materials used, not only in the payload but also in the spacecraft as a whole, should have low outgassing property. Particulate contaminants on optical surfaces can scatter light causing loss of signal and enhanced background. A single speck of particle on a CCD cover can cast a shadow on the active surface. Therefore, the assembly and test of electro-optical sensors are carried out in a controlled environment. For this purpose, specially designed clean rooms are used wherein the concentration of airborne particles is controlled as well as temperature and humidity. Clean rooms are planned and constructed using strict protocol and methods. However excellent the clean room design is, the use of the clean room requires strict discipline. For example, the entry and exit to the clean room should be through airlocks and air showers, and the personnel must wear special clothing. Clean rooms are classified according to the number and size of particles permitted per volume of air. A "class 1000" denotes that the number of particles of size 0.5 μm or larger permitted per cubic foot of air is not more than 1000. The whole laboratory cannot be maintained at the required cleanliness level. Therefore, clean air tents/work stations are used where cleanliness of class 100 is required. Though the classification mentioned here has been popularly used, the International Organization of Standards (ISO) has come up with a new standard—ISO 14644-1 (Whyte 2010).

10.6   Reviews

Formal reviews are one of the important steps for reducing in-orbit failures. The technical review is an evaluation of the products and processes by an independent group of experts by critically assessing the design and fabrication based on the documents and test results presented to them. The designers should appreciate the role of the reviewers as they are trying to ensure that the payload functions reliably in orbit. The best result from a review can happen only when designers are frank and transparent in presenting their failure and concerns. This attitude should prevail at all levels. The author, when in charge of camera development, while welcoming the review committee, used to conclude by saying "beat us now so that you can pat us on the back after the launch."

We discussed in Section 10.4 three formal reviews at different stages of the development of the payloads. We shall here elaborate their role.

10.6.1   Baseline Design Review

Baseline design review (BDR) is also known as conceptual design review. This is carried out during the concept phase. BDR reviews various possible configurations of the subsystems to meet the mission goal and the feasibility of the suggested system configuration to realize the payload within the schedule and cost.

10.6.2   Preliminary Design Review (PDR)

Preliminary design review (PDR) is conducted at the end of the initial development phase when the block diagrams and the specifications of major elements are finalized. The review critically examines each design to ensure that the proposed basic design/fabrication approach can realize a payload meeting the required specifications. At this stage, if the realizability of any submodule is doubtful in terms of schedule or performance, alternate options to build/buy should be investigated. The compatibility with the spacecraft is also assessed during this stage. The action items generated during the BDR need to be resolved and an action closeout document shall be provided to the PDR committee.

10.6.3   Critical Design Review

Critical design review (CDR) is conducted after hardware is generated to form, fit, and function. The closeout document of the actions generated during the PDR shall be presented during the review. Any suggestions made during PDR that could not be met, and the reason and impact on the reliability and performance on the payload shall be presented to the CDR team. The

CDR committee reviews the test results of the hardware realized and its ability to function properly in the expected environment. This is the last formal review before hardware fabrication. From now onward all the fabrication documents shall be formally reviewed by the R&QA team and approved by a competent authority.

10.6.4   Preshipment Review

The payload, after undergoing satisfactorily all environmental tests, is subjected to a preshipment review (PSR) to authorize that the payload can be integrated with the FM spacecraft. During this review, all the environmental test results are reviewed and if there is any deviation from the expected behavior its implication is analyzed. Any failure that occurred during the development of the FM is also reviewed. Once the PSR committee clears, the payload is authorized for integration with the satellite.

Before each review, a proper review document should be prepared and shall be made available to the members well in advance for their perusal. These three reviews are conducted by formal expert teams and it is desirable the same team conducts all three reviews. Considering the diverse specialties, it is desirable that initially separate reviews are conducted for each subsystem such as optics, focal plane array, mechanical systems, and electronics/electrical systems before a total payload-level review is done. In addition to the formal reviews, the project manager may conduct peer reviews at the circuit/subsystem level as the design progresses, so that midcourse corrections can be taken if found necessary.

In the course of development of the payload, there is a need to monitor and take corrective measures for failures and deviation of performance observed. For this purpose, a number of internal monitoring mechanisms are set up during the project. Some of the important ones are given in Sections 10.6.5 and 10.6.6.

10.6.5   Configuration Change Control and Nonconformance Management

Configuration control is important so that the system is not altered without understanding the impact of the change on other subsystems within the payload and with the satellite, as well as on the validity of any test or qualification already carried out. Therefore, practices and procedures should be evolved to handle changes systematically so that the system maintains its intended functional integrity over time. A review team including the designer, R&QA representative, and peer experts shall be constituted for this purpose. The designer generates a *configuration change* request and submits it to the committee with the reason for change and his or her analysis of the implication of the change. The committee evaluates the proposed change, its desirability, and its impact on the mission and schedule and makes its recommendation. If the configuration change affects the basic performance requirement of the payload or calls for any change in electrical/mechanical or thermal interfaces with spacecraft/ other subsystems, then it requires further consultation with the spacecraft team and end user. All the changes have to be formally approved by this review team, and it is necessary to maintain an inventory of such changes.

Even if there is no change in the configuration during the FM realization phase, it is possible that some of the products do not exactly confirm to the original projection. For example, it could happen that a mechanical item after fabrication has some dimensions that are not within the tolerance specified. Under similar situations, it is necessary to assess whether the part can be accepted as it is or whether it needs to be refabricated. All nonconformances related to components, materials, fabrication processes, or workmanship noticed or observed once the configuration is finalized must be reviewed and dealt with. The earlier or any other duly constituted committee shall assess whether the nonconformances put the mission at risk, and whether some additional validation is required to ensure its acceptability or whether the parts can be accepted as they are. On the basis of these assessments, the committee makes a suitable disposition. The disposition can be in the form of some remedial rework, a total rejection of the item, or use as is in cases where the nonconformance is minor. The nonconformance management procedure must be implemented even if the deviations appear trivial. All such nonconformance and its disposition shall be properly recorded. This will be useful for analysis if, in the future, some failure/non-nominal performance occurs.

10.6.6   Failure Review Committee

During the realization of the space hardware, failures do happen due to various reasons. It could be a design inadequacy or mere accident. A part or a system encompassing hardware or software is considered to have failed under one of the following three conditions:

1. When the system becomes totally inoperable
2. When the deterioration has made the performance of the system erratic/unreliable
3. When it is still operable but is no longer able to perform its planned functions satisfactorily

Conditions (1) and (2) are unambiguous and are easily identified. However, the last condition involves a certain amount of subjective assessment as to what is considered satisfactory functioning. In such cases, functional specification and the operational tolerances of the system should be the basis for judging the malfunction observed (ISRO 2005). A standing Failure Analysis Board (FAB) shall be constituted by the project director for each project or by the head of the institution where the payload is fabricated. To get the best feedback, separate boards may be necessary for different disciplines. Once the configuration is finalized, a failure, however trivial it may appear, shall be referred to FAB. As electronics component count is the largest in the payload, most of the failures are related to electronics components. The designers must be transparent and bring all the failures to the notice of FAB in a systematic formal way. *Never ever replace a failed component unless the cause is understood.* If the failure is due to design inadequacy, a thorough review is to be carried out on whether similar design is elsewhere in the system and the design needs to be modified. In the case of failure due to the generic nature of the component, all the components used from the batch may have to be replaced.

10.7   Parts/Components Procurement

The quality of the parts selected is an important factor in realizing reliable hardware. The components meeting the standards required for space use are manufactured only by a select few vendors and some parts may have to be custom-built. Therefore, proper product selection at an early stage of the project allows for its availability as per the project schedule and at optimal competitive prices while maintaining reliability and performance. For space-borne cameras, as parameters of optics are specific to a mission, they are custom-built. In many cases, focal plane arrays are also designed as per the specific configuration. Generating the performance parameters/specifications, test requirements, and the quality criteria needs special attention. To procure custom-built parts, it is a good practice to send letters to manufacturers in the field giving broad requirements to assess their interest in fabricating the product. This helps to identify potential vendors in advance of finalizing the specifications.

Electronic, electric, and electromechanical (EEE) parts are the basis of all electronics. They significantly determine the quality and reliability of electronic systems, and hence selection of active and passive EEE parts requires special attention concerning, packaging, temperature range of operation, and radiation tolerance/susceptibility (NASA 2003). The number of types is higher for EEE parts compared to any other components in a camera and have to be procured from different manufacturers; hence their availability can be in the critical path of the payload fabrication. In addition, on a parts count basis the failure probability is highest for EEE components. Failure of a single component may lead to poor performance of the system or could be the cause for losing the complete mission. Therefore, EEE parts procurement should follow a strict product assurance plan. High-reliability (Hi-Rel) Class-S space-qualified components are generally procured meeting the specifications of MIL-PRF-38534 for camera electronics. These components undergo all the screening tests at the source.

In the case of nonavailability of any component in this grade, alternative parts of either MIL-STD-883H or industrial grade are procured and qualification/screening tests are done in-house before their use in the FM.

One of the important criteria for EEE parts and also for mechanical part selection is derating. Derating of a part is the intentional reduction of its applied stress, with respect to its rated stress, so that there is a margin between the applied stress and the demonstrated limit of the part's capabilities. Derating decreases the degradation/failure rate, thereby ensuring a reliable operation during the designed life time. Therefore, the circuit design should incorporate derating requirements at the design stage itself.

Generally, each space agency has a preferred part list (PPL). The PPL consists of a list of the parts approved for use in a satellite consistent with reliability and quality requirement. The parts for use in space systems are generally selected from the PPL. If a certain required part is not available in the PPL, the component has to be specially selected and qualified as per established procedure before being used in the system. As a general guideline, it is desirable to reduce the number of types of EEE components to be used in the payload, which will help in reducing cost and procurement hassle. NASA (2003) gives baseline criteria for selection, screening, qualification, and derating of EEE parts for use on NASA GSFC space flight projects. More information on technical requirements for electronic parts, materials, and processes (electronic PMP) used in the design, development, and fabrication of space and launch vehicles can be obtained from Robertson et al. (2006). Once parts arrive after undergoing mandatory tests, they have to be stored in a place with a controlled environment, referred to as a bonded store, with access only to authorized personal.

Another important aspect when using optical payload is outgassing. As a general guideline, the materials used should exhibit a total mass loss (TML) of not more than 1.0% and a collected volatile condensable material (CVCM) of not more than 0.1%. Because the payload is part of the satellite, the payload manager should insist early in the program for a Contamination Control Plan for the whole satellite.

10.8 Reliability and Quality Assurance

A reliability and quality assurance program plays a very significant role in any industry. In building space hardware, it is very vital as we have very little flexibility to correct any failure happening after launch. The R&QA team gives an independent assessment that the system actually produced and delivered shall function properly as per the design goals over its intended life. The quality and reliability philosophy is followed in a systematic manner throughout the project/payload development cycle through a product assurance program. A document called the "Product Assurance (PA) Plan" is generated early in the project life cycle. The PA plan describes quality- and reliability–related activity to be followed during the realization of the subsystems and integrated payload and also lists documents to be generated and reviewed, for each activity. The PA plan is tailored for the product assurance activities for a particular project. For example, an operational communication satellite with a 15-year mission life requires that the components used are of the highest quality, whereas an experimental mission may use commercial off-the-shelf components (COTS) to save money and time.

Thus the R&QA aspects encompass almost all activities of satellite realization from concept to completion. For that reason, it is essential that an R&QA team with a leader (say, a project manager [PM]) be constituted right at the initiation of the payload project. The R&QA team is an independent body and the PM (R&QA) reports to the head of the institution where the payload is made. The R&QA responsibility spans a large spectrum from analyzing the design to evolving the final test philosophy. Some of the tasks carried out under the R&QA role include parts and materials selection guidelines, fabrication process control guidelines, test and evaluation guidelines, and design analysis, as well as comprehensive nonconformance management, statistical reliability estimates, and analysis such as Failure Modes Effect and Criticality Analysis (FMECA), Worst Case Circuit Analysis (WCCA), and component derating criteria. Because all components may not be available from vendors whose production line is approved for space hardware production, qualifying new vendors is another essential activity of R&QA. Vendor qualification has several key issues such as certification of production line, training and certification of manpower, and periodical auditing of vendor's facilities. With many systems having embedded software, software QA is also crucial.

These efforts initiated at the start of projects should continue during the entire duration of the project. The R&QA activities provide the basis for management to take appropriate technomanagerial decisions to ensure that the payload performs as expected with the required reliability goals. Needless to stress the fact that PM (R&QA) should work in tandem with the payload manager to ensure that not only we have a reliable payload, but it is also possible to realize it on schedule and within the budget.

What is given previously is not an exhaustive function of R&QA, but only to point out the need for a robust reliability and quality assurance program to realize a payload meeting all mission goals.

Those who develop space hardware—beware of Murphy's Law
"If anything can go wrong, it will!!"
So do not leave anything for chance.

References

Åhlén, L. 2005. Grounding and Shielding, lecture notes. http://www.ltu.se/ cmsfs/1.59472!/gnd_shi_rym4_handout_a.pdf/ (accessed on June 25, 2014).

Arenas, J. P. and R. N. Margasahayam. 2006. Noise and vibration of spacecraft structures. Ingeniare. Revista chilena de ingeniería, 14(3):251–226. http://www.scielo.cl/pdf/ingeniare/v14n3/art09.pdf (accessed on May 14, 2014).

Clawson, J. F., G. T. Tsuyuki, B. J. Anderson et al. 2002. *Spacecraft Thermal Control Handbook*. ed. David G. Gilmore. 2nd Edition. American Institute of Aeronautics and Astronautics, Inc. The Aerospace Press, CA.

Dasseau, L. and J. Gasiot. 2004. *Radiation Effects and Soft Errors in Integrated Circuits and Electronics Devices*. World Scientific Publishing Company, Singapore. ISBN 981-238-940-7.

ECSS. 2013. *Requirements & Standards Division Spacecraft Mechanical Loads Analysis Handbook*. ECSS-E-HB-32-26A. http://www.ltas-vis.ulg.ac.be/cmsms/uploads/File/ECSS-E-HB-32-26A_19February2013_.pdf (accessed on May 15, 2014)

Farmer, V. R. 1992. Review of reliability and quality assurance issues for space optics systems. *Proc. of SPIE*. 1761:14–24.

Hariharan, V. K., P. V. N. Murthy, A. Damodaran et al. 2008. Satellite EMI/ESD control plan. *10th International Conference on Electromagnetic Interference & Compatibility. 2008. INCEMIC*: 501–507.

ISRO. 2005. Failure Reporting, Analysis and Corrective Action Procedures. ISRO-PAS-100. Issue 2.

Janesick, J. R. 2001. *Scientific Charge-Coupled Devices*. SPIE Press, ISBN: 9780819436986.

Joint Electron Device Engineering Council. http://www.jedec.org/ (accessed on May 15, 2014).

Merstallinger, A., E. Semerad, B. D. Dunn and H. Störi. 1995. Study on adhesion/cold welding under cyclic load and high vacuum. *Proceedings of the Sixth European Space Mechanisms and Tribology Symposium*. Zürich, Switzerland. ESA SP-374.

Merstallinger, A., E. Semerad and B. D. Dunn. 2009. Influence of coatings and alloying on cold welding due to impact and fretting. *ESA STM-279*. http://esmat.esa.int/Publications/Published_papers/STM-279.pdf (accessed on May 15, 2014).

NASA-HDBK-4001. 1998. Electrical grounding architecture for unmanned spacecraft. https://standards.nasa.gov/documents/detail/3314876 (accessed on June 25, 2014).

NASA. 1999. Environmental Test Sequencing. *Public Lessons Learned Entry: 0779*. http://www.nasa.gov/offices/oce/llis/0779.html (accessed on May 15, 2014).

NASA. 2003. Instructions for EEE Parts Selection, Screening, Qualification and Derating. NASA/TP-2003-212242. http://snebulos.mit.edu/projects/reference/NASA-Generic/EEE-INST-002.pdf (accessed on May 15, 2014).

NASA/SP-2007-6105. 2007. *NASA Systems Engineering Handbook*. http://www.acq.osd.mil/se/docs/NASA-SP-2007-6105-Rev-1-Final-31Dec2007.pdf (accessed on May 15, 2014).

Petkov, M. P. 2003. The Effects of Space Environments on Electronic Components. http://trs-new.jpl.nasa.gov/dspace/bitstream/2014/7193/1/03-0863.pdf (accessed on May 15, 2014).

Robertson, S. R., L. I. Harzstark, J. P. Siplon, et al. 2006. Technical requirements for electronic parts, materials, and processes used in space and launch vehicles. *Aerospace Report No. TOR-2006(8583)-5236*. https://aero1.honeywell.com/thermswtch/docs/TOR-2006%288583%29-5236%2012-11-06.pdf (accessed on May 15, 2014).

Whyte, W. 2010. *Clean Room Technology: Fundamentals of Design, Testing and Operation*, 2nd edition. John Wiley & Sons, West Sussex, United Kingdom.

Appendix: Representative Imageries

We shall present here representative imageries from some of the Earth observation cameras. Imageries can be represented as *black and white* when only one spectral band is used or as a color picture by combining two or more spectral bands. In a black-and-white picture, different digital values are seen in different gray tones. A color picture is made of three primary colors: red, green, and blue. When we have a mixture of different proportions of these primary colors, we get different color shades. In principle, imagery from any three spectral bands can be combined to generate color pictures. When the imagery taken in the red band is assigned red color, the imagery taken in the green band is assigned green color, and the data from the blue band is assigned blue color, we have a natural color composite (NCC). Here the colors appear as we see them, that is, green vegetation appears as green, a red apple appears red, and so on. Any three bands can be combined to generate a color picture. However, they do not represent colors as we see, and they are termed a false color composite (FCC). The most popular band combination to produce FCC is the following:

Near infrared (NIR) is assigned red.

Red is assigned green.

Green is assigned blue.

In this combination, the vegetation appears in different shades of red. I leave it to the reader to reason out in this color combination why the vegetation looks red.

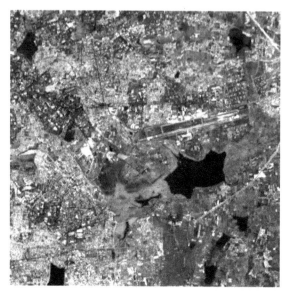

Natural color

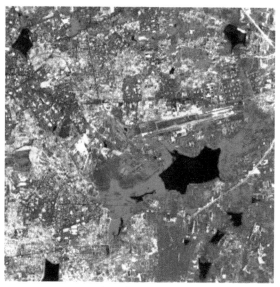

False color composite

FIGURE A.1
Figure A.1 gives a scene represented in natural color composite and false color composite. Natural color is generated using the primary colors: blue, green, and red. The green vegetation appears green. The FCC image uses green, red, and NIR bands as blue, green, and red, respectively. Here, the vegetation appears in different hues of red. (Reproduced with permission from Joseph G., *Fundamentals of Remote Sensing*, 2nd Edition, Universities Press (India) Pvt. Ltd., Hyderabad, Telangana, India, 2005, Plate 1.2.)

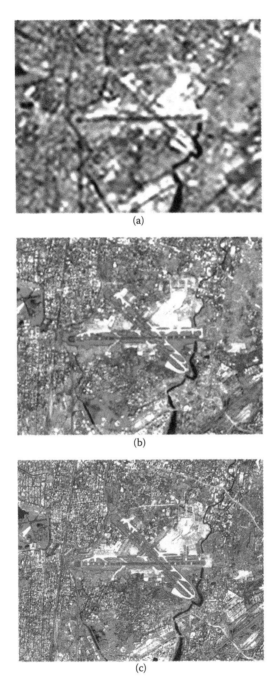

(a)

(b)

(c)

FIGURE A.2

The same scene is shown in three resolutions: (a) AWiFS 56 m, (b) LISS3 23 m, and (c) LISS4 5 m. As resolution increases, the discernible information content also increases. (Credit: Indian Space Research Organisation.)

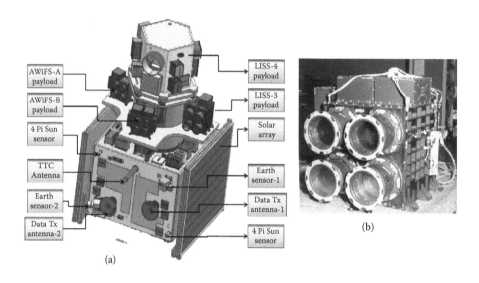

AWiFS-A payload

AWiFS-B payload

4 Pi Sun sensor

TTC Antenna

Earth sensor-2

Data Tx antenna-2

LISS-4 payload

LISS-3 payload

Solar array

Earth sensor-1

Data Tx antenna-1

4 Pi Sun sensor

(a)

(b)

(c)

(d)

FIGURE A.3
(a) Artist's view of Resourcesat with solar panel stowed showing accommodation of various subsystems. (b) Photograph of LISS-3 camera. (c) Photograph of AWiFS camera. (d) Photograph of LISS-4 camera. (Credit: Indian Space Research Organisation.)

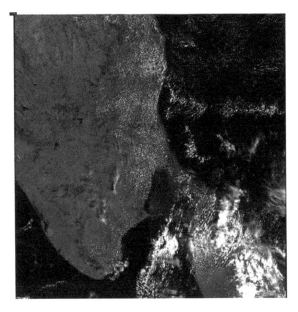

FIGURE A.4
Resourcesat AWiFS false color composite imagery. IGFOV 56 m; swath 740 km. (Credit: Indian Space Research Organisation.)

FIGURE A.5
Resourcesat LISS-3 false color composite over an agricultural area over India. IGFOV 23.5 m; area covered is about 19 × 18 km. Size of one of the fields is shown on the image. (Credit: Indian Space Research Organisation.)

FIGURE A.6
Resourcesat LISS-4 false color composite over an urban area. IGFOV 5.8 m: the image size is about 5.1 × 5.5 km. (Credit: Indian Space Research Organisation.)

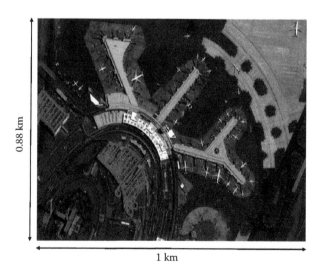

FIGURE A.7
Carto2 imagery (IGFOV 0.8 m) over airport. (Credit: Indian Space Research Organisation.)

FIGURE A.8
WorldView-2 imagery over Dubai. Color composite of NIR, red, and green. IGFOV 2 m. (Credit: DigitalGlobe.)

FIGURE A.9
Quickbird PAN. IGFOV 0.61. Part of Kuala Lumpur. (Credit: DigitalGlobe.)

FIGURE A.10
Pleadis imagery. Three-band natural color. 50 cm PAN sharpened. (Credit: CNES-Distribution
Astrium Services/Spot Image S.A.)

FIGURE A.11
Imagery from geostationary satellite INSAT3D in visible (0.5–0.75 μ) band. Spatial resolution
at nadir 1 km. (Credit: IMD/MOES.)

Index

A

Aberrations, 9, 17–18
  spherical, 18–19
Absolute accuracy, 88
Absorbed dose, 307
Acceptance tests, 318
Accuracy, 88
Across-track stereoscopy, 285–287
Active pixel linear array, 151
Active pixel sensor (APS), 298
Active thermal control techniques, 305
Adaptive differential pulse-code
        modulation (ADPCM), 239
ADC, *see* Analog-to-digital converter
Adjacency effect, 246
  of pixels, 89
ADPCM, *see* Adaptive differential
        pulse-code modulation
Advanced Land Observation Satellite
        (ALOS), 294–295
Advanced wide field sensor (AWiFS), 190
  false color composite imagery, 331
Aerial data, 160
Aerial image, 1
Airborne visible/infrared imaging
        spectrometer, 274
Air Force tri-bar target frequencies, 82
Airy pattern, 25, 77
  intensity distribution of, 26
All-SiC telescope technology at
        EADS-Astrium, 198
Along-track stereoscopy, 287–289
  by satellite tilt, 299
  using multiple telescopes
        ALOS, 294–295
        IRS CARTOSAT 1, 289–292
        SPOT high-resolution stereo
            camera, 292–294
ALOS, *see* Advanced Land Observation
        Satellite
Aluminum mirrors, 58
Analog-to-digital converter (ADC), 153
Antrix Corporation Limited (Antrix), 6

Aperture stop, 16
Aplanatic system, 20
Application-specific integrated circuit
        (ASIC), 159
Application Technology Satellite-1
        (ATS-1), 107
APS, *see* Active pixel sensor
Array configuration, staggered, 225–227
Aryabhata, 6
ASIC, *see* Application-specific integrated
        circuit
Astigmatism, 20
  misalignments, 69, 70
Asynchronous imaging, 223–225
Atmospheric effect of pixels, 89
"Atmospheric window" regions, 3
Atmospheric windows, 31
ATS-1, *see* Application Technology
        Satellite-1
AT spectral band separation, 200
AWiFS, *see* Advanced wide field sensor
Axial chromatic aberration, 22

B

Ball Aerospace, 239
Band-limiting filters, 133
Band-to-band registration (BBR),
        172, 173
Baseline configuration, 310
Baseline design review (BDR), 320
BBM, *see* Breadboard model
BBR, *see* Band-to-band registration
BDR, *see* Baseline design review
Beam splitter, 114
Bendix multispectral scanner, 160
Beryllium, 57
Bhaskara missions, 6
Bias voltages, 154
Bidirectional Reflectance Distribution
        Function (BRDF), 90
Binocular vision, principle of, 281
Black body, 12

Blooming, 149
Bolometers, 120
Bonding pad design, 67
BRDF, *see* Bidirectional Reflectance
    Distribution Function
Breadboard model (BBM), 310
Bridge amplifiers, 132
Building space hardware, 303

C

Camera obscura, 1
Camera radiometric transfer function, 94
Camera system, focusing of, 169–172
Carbon-fiber-reinforced plastic (CFRP),
    243
Carbon fiber reinforced polymer
    (CFRP), 310
CARTO1, 290–292
Cartographic satellites, 6
Carto2 imagery, 332
CARTOSAT2, 213
Cartosat camera, in-orbit measured
    values of, 82
CARTOSAT1, IRS, 289–292
Cartosat-2 series satellites, 243–244
Cassegrain telescope, 16, 44–46, 49, 265
    stray light paths in, 53
Cassegrain telescope beam, 210
Cassegrain-type telescopes, 48
    diffraction-limited MTF for, 50
Catadioptric system, increasing field,
    view of telescopes, 47–48
CAVIS, 242–243
Cavities, 86
CCDs, *see* Charge-coupled devices
CDR, *see* Critical design review
CDS process, *see* Correlated double
    sampling process
Centered optical system, 14
Centre of gravity (CG), 167
CFRP, *see* Carbon-fiber-reinforced
    plastic; Carbon fiber reinforced
    polymer
CG, *see* Centre of gravity
Chandrayaan1 TMC, 298–299
Charge-coupled devices (CCDs), 41,
    147–150, 228, 291, 295, 297
    arrays, 2, 7, 20, 145

linear arrays, 221
output signal, 154–155
signal generation and processing,
    153–159
Charge transfer efficiency (CTE), 149
Chemical-vapor-deposited
    (CVD)-coated SiC, 57
Chief ray, 16
Chromatic aberration, 20–22
Circle of least confusion, 18
Civilian Earth imaging system, 3–5
Civilian remote sensing satellite system,
    222
Clamp-sample approach, 157
Clamp switch, 157
CMOS photon detector array, *see*
    Complementary metal oxide
    semiconductor photon detector
    array
Coefficient of thermal expansion (CTE),
    55
Cold welding, 305–306
Color pictures, 327
Coltman formula, 30
Coma, 19–20
    misalignments, 69, 70
Commercial Earth observation system, 8
Common path interferometer, 266
Complementary metal oxide
    semiconductor (CMOS) photon
    detector array, 150–152
Compression ratio (CR), 232
Computer-aided optical design tools, 24
Conceptual designs, 310
Configuration change control, 321–322
Conical scanning, 111, 112
Contamination control, 319
Contoured back mirrors, 60
Contrast transfer function (CTF), 30
Control Moment Gyro (CMG) assembly,
    as actuators, 198, 242
Conventional photographic camera, 92
Conventional spectrometers, 162
Corona systems, 2
Correlated double sampling (CDS)
    process, 156–157
"Coupled system," 6
CR, *see* Compression ratio
Critical angle, 13

Critical design review (CDR), 311, 320–321
CRYSTAL, 2
CTE, *see* Charge transfer efficiency; Coefficient of thermal expansion
CTF, *see* Contrast transfer function
Curvature of field, 20
Cutoff frequency, 30
CVD-coated SiC, *see* Chemical-vapor-deposited-coated SiC

D

Dall-Kirkham (DK) configuration, 46
Dark current, 126–127, 135
Data compression, 232–233
Data rate, 135
Data transmission, 229–233
DDS, *see* Digital double sampling
Deconvolution, 226–227
Defocus, misalignments, 69, 70
DEM, *see* Digital elevation model
Detection, identification, and recognition (DIR) of military objects, 82
Detection of photons, CCDs, 147–148
Detectors, 291, 293
    figure of merit, 117–120
    operating temperature, 126–129
    photon detectors, 121–125
    signal processing, 130–133
    thermal detectors, 120–121
D*, figures of merit, 119
Dichroic beam splitters, 114, 115, 276
Dielectric constant, 10
"Differential-in-time amplifier," 157
Diffraction, 13–14
Diffraction-limited system, 225
Diffraction orders, 259
Diffraction pattern of image, 77
Digital double sampling (DDS), 157
Digital elevation model (DEM), 281
    extraction of, 232
Digitizer, 157–158
Direct current (DC) converters, 154
Direct hit rays, 52
DIR of military objects, *see* Detection, identification, and recognition of military objects

Discrete sampling device, 225
Dispersion, 11
Dispersive spectrometers, 258–261, 272
Dispersive system, 113–117
Distortions, 9, 20
Distributed satellite systems, concept of, 7
"Divoli," 196, 197
DK configuration, *see* Dall-Kirkham configuration
Dose, 307
Double Gauss lens, layout of, 38
DPCM algorithm, 192
Dual-axis flexure, 66
Dwell time for pushbroom scanner, 146
Dynamic range, figures of merit, 119

E

Earth imaging cameras, 16, 52
Earth observation cameras, 75–76
    imaging modes, 92–93
    on-orbit performance evaluation, 93–98
    performance specification, 91–92
    radiometric resolution, 87–88
        radiometric quality, 88–90
    spatial resolution, 76–82
    spectral resolution, 82–86
        interference filter, 86–87
    spectral selection schemes of, 253
    temporal resolution, 90–91
Earth Observation 1 (EO-1), NASA, 58
Earth observation satellites, 219
Earth observation systems, 6–8, 33, 107
    of ISRO, 160
Earth observation technology, 8
Earth resources survey, 206
Earth Resources Technology Satellite (ERTS), 4
Earth Resources Technology Satellite Landsat 1, 138
Earth's rotation, 289–290, 294–295
ECL devices, *see* Emitter-coupled logic devices
Edge-spread function (ESF), 95, 97
EEE, *see* Electronic, electric, and electromechanical
Effective focal length (EFL), 15–16, 220

Effective instantaneous field of view (EIFOV), 81
Effective refractive index, 86
EFL, *see* Effective focal length
EGSE, *see* Electrical ground support equipment
EIFOV, *see* Effective instantaneous field of view
Electrical ground support equipment (EGSE), 318, 319
Electrodynamic shaker, 315
Electromagnetic (EM) energy, 10
Electromagnetic frequency spectrum, use of, 231
Electromagnetic interference/compatibility tests, 317
Electromagnetic radiation, 10, 258
    diffraction, 13–14
    for imaging, source of, 31
    from one medium to another, propagation of, 12–13
    quantum nature of, 11
Electromagnetic spectrum, 257
Electronic circuit, 308
Electronic, electric, and electromechanical (EEE), 323–324
Electronics, 168
Electro-optical assembly, 276
Electro-optical design, 139
Electro-optical digital imaging, 2
Electro-optical modules, 190
Electro-optical sensors, 78, 319
    output of, 88
Electro-optical system, 79
Electro-optics module, 166
EM energy, *see* Electromagnetic energy
Emissivity, 12
Emitter-coupled logic (ECL) devices, 154
Encircled energy, 25
Enhanced thematic mapper plus (ETM+), 138–142, 199
Entrance pupil, 16
Environmental tests
    contamination control, 319
    ground support equipment/facilities, 318–319
    levels, 317–318
    mechanical tests, 314–316
    TV test, 316–317

ENVISAT, 7
EO module, CARTO1, 290, 291
EOS A-Train system, 7
ERTS, *see* Earth Resources Technology Satellite
ESA GEO-HR system, 209–212
ESF, *see* Edge-spread function
ETM+, *see* Enhanced thematic mapper plus
Exercising random vibration, 315
Exit pupil, 16
External stable calibration sources, 94

F

Failure Analysis Board (FAB), 322–323
Failure review committee, 322–323
Fairchild 143A CCD packaging outline, 162, 163
False color composite (FCC), 327, 328
Fast Fourier Transform (FFT), 268
"Fat zero," 187
FCC, *see* False color composite
FCO, *see* Field correcting optics; Focal plane corrector
FEM, *see* Finite element method
FFT, *see* Fast Fourier Transform
Field correcting optics (FCO), 243
Field curvature, 9
Field-effect transistor switch, 154
Field of view (FOV), 16, 161
    of instrument, 75, 79, 86
Field programmable gate array (FPGA) devices, 154
Fieldstop, 16
Figure of merit (FM)
    of camera, 80
    detectors, 117–120
Filter-based systems, 270–273
Finite element method (FEM), 63
First-order optics, 18
First-order theory, 20
Flight model (FM), 310, 311
FM, *see* Figure of merit
f/4 multielement telecentric lens assembly, 277
f-number, 17
Focal length, 15
Focal plane, 15

Focal plane area array, 277
Focal plane assembly (FPA), 187
Focal plane corrector (FCO), 48
Focal plane detector system, 196
Focal plane layout, 162
    multiple lens option, 165–166
    optomechanical scanners, 113–117
    single collecting optics scheme,
        162–165
Focal point, 14
Focusing of camera system, 169–172
Fold mirror, 194
Four differential equations, 10
Fourier lens, 268
Fourier transform
    of LSF, 95
    of PSF, 27
Fourier transform hyperspectral imager
        (FTHSI), 274, 275
Fourier transform spectrometers (FTSs),
        261–263
    Michelson interferometers, 263–266
    Sagnac interferometers, 266–269
FOV, *see* Field of view
FPA, *see* Focal plane assembly
FPGA devices, *see* Field programmable
        gate array devices
Frame by frame imaging mode, 92
Fraunhofer diffraction, 25
"Fraunhofer lines," 257
French satellite SPOT-1, 159
French SPOT-1 system, 6
Fresnel reflection, 12
FTHSI, *see* Fourier transform
        hyperspectral imager
FTSs, *see* Fourier transform
        spectrometers
Full width at half maxima (FWHM),
        84, 85
Functional block diagram of payload
        evaluation system, 179
FWHM, *see* Full width at half maxima

G

Galactic cosmic rays (GCR), 306
Gaussian optics, 18
GCPs, *see* Ground control points
GCR, *see* Galactic cosmic rays

Generation-recombination (G-R) noise,
        136
GEO, *see* Geosynchronous Earth Orbit
GeoEye-1, 240
GeoEye camera, in-orbit measured
        values of, 82
GEO high-resolution imaging systems,
        208–209
Geo-Oculus, spectral channels
        proposed for, 210
Georeferencing function, 93
GEO-resource survey systems
    ESA GEO-HR system, 209–212
    geostationary hyperspectral imaging
        radiometer, 214
    GOCI, 212–213
    ISRO GISAT, 213–214
Geostationary hyperspectral imaging
        radiometer (NASA), 214
Geostationary ocean color imager
        (GOCI), 212–213
Geostationary satellite, imagery from,
        334
Geosynchronous Earth Orbit (GEO), 307
Geosynchronous observation
        systems, 90
"Ghost interferograms," 267
GOCI, *see* Geostationary ocean color
        imager
Graphite-epoxy laminate, 68
Grating imaging spectrometer, 261
Gray body, 12
Gregorian telescope, 44, 45
G-R noise, *see* Generation-recombination
        noise
Ground control points (GCPs), 234
Ground sampling distance (GSD), 81
Ground support equipment (GSE), 310,
        311, 318–319
Gyro-stabilization, 3

H

Hard vacuum environment, 305
Haute Résolution Stéréoscopique (HRS),
        292–294
Haute Resolution Visible (HRV), 194
Height error, 282, 283
Herschel IR telescope, 57–58

High-resolution imagery, 236
High Resolution Imaging Science
    Experiment (HiRISE) camera,
    49–50
High-resolution imaging system,
    considerations for realizing,
    220–221
High-resolution imaging systems, 8
High-resolution system, 219
HiRISE camera, *see* High Resolution
    Imaging Science Experiment
    camera
HRS, *see* Haute Résolution
    Stéréoscopique
HRV, *see* Haute Resolution Visible
HRVIR focal plane arrangement, 196,
    197
Hubble Space Telescope, 208
Hybrid arrays, 152–153
Hybrid scanner, 205–208
    GEO high-resolution imaging
        systems, 208–209
    GEO-resource survey systems
        ESA GEO-HR system, 209–212
        geostationary hyperspectral
            imaging radiometer, 214
        GOCI, 212–213
        ISRO GISAT, 213–214
Hyperion, 275–276
Hyperspectral imagers (HISs), 213,
    254, 275
Hyperspectral imager SWIR
    (HySI-SWIR), 213
Hyperspectral imager VNIR
    (HySI-VNIR), 213
Hyperspectral imaging spectrometer
    development, 274
Hyperspectral imaging systems,
    251–254
    configuration, 254
        pushbroom approach, 256–257
        scanner approach, 255–256
    instruments, 274–275
        hyperion, 275–276
        Indian space research
            organization hyperspectral
            imager, 277–278
        MightySat II, 275
        NASA EO-1, 276–277

spectrometers, 257–258
    dispersive, 258–261
    filter-based systems, 270–273
    Fourier transform spectrometers,
        261–269
    "Smile" and "Keystone" effects,
        273–274
Hyperspectral sensors, 252
HySI-SWIR, *see* Hyperspectral imager
    SWIR
HySI-VNIR, *see* Hyperspectral imager
    VNIR

I

Ideal aberration-free system, 22
Ideal imaging system, 9
    image formation by, 23
Identical CCD linear arrays, 225
IFC system, *see* In-flight calibration
    system
IFOV, *see* Instantaneous field of view
IGFOV, *see* Instantaneous geometric
    field of view
IKONOS, 8, 219, 222, 237–239
    camera, in-orbit measured values
        of, 82
Image formation
    aberrations, 17–18
        chromatic aberration, 20–22
        coma, 19–20
        spherical aberration, 18–19
    electromagnetic radiation, 10–14
        for imaging, source of, 31
    image quality evaluation, 25–27
    imaging systems
        basic conditions of, 9
        useful terminologies of, 14–17
    modulation transfer function, 27–30
    radiometric consideration, 31–33
    wave optics, 22–24
Image matching, 284
Image quality evaluation, 25–27
Image, radiometric quality of, 88–90
Imagery, 327–331
Image space scanning, 104
Imaging geometry of optical system, 32
Imaging modes, 92–93
    with agile satellite, 237

Imaging optics
  catadioptric, 35
  reflective telescope, *see* Reflective
    telescopes
  refractive optics, *see* Refractive optics
  stray light control and baffling, 52–54
Imaging spectrometer
  concept of, 256
  in optomechanical scanner
    configuration, concept of, 255
Imaging systems, 77, 180, 254
  basic conditions of, 9
  history of, 1
  useful terminologies of, 14–17
Indian Earth observation program, 5–6
Indian Remote Sensing (IRS) cameras,
    40, 86
  alignment and characterization,
    168–169
    flat field correction, 174–178
    focusing of camera system,
      169–172
    image format matching and BBR,
      172–174
  electronics, 168
  focal plane layout, 162
    multiple lens option, 165–166
    single collecting optics scheme,
      162–165
  LISS-1 and -2, 159–162
  mechanical design, 166–168
  qualification, 179–180
Indian Remote Sensing (IRS) Satellite,
    219
  high-resolution imaging systems,
    243–244
  IRS-1A, 5
  LISS-4 camera, 149
Indian Space Research Organization
    (ISRO), 6–7, 243, 275
  Earth observation system of, 160
  Geostationary Imaging Satellite
    (GISAT), 213–214
  hyperspectral imager, 277–278
  IRS-1A in 1988, 159
  satellite INSAT-2E, 205
Indium gallium arsenide (InGaAs), 125
  linear array, 153, 181
In-flight calibration (IFC) system, 93, 168

Infrared astronomy satellite (AKARI)
    telescope, 58
Infrared (IR) detector array, 152
InGaAs, *see* Indium gallium arsenide
Instantaneous field of view (IFOV), 220
  of camera, 78, 79
    in-orbit measured values of, 82
Instantaneous geometric field of view
    (IGFOV), 78, 79, 101, 145, 146,
    223, 236, 255, 256
Instruments, hyperspectral imaging,
    274–275
  hyperion, 275–276
  Indian space research organization
    hyperspectral imager, 277–278
  MightySat II, 275
  NASA EO-1, 276–277
Integration time, 221
  array configuration, staggered,
    225–227
  asynchronous imaging, 223–225
  time delay and integration, 222–223
Interband registration, 172
Interference filter, 86–87
Interference fit, 39
Interferogram, 263
  generation by temporal imaging
    Fourier transform
    spectrometer, 265
International laws, 1
International Society for
    Photogrammetry and Remote
    Sensing (ISPRS), 91
International Telecommunication Union
    (ITU), 231
*International Workshop on Radiometric and
    Geometric Calibration* (2003), 91
Invar, 68
Ionizing radiation, 36–37
IRS-1A satellite, 161, 180
IRS-1B satellite, 180
IRS CARTOSAT 1, 289–292
IRS-1C/D camera, 180–181
  LISS-3 design, 181–182
  PAN camera, 184–187
    payload steering mechanism,
      187–188
  wide field sensor, 182–183
IRS-1C/1D PAN camera, 78, 286, 288

IRS-1C LISS-III camera, 90
IRS-1C satellite, 188
IRS LISS cameras, 94
IRS PAN cameras, 78
IRS Satellite, *see* Indian Remote Sensing
    Satellite
ISPRS, *see* International Society for
    Photogrammetry and Remote
    Sensing
ISRO, *see* Indian Space Research
    Organization
ITU, *see* International
    Telecommunication Union

J

James Webb Space Telescope (JWST),
    57, 208
Japan Earth Resources Satellite (JERS-1),
    4, 289
Jet Propulsion Laboratory, 274
Jitter modulation transfer functions, 234
Johnson noise, 135–136
JPL convex grating design, 276
JWST, *see* James Webb Space Telescope

K

"Keystone" effects, 273–274
Knife-edge method, 97
Korean Geostationary Ocean Color
    Imager (GOCI), 207
Korean multifunctional geostationary
    satellite, 212

L

LAC, *see* Linear Etalon Imaging Spectral
    Array Atmospheric Corrector
Landsat 8, 94, 198–199
    OLI, 199–202
    TIRS, 202–204
Landsat 1 BBV camera, 38
Landsat Data Continuity Mission
    (LDCM), 198–204
Landsat MSS system, 4, 160
    bands, 85
    detectors, projection of, 116
    scan mirror, 57

Landsat Thematic Mapper (TM), 160
    scan mechanism, 107
    telescope structure, 68
Large-format camera (LFC), 36
Lateral chromatic aberration, 22
Launch environment, 303–304, 316
LDCM, *see* Landsat Data Continuity
    Mission
"Leapfrog" configuration, 269
LEDs, *see* Light emitting diodes
LEO environment, *see* Low Earth orbit
    environment
LFC, *see* Large-format camera
Light emitting diodes (LEDs), 94, 168
Linear arrays
    CCDs, 161
    for pushbroom scanning, 147
        CCDs, 147–150
        CMOS photon detector array,
            150–152
        hybrid arrays, 152–153
Linear Etalon Imaging Spectral Array
    Atmospheric Corrector (LAC),
    276
Linearity of mirror motion, 110
Linear transfer function, 87
Linear variable filter (LVF), 270
Line by line imaging mode, 92
Line of sight (LOS) disturbance, 234
Line-spread function (LSF), 95, 97
LISS-3, 189
    design of, 181–182
    EO module, mechanical structure
        of, 182
    false color composite imagery, 331
LISS-4
    false color composite imagery, 332
    multispectral camera, 190–194
LISS lenses, 38–40
Longitudinal spherical aberration,
    18
Low Earth Orbit (LEO) environment,
    305
Low-level random vibration test,
    315–316
Low-level sinusoidal excitation,
    314–315
LSF, *see* Line-spread function
LVF, *see* Linear variable filter

M

Marginal ray, 16
Marine Observation Satellite (MOS-I), 4
Mars Reconnaissance Orbiter
    spacecraft, 49–50
Master reference cube (MRC), 172
Maturity levels, 313
Maxwell's equations, 10
MCT, *see* Mercury cadmium telluride
Mechanical design, 310
    IRS cameras, 166–168
Mechanical ground support equipment
    (MGSE), 318
Mechanical tests, 314–316
MEOSS, *see* Monocular Electro-Optical
    Stereo Scanner
Mercury cadmium telluride (MCT), 123
Meridional plane, 20
Mersenne afocal reflective telescope, 113
MFDs, *see* Mirror fixing devices
MGSE, *see* Mechanical ground support
    equipment
Michelson interferometers, 263–266
Microlens increases system complexity,
    151
MightySat II, 275
Mineral halloysite, reflectance spectra
    of, 252
Miniaturized designs, 154
Minimum resolvable wavelength
    difference, 257
Mini-TES for Mars observation, 265
Mirror fixing devices (MFDs), 50, 290
Mirror mounts, 64–65
    bipod mounts, 65–67
Mirrors
    alignment of, 67–71
    lightweighting of, 59–63
Mirror support system (MFD), 63, 64
Misalignments, types of, 69–70
MLG, *see* Multilinear gain
Model philosophy, 313
Modular Optoelectronic Multispectral
    Scanner (MOMS), 159
Modulation transfer function (MTF),
    27–30, 79–81, 89, 149, 171, 223
    estimation technique, 97
    measurement, 95

Modulation transfer functions, 149
"Moments" method, 83, 85
MOMS, *see* Modular Optoelectronic
    Multispectral Scanner
Monocular Electro-Optical Stereo
    Scanner (MEOSS), 295–297
Monolithic detector arrays, 152
MRC, *see* Master reference cube
M-shaped target for BBR evaluation,
    172–174
MSS, *see* Multispectral scanners;
    Multispectral scanner system
MTF, *see* Modulation transfer function
MTF degradation, 228
Multidisciplinary global approach, 6
Multielement lens, 15
Multiface prism scanner, 105
Multilinear gain (MLG), 192
Multiple lens option, 165–166
Multiresolution data, 98
Multispectral camera, 180
Multispectral (MXL) channels, 241
Multispectral classification, 81
Multispectral imagery, 233
Multispectral imaging cameras, 82
Multispectral remote sensing, 82
Multispectral scanners (MSS), 103,
    162, 270
    scan mechanism, 106–107
Multispectral scanner system (MSS), 4,
    138, 219
Multispectral sensors, 252
Multispectral VNIR (MX-VNIR), 213
MX-VNIR, *see* Multispectral VNIR

N

NASA, *see* National Aeronautics and
    Space Administration
NASA's Ocean Biology and
    Biogeochemistry Working
    Group (OBBWG), 214
National Aeronautic and Space
    Administration Earth
    Observation 1 (NASA's EO-1),
    275
National Aeronautics and Space
    Administration (NASA), 4, 275
EO-1, 58, 276–277

JWST, 57
Mars Reconnaissance Orbiter
    spacecraft, 49–50
project life-cycle phases, 312
National Map Accuracy Standards
    (NMAS), 282
Natural color composite (NCC), 327, 328
Near infrared (NIR), 1
Near-Infrared Mapping Spectrometer of
    the Galileo mission, 274
NEΔL, *see* Noise equivalent differential
    radiance
NEM, *see* Noise equivalent modulation
NEP, *see* Noise equivalent power
Neural network classifiers, 233
Newtonian telescope, 44
NMAS, *see* National Map Accuracy
    Standards
Noise equivalent differential radiance
    (NEΔL), 80, 87
Noise equivalent modulation (NEM), 80
Noise equivalent power (NEP), 118
Nonconformance management, 321–322
Nonimaging spectrometers,
        conventional configuration
        of, 259
Normalization process, 177
Notching, 315
Nyquist frequency, 170
Nyquist noise, 135–136
Nyquist sampling criteria, 264

O

OBBWG, *see* Ocean Biology and
        Biogeochemistry Working
        Group
Object plane scanning, 105
Object space scanning, 104–105
Object space telecentric lenses, 41
Obscured TMA (OTMA) telescope, 49
Ocean Biology and Biogeochemistry
        Working Group (OBBWG), 214
Ocean Color Monitor (OCM)
    lens, 41, 43
    sensors, 85
Off-axis parabolic mirror, 169
Offner convex-grating spectrometers, 261

Offner spectrometer, 261
Off-the-chip signal processing,
        155–159
OIR sensors, *see* Optical-infrared
        sensors
OLI, *see* Operational land imager
Onboard calibration system, 93, 94
Onboard Solid State Recorder, 192
1/f noise, 136
OPD, *see* Optical path difference
Open back mirror configuration, 61, 62
Operating temperature, detectors,
        126–129
Operational land imager (OLI),
        199–202
OPS, *see* Optical sensor
Optical axis, 14, 223
Optical-infrared (OIR) sensors, 75
Optical path difference (OPD), 23, 262
Optical sensor (OPS), 289
Optical system design, 24
Optical telescope, 229
Optical transfer function (OTF), 27, 95
Optimal parameters of camera, 91
Optimizing lightweight mirror
        structure, 63–64
Optomechanical scanners, 254
    configuration, concept of imaging
        spectrometer in, 255
    detectors, *see* Detectors
    dispersive system and focal plane
        layout, 113–117
    enhanced thematic mapper plus,
        138–142
    operation principle, 101–103
    optics for, 112–113
    scanning systems, *see* Scanning
        systems
    system design considerations,
        133–138
Order sorting filters, 260
Organization of linear CCD, 149
OTF, *see* Optical transfer function
OTMA telescope, *see* Obscured TMA
        telescope
Outgassing, 305
Out-of-band radiation, 271
Out-of-field stray light, 52

P

Paintbrush mode of operation, 236
Panchromatic (PAN) array, 242
Panchromatic (PAN) band, 251
  FPA, 199
Panchromatic (PAN) camera, 184–187
  payload steering mechanism, 187–188
Panchromatic (PAN) channels, 241
Panchromatic (PAN) electro-optics
  module, 186
Panchromatic (PAN) imagery, 219
Panchromatic Remote Sensing
    Instrument for Stereo Mapping
    (PRISM), 294–295
PA Plan, *see* Product Assurance plan
Parallax, 281
  determination of, 284
Parallel rays, 22
Paraxial optics, 18
Passive coolers, 129
Passive remote sensing, 3
Passive sensors, 75
Passive thermal control techniques, 305
Payload data acquisition system, 179
Payload evaluation system, functional
    block diagram of, 179
Payload status indicator, 179
Payload steering mechanism, 187–188
PDR, *see* Preliminary design review
Peak to valley (P-V) wave front error, 23
Peltier effect, 128
Perkin-Elmer Corporation, 205
PFM, *see* Proto-flight model
Phase transfer function (PTF), 27
Philosophy model, 313
Photoconductive detectors, 123, 131
Photodiodes, 125, 148
Photoemissive detectors, 122–123
Photomultiplier tubes (PMTs), 122
Photon, 11
Photon detectors, 121–122
  photoemissive detectors, 122–123
  photovoltaic detectors, 123–125
Photoresponse nonuniformity (PRNU),
    174, 175
Photovoltaic detectors, 123–125, 130
Photovoltaic effect, 123

Pixel, 78
Pixel by pixel imaging mode, 92
Planck's constant, 11
Planck's law, 12
Plane elliptical mirror, 194
Plane reflection grating, 261
Planimetric error, 282
Pleadis imagery, 334
Pleadis telescope, 68
PMTs, *see* Photomultiplier tubes
Point-spread function (PSF), 26, 95
Polychromatic radiation, 258
Power distribution system, 180
Power package, 166
Power spectral density (PSD), 315
PPL, *see* Preferred part list
Practical optical system, 23
Preamplifier noise, 136
Preferred part list (PPL), 324
Preliminary design review (PDR),
    311, 320
Preshipment review (PSR), 321
Principal axis, 14
Principal planes, 15, 16
Principal points, 15
Principal surface, 15
Principles of interferometers, 261
PRISM, *see* Panchromatic Remote
    Sensing Instrument for Stereo
    Mapping
Prism spectrometer, 258, 259
PRNU, *see* Photoresponse
    nonuniformity
Product Assurance (PA) plan, 324–325
Proto-flight model (PFM), 313
PSD, *see* Power spectral density
PSF, *see* Point-spread function
PSR, *see* Preshipment review
PTF, *see* Phase transfer function
Pushbroom approach, 256–257
Pushbroom imagers, principle of
    operation, 145–147
Pushbroom scanners, 254
  integration time for, 229
Pushbroom scanning technique, 92,
    146, 256, 270
  linear array for, *see* Linear arrays,
    for pushbroom scanning

P-V wave front error, *see* Peak to valley
    wave front error
Pyroelectric detectors, 120–121
Pyrotechnic devices, ignition of, 304

Q

Qualification tests, 317
Quantum detectors, *see* Photon
    detectors
Quantum efficiency (QE), figures of
    merit, 118
Quantum theory, 11
Quantum well infrared photodetectors
    (QWIPs), 126, 127, 202–203
Quickbird-2, 239–240
Quickbird PAN imagery, 333
Quilting effect, 61–62

R

Radar Imaging Satellite, 6
Radiance domain, Earth observation
    camera, 92
Radiant flux, 31
Radiation environment, 306–309
Radiation incident on imaging
    system, 16
Radiative transfer model, 94
Radiometric accuracy, 88
    MTF effect on, 89
Radiometrically accurate IFOV
    (RAIFOV), 81, 95
Radiometric calibration look-up table
    (RADLUT), 177
Radiometric errors, 89–90
Radiometric quality of image, 88–90
Radiometric resolution, sensor
    parameters, 87–88
Radiometric sensitivity, 93
Radio regulations, 231
RADLUT, *see* Radiometric calibration
    look-up table
RAIFOV, *see* Radiometrically accurate
    IFOV
Random refractive index variation, 245
Random vibration, 315
RapidEye Earth Imaging System
    (REIS), 58

Ray, 12
Rayleigh criterion, 77
RBV, *see* Return Beam Vidicon
RC design, *see* Ritchey-Chretien design
RC telescope, *see* Ritchey-Chretien
    telescope
Readout integrated circuit (ROIC), 151
Readout register, 148
Real optical system, 17
Reconnaissance satellite, 159
Reflectance spectra of mineral
    halloysite, 252
Reflecting cube, 172
Reflecting property, of mirror, 60
Reflective telescopes, 54–55
    drawback of, 50
    mirror fabrication, 58–64
    mirror material selection, 55–58
    mirror mounts, 64–67
    mirrors alignment, 67–71
    types of, 43–46
Refracted ray, 12–13
Refraction, 12
Refractive index (RI), 11
Refractive optics, 36–40
    telecentric lenses, 40–43
REIS, *see* RapidEye Earth Imaging
    System
Relative accuracy, 88
Reliability and Quality Assurance
    (R&QA), 324–325
Reliable test, 179
Remote sensing, 2–3, 6, 8
    applications, 251
Remote sensors, 75
    classification of, 76
Repetivity, 90
Resolution limit, 77
RESOURCESAT series, 188–189
    AWiFS, 190
    LISS-3, 189
    LISS-IV multispectral camera, 190–194
    view of, 330
RESOURSESAT-2 cameras, parameters
    of, 193
RESOURSESAT-2 LISS3 bands, 83
Responsivity, figures of merit, 117
Return Beam Vidicon (RBV), 4
    camera, 219

Revisit capability, 90
RI, *see* Refractive index
Ribs, 185
Rib-stiffened lightweight mirror, 62–63
Ritchey-Chretien (RC) design, 169
Ritchey-Chretien (RC) telescope, 46,
    243, 275
RMS, *see* Root mean square
Rocket-propelled camera systems, 3
ROCSat-2, 58
ROIC, *see* Readout integrated circuit
Root mean square (RMS), 315
    wave front error, 24
R&QA, *see* Reliability and Quality
    Assurance

S

Sagittal plane, 20
Sagnac interferometers, 266–269
Sample-scene phasing effect, 96
Sandwich mirrors, 60–61
SAR, *see* Synthetic aperture radar
Satellite, 294
    temperature of, 304
Satellite-based imaging system, 4
Satellite-based remote sensing
    systems, 6
Satellite-borne sensors, 128
Satellite imagery, potential of, 4
Satellite Interface Simulator (SIS), 319
Satellite orbit, 305
Satellite program in India, 5
Satellite remote sensing, 90
Saturation radiance (SR), 87
SCA, *see* Sensor chip array
Scan geometry, 110–112
Scan line corrector (SLC), 140–142
Scan mirror, 92
Scanner approach, 255–256
Scanning systems, 103–110
    scan geometry and distortion,
    110–112
Scene simulator, 169
Schmidt camera, 194
Schmidt-Cassegrain, 47
Schmidt corrector, 47
Schmidt telescope, 47
Schott Catalogue, 86

SeaWiFS, 85, 90
SEB, *see* Single-event burnout
SEE, *see* Single-event effect
Seidel aberrations, 18
SEL, *see* Single-event latch-up
Semi-open back mirror configuration,
    61, 62
Sense capacitor, 154
Sensor chip array (SCA), 199
Sensor parameters
    radiometric resolution, 87–88
        radiometric quality, 88–90
    spatial resolution, 76–82
    spectral resolution, 82–86
        interference filter, 86–87
    temporal resolution, 90–91
Sensors, 75
    parameters, *see* Sensor parameters
Sentinel-2 satellite, 58
Set partitioning in hierarchical trees
    (SPIHT) coding, 232
SEU, *see* Single event upset
Shannon's criteria, 133
Shannon's sampling criteria, 225
Shock tests, 316
Shortwave infrared (SWIR) detection,
    152, 181, 189, 197
Shortwave infrared (SWIR) spectral
    region, 251
Shot noise (photon noise), 136
Shower curtain effect, 245
"Shutter Control," 7
Shuttle Mission STS-7, 159
Signal processing, detectors, 130–133
Signal to noise ratio (SNR), 147, 220
Silicon-based devices, 152
Silicon carbide (SiC) mirrors, 57–58
Silicon linear array, 205
Single-band CCD camera, architecture
    of, 153
Single collecting optics scheme, 162–165
Single-event burnout (SEB), 309
Single-event effect (SEE), 308
Single-event latch-up (SEL), 308
Single event upset (SEU), 308
Single performance parameter, 81
Single scan mirror, 105
Single-structural thermal model (STM),
    313

SIS, *see* Satellite Interface Simulator
SLC, *see* Scan line corrector
Slow systems, 17
Smear modulation transfer functions, 234
"Smile" effects, 273–274
Snell's equation, 18
Snell's law, 13, 18
    of refraction, 258
SNR, *see* Signal to noise ratio
Solar diffuser-based calibration, 200
Solar diffusers, 94
Solar radiation, 31
Solid-state image sensors, concept of, 147
Spaceborne hyperspectral sensor, 252
Spaceborne imaging spectrometers, 274
Spaceborne pushbroom cameras, 159
Spacecraft, 303–304
Spacecraft drift/jitter, 208–209
Spacecraft interface simulator, 179
Spacecraft level alignment
        measurements, 172
Spacecraft thermal control (STC), 304
Space environment, 304–309
Space hardware realization approach, 309–313
Space radiation, 36
Space-to-Earth data transmission, 231
Spatial domain, Earth observation
        camera, 91
Spatial resolution, 244–246
    imaging systems, 234
    sensor parameters, 76–82
Specific detectivity, figures of merit, 119
Spectral bands, 82, 327
    selection, 212
Spectral bandwidth definitions, 84
Spectral channels proposed for Geo-
        Oculus, 210
Spectral domain, Earth observation
        camera, 91
Spectral keystone, 273
Spectral resolution, sensor parameters,
        82–86
    interference filter, 86–87
Spectral response, figures of merit, 118
Spectral separation, techniques of, 258
Spectral smile, 273

Spectrometers, hyperspectral imaging, 257–258
    dispersive, 258–261
    filter-based systems, 270–273
    Fourier transform spectrometers, 261–269
    "Smile" and "Keystone" effects, 273–274
Spectroscopy, 254, 257
Spherical aberration, 18–19
SPIHT coding, *see* Set partitioning in
        hierarchical trees coding
Spinning satellite, 108, 109
Spin scan system, 108
SPOT Earth observation camera, 194–198
SPOT Haute Resolution Visible
        (HRV), 87
    camera, 195
SPOT satellite, 187
SPOT-5 satellite, 198
SPOT stereo pairs, 286
"Square wave" pattern, 89
Square wave response (SWR) of system, 170–171
SR, *see* Saturation radiance
Stairing mode, imaging, 92
STC, *see* Spacecraft thermal control
Steering mirror optical performance, 187
Stereo imaging, 284
    capacity, 236
Stereo pair
    generation of, 284–289
    with single optics, MEOSS, 295–297
Stereoscopy, 281
    across-track, 285–287
    along-track, 287–289
Stirling cycle coolers, 129
STM, *see* Single-structural thermal
        model
Stray light, control and baffling of, 52–54
Strehl ratio, 25
Structural analysis, 167
Structural hardware model, 313
Submeter class imaging systems, 221
Submeter imaging, 219
    constraints on satellite, 234–237
    data transmission, 229–233

faster optics, choosing, 227–229
high-resolution imaging system,
    considerations for realizing,
    220–221
integration time, increasing, 221
    array configuration, staggered,
    225–227
    asynchronous imaging, 223–225
    time delay and integration,
    222–223
spatial resolution, 244–246
submeter resolution, imaging
    cameras with
    GeoEye-1, 240
    iKONOS, 237–239
    Indian remote sensing satellite
    high-resolution imaging
    systems, 243–244
    Quickbird-2, 239–240
    WorldView imaging systems,
    240–243
systems, 236, 243
Submeter telescope, 229
Subsatellite velocity vector, 223
Successive frames, 212
    image, 92
Successive scan lines, 92
Sun, 306–307
Supermode, 226
"Synchronized strategy," 6
"Synchronous Earth Observation
    Satellite," 205
Synchronous mode, 223
Synthetic aperture radar (SAR), 4–5

T

Tangent plane, 20
Target pixels, 97
Taylor series, 18
TC system, *see* Thermal control system
TDI, *see* Time delay and integration
Technical review, 320
Technology Experiment Satellite (TES),
    221, 243
Telecentric lenses, refractive optics,
    40–43
Telescopes
    design, 290

increasing field of view, 46
metering structure of, 68
reflective wide field of view, 48–52
use of, 77
Television Infrared Observation
    Satellite-1 (TIROS-1), 4
Television Infrared Observation Satellite
    (TIROS), 6
Temporal frequency, 90
Temporal imaging Fourier transform
    spectrometer, interferogram
    generation by, 265
Temporal resolution, sensor parameters,
    90–91
Terrain Mapping Camera (TMC), 297
    Chandrayaan1, 298–299
TES, *see* Technology Experiment
    Satellite; Thermal Emission
    Spectrometer
Thematic Mapper, 219
THEOS, 58
Thermal control (TC) system, 129
Thermal detectors, 120–121
Thermal Emission Spectrometer (TES),
    265
Thermal enclosure temperature, 294
Thermal environment, 304–305
Thermal infrared sensor (TIRS), 202–204
Thermal noise, 135–136
Thermal radiation, 12
Thermocouple, 120
Thermoelectric coolers, 128
Thermopile, 120
Thermovacuum (TV) test, 316–317
Third-order optics, 18
Three-mirror anastigmatic (TMA), 237
    telescope, 49, 261
Three-mirror telescopes, 49
TIA, *see* Transimpedance amplifier
TID, *see* Total ionizing dose
Tilt, misalignments, 70
Time constant, figures of merit, 119–120
Time delay and integration (TDI), 8,
    221–223
Time domain FTS, 263
TIRS, *see* Thermal infrared sensor
TMA, *see* Three-mirror anastigmatic
TMC, *see* Terrain Mapping Camera
Total internal reflection, 13

Total ionizing dose (TID), 308, 309
Transfer gate, 148
Transimpedance amplifier (TIA), 124, 130
Transmission degradation, model simulation of, 37
Transport shift registers, 148, 229
Transverse spherical aberration, 18
Tunable filter system, 270
Tungsten halogen lamps, 175
Tungsten lamp, 200
TV test, *see* Thermovacuum (TV) test
2D detector array, 272
Two-image stereo reconstruction, 284
Two-mirror telescope system, 44–46, 68–69
2048-element linear array CCD, 183
Typical spectral response of silicon CCD, 152

U

United Nations (UN), 231
Unobscured three-mirror anastigmatic (UTMA) telescope, 51–52
U.S. Air Force Research Laboratory's satellite, 275
USDA National Agricultural Statistics Service, 194
UTMA telescope, *see* Unobscured three-mirror anastigmatic telescope

V

Vacuum environment, 305–306
Van Allen radiation belts, 306–307
Variable filter system, 270
Veiling glare, 52
Vertical temperature profiling radiometer (VTPR), 75
Very near infrared (VNIR), 289
 band, 181, 189, 275

region, 43
spectrometer, 276
Vibration load, 314
Vicarious calibration, 94
Visible and near-infrared (VNIR), 85
 region, 148
Visible-infrared (IR) region, 251
VNIR, *see* Very near infrared; Visible and near-infrared
VTPR, *see* Vertical temperature profiling radiometer

W

Water vapor, 305
Wave front, 22
Wave front error (WFE), 51, 63–64
Wavelength selection, 86
Wave optics, 22–24
Wedge filter camera system, 272
Wedge filter spectrometer, concept of, 270
Well-corrected optical system, 15
WFE, *see* Wave front error
Whiskbroom scanners, 101, 202
Whiskbroom scanning, 92
Wide field sensor (WiFS), 182–183
 camera, 180
WorldView-2 (WV-2), 234
 imagery, 333
WorldView-3 (WV-3), 242
WorldView imaging systems, 240–243
WV-2 satellite, 241

Z

Zenit, 2
Zernike polynomials, 24
Zero-biased photovoltaic detectors, 130
Zerodur, 239, 243
Zero path difference (ZPD), 264

Milton Keynes UK
Ingram Content Group UK Ltd.
UKHW022036141024
449569UK00014B/632